FLORA ZAMBESIACA

Flora terrarum Zambesii aquis conjunctarum

VOLUME TEN: PART FOUR

FLORA ZAMBESIACA

MOZAMBIQUE

MALAWI, ZAMBIA, ZIMBABWE

BOTSWANA

VOLUME TEN: PART FOUR

Edited by
G.V. POPE & E.S. MARTINS

on behalf of the Editorial Board:

S.J. OWENS
Royal Botanic Gardens, Kew

M.A. DINIZ
*Centro de Botânica, Instituto de Investigação
Científica Tropical, Lisboa*

G.V. POPE
Royal Botanic Gardens, Kew

Published by the Royal Botanic Gardens, Kew,
for the Flora Zambesiaca Managing Committee
2002

Typesetting and page make-up by Media Resources, Information Services Department,
Royal Botanic Gardens, Kew

Printed in the European Union by
The Cromwell Press

ISBN 1 84246 053 6

CONTENTS

INCLUDED IN VOLUME X, PART 4

GRAMINEAE

Tribe Andropogoneae

LIST OF NEW NAMES PUBLISHED IN THIS PART

The tribal arrangement, as set out in Volume **10** part 1 (1971), on page 7 and in the key on page 13, has been modified in the light of subsequent research (see Clayton & Renvoize, Genera Graminum (1986)).

Tribe Arundineae has been enlarged to incorporate the Danthonieae and becomes tribe XIX. In its original concept, as arranged and keyed out in volume **10** part 1, the tribe Arundineae consisted of a single genus *Phragmites* and was dealt with in volume **10** part 1; the rest of the tribe (i.e. the 'Danthonieae') was dealt with in volume **10** part 2.

Tribe Leptureae becomes No. XXI.

Tribe Chlorideae becomes Tribe Cynodonteae (tribe No. XXII), which now incorporates the Zoysieae (there is no longer a tribe numbered XXIII).

Tribes Paniceae and Arundinellae become Nos. XXIV and XXVI respectively.

Tribe Maydeae is now incorporated in the Andropogoneae (tribe No. XXVII).

The tribes now recognized as being represented in the Flora Zambesiaca area are as follows:

I.	BAMBUSEAE	XIV.	AGROSTIDEAE
II.	PHAREAE	XV.	ARUNDINEAE
III.	OLYREAE	XVI.	ARISTIDEAE
IV.	ORYZEAE	XVII.	CENTOTHECEAE
V.	PHYLLORHACHIDEAE	XVIII.	PAPPOPHOREAE
VI.	EHRHARTEAE	XIX.	ARUNDINEAE (in part)
VII.	POEAE	XX.	ERAGROSTIDEAE
VIII.	BROMEAE	XXI.	LEPTUREAE
IX.	BRACHYPODIEAE	XXII.	CYNODONTEAE
X.	TRITICEAE	XXIV.	PANICEAE
XI.	MELICEAE	XXV.	ISACHNEAE
XII.	AVENEAE	XXVI.	ARUNDINELLEAE
XIII.	PHALARIDEAE	XXVII.	ANDROPOGONEAE

GRAMINEAE

XXVII. ANDROPOGONEAE Dumort.

By T.A. Cope*

Andropogoneae Dumort., Observ. Gramin. Belg.: 84 (1824).

Inflorescence of fragile (or very rarely tough) racemes, these sometimes in a large panicle, but usually single, paired or digitate, terminating the culm or axillary and numerous, in the latter case each true inflorescence subtended by a modified leaf-sheath (spatheole) and often aggregated into a leafy false panicle. Racemes bearing the spikelets in pairs (rarely singly or in threes, but usually terminating in a triad), nearly always with one sessile and the other pedicelled, these sometimes alike but usually dissimilar, the sessile being bisexual and the pedicelled male or barren (very rarely the sexes reversed); occasionally with 1 or more of the lowermost pairs in the raceme (homogamous pairs) alike, infertile and persisting for some time after the other spikelets have fallen. Fertile spikelet usually dorsally compressed, 2-flowered without rhachilla extension, falling entire at maturity with adjacent pedicel and internode (unless the raceme tough). Glumes 2, indurated, the inferior facing outward and very variable in shape and ornamentation, the superior usually boat-shaped and fitting between internode and pedicel. Inferior floret male or barren, the lemma hyaline to membranous and awnless, the palea usually suppressed when floret barren. Superior floret bisexual, its lemma hyaline or membranous, with or without a geniculate awn with spirally twisted column, its palea short or absent. Lodicules 2, fleshy. Stamens usually 3. Stigmas 2. Caryopsis ellipsoid with large embryo; hilum punctiform. Pedicelled spikelet sometimes similar to the sessile, but commonly male to barren, awnless, and smaller or even vestigial (though occasionally large and colourful); rarely the pedicel absent or fused to the internode. Chromosomes small, basic number usually 10.

A tribe comprising 85 genera. Distributed throughout the tropics and extending into warm temperate regions.

Paired spikelets borne on a fragile rhachis are typical of the tribe, but they are often much modified by suppression of sexuality in the pedicelled spikelet, and by transformation of internode and pedicel into a protective shield around the fertile sessile spikelet. Protection of the fertile spikelet may be reinforced by an involucre of homogamous pairs, so that three distinct spikelet morphs are sometimes present. The genera *Cleistachne* and *Oxyrhachis* and some species of *Arthraxon* have no pedicel or pedicelled spikelet; their inclusion in the tribe depends on their similarity to adjacent genera.

Another characteristic of the tribe is its tendency towards reduction of the inflorescence to 1 or 2 short racemes, coupled with axillary branching and modification of the subtending leaf into an inflated laminaless spatheole. The result is a leafy branch system which, when crowded towards the top of the culm, can closely imitate a panicle.

Zea mays L. (maize) is widely cultivated throughout the region. It is monoecious, having a terminal male 'tassel' of digitate racemes, and highly modified female inflorescences taking the form of axillary 'ears'.

Coix lacryma-jobi L. (Job's tears) is occasionally grown as an ornamental. It is monoecious, with the female inflorescence enclosed in a hardened flask-like spatheole from which necklaces and other decorative articles can be made.

---

* With subtribes *Saccharinae* (genera 1 to 8) by W.D. Clayton, *Sorghinae* (genera 9 to 16) by M.P. Setshogo and *Andropogoninae* (genera 21 to 25) by Fatíma Sales.

1. Rhachis internodes and pedicels slender, filiform or linear, rarely thickened upwards and
    then the superior lemma of fertile spikelet awned or the callus fulvously barbate · · · · · ·2
 − Rhachis internodes and pedicels stout, thickening upwards; superior lemma awnless · · 36
2. Superior lemma awned from low down on the back · · · · · · · · · · · · · · · · 24. **Arthraxon**
 − Superior lemma awnless, awned from the apex, or awned from the sinus of a bilobed apex
    ·3
3. Spikelets all alike or nearly so · 4
 − Spikelets of each pair significantly different, sometimes the pedicelled rudimentary or
    represented only by its pedicel · 12
4. Inflorescence a panicle with elongated central axis · 5
 − Inflorescence of 1–many subdigitate racemes · 10
5. Panicle branches distally fragile; one spikelet of each pair sessile · · · · · · · · · · · · · · · 6
 − Panicle branches tough; all spikelets pedicelled · 8
6. Callus hairs rufous or tawny; racemes short, dense · · · · · · · · · · · · · · · · · 2. **Eriochrysis**
 − Callus hairs silvery; racemes long, loose · 7
7. Callus hairs much longer than the spikelet · 1. **Saccharum**
 − Callus hairs shorter than the spikelet · 10. **Sorghastrum**
8. Callus pubescent, the hairs less than $^1/_4$ the length of the spikelet · · · · · · 11. **Cleistachne**
 − Callus conspicuously barbate, the hairs at least $^1/_4$ the length of the spikelet · · · · · · · · · 9
9. Panicle loose, the component racemes distinct · 3. **Miscanthus**
 − Panicle spiciform · 4. **Imperata**
10. Callus acutely conical to pungent; rhachis fragile (if tough see 8. *Trachypogon*) · · · · · · · ·
    · 6. **Homozeugos**
 − Callus obtuse ·11
11. Inferior glume convex or concave, often villous ·5. **Eulalia**
 − Inferior glume medianly grooved, glabrous or ciliate only at the apex (in Flora Zambesiaca
    area) · 7. **Microstegium**
12. Sessile spikelet male or barren (rarely fertile), the pedicelled fertile · · · · 8. **Trachypogon**
 − Sessile spikelet fertile, the pedicelled spikelet male, barren or suppressed · · · · · · · · · 13
13. Inflorescence a panicle with an elongated central axis, not supported by spatheoles · · 14
 − Inflorescence of single, paired or subdigitate racemes, often supported by spatheoles · · 18
14. Sessile spikelets laterally compressed or terete · · · · · · · · · · · · · · · · · · · 12. **Chrysopogon**
 − Sessile spikelets dorsally compressed · 15
15. Pedicels barren · 10. **Sorghastrum**
 − Pedicels all bearing spikelets · 16
16. Rhachis internodes and pedicels solid, without a thinner median line · · · · · · 9. **Sorghum**
 − Rhachis internodes and pedicels with a thinner translucent or balsmiferous, sometimes
    purple-tinged, median line ·17
17. Racemes composed of 1–2 sessile spikelets · 14. **Capillipedium**
 − Racemes composed of 10 or more sessile spikelets · · · · · · · · · · · · · · · 15. **Bothriochloa**
18. Fertile lemma awned from its entire apex · 19
 − Fertile lemma awned from the sinus of its bilobed apex, rarely awnless (lemma entire in
    Andropogon brazzae and *Schizachyrium lopollense*) · 23
19. Racemes composed of a single awned sessile spikelet and 2 pedicelled spikelets, enclosed
    by an involucre of 4 sterile spikelets · 33. **Themeda**
 − Racemes composed of many pairs of spikelets · 20
20. Callus pungent · 32. **Heteropogon**
 − Callus obtuse (but if inflorescence a dense spatheate head see *Cymbopogon densiflorus*) · · 21
21. Rhachis internodes and pedicels solid, without a thinner median line · · 13. **Dichanthium**
 − Rhachis internodes and pedicels with a thinner translucent or balsamiferous, sometimes
    purple-tinged, median line · 22
22. Racemes erect or divergent, without homagamous pairs · · · · · · · · · · · ·15. **Bothriochloa**
 − Racemes nodding, with 1–3 homogamous pairs at the base · · · · · · · · · · · · 16. **Euclasta**
23. Inferior glume of sessile spikelet transversely rugose; pedicelled spikelet represented only
    by a narrow curved pedicel · 18. **Thelepogon**
 − Inferior glume of sessile spikelet smooth, rarely rugose and then the pedicelled spikelet
    well developed · 24
24. Callus of sessile spikelet inserted into the crateriform or cupuliform apex of the internode,
    at least the rim of the internode lapping over and concealing the tip of the callus; inferior
    glume of sessile spikelet 2-keeled or with the margins sharply inflexed and usually

depressed between the keels, rarely the flanks abruptly rounded but then deeply grooved between them · 25

− Callus of sessile spikelet applied obliquely to the apex of the internode with its tip free, usually acute to pungent; inferior glume of sessile spikelet convexly rounded on the back without keels (rarely with a median groove); internodes and pedicels linear · · · · · · · · 31

25. Inferior floret of sessile spikelet male, with a well developed palea · · · · · · · · · · · · · · 26

− Inferior floret of sessile spikelet barren and reduced to a lemma · · · · · · · · · · · · · · · 28

26. Racemes paired, digitate or subdigitate (rarely single but then the superior glume neither awned nor crested) · 17. **Ischaemum**

− Racemes single · 27

27. Glumes inconspicuously winged, the superior awned · · · · · · · · · · · · · · · · · · 19. **Sehima**

− Glumes with a prominent wing-like crest, muticous · · · · · · · · · · · · · · · 20. **Andropterum**

28. Callus of sessile spikelet acute to pungent, 1–5 mm long · · · · · · · · · · · 25. **Diheteropogon**

− Callus of sessile spikelet obtuse, usually very short · 29

29. Racemes single; inferior glume of sessile spikelet with several intercarinal nerves · · · · · ·
· 23. **Schizachyrium**

− Racemes paired or digitate, rarely single and then the inferior glume of the sessile spikelet nerveless between the keels · 30

30. Racemes not deflexed, borne upon unequal terete raceme-bases; leaves not aromatic · · · · ·
· 21. **Andropogon**

− Racemes deflexed at maturity, borne upon subequal flattened raceme-bases, seldom exceeding the spatheole in length; rhachis internodes and pedicels linear; leaves aromatic; panicle dense, decompound · 22. **Cymbopogon**

31. Racemes paired, rarely single, but then pedicelled spikelet without a callus · · · · · · · · 32

− Racemes single; pedicelled spikelet with distinct narrowly oblong to linear callus · · · · 35

32. Inferior glume of sessile spikelet with a median groove · · · · · · · · · · · · · 28. **Hyperthelia**

− Inferior glume of sessile spikelet convex on the back · 33

33. Superior glume of sessile spikelet awned · 29. **Elymandra**

− Superior glume of sessile spikelet awnless · 34

34. Superior raceme-base up to 10 mm long, but usually much shorter · · · · 26. **Hyparrhenia**

− Superior raceme-base 15–25 mm long; homogamous pairs 2 at the base of each raceme, forming an involucre · 27. **Exotheca**

35. Spatheoles linear to narrowly lanceolate; racemes loose, the internode visible · · · · · · · · ·
· 30. **Anadelphia**

− Spatheoles cymbiform; racemes dense, the internodes concealed · · · · 31. **Monocymbium**

36. Pedicels distinct · 37

− Pedicels fused to the internode, rarely absent · 43

37. Inferior glume of sessile spikelet produced into a long flattened tail; spikelets similar · · ·
· 38. **Vossia**

− Inferior glume of sessile spikelet without a herbaceous tail · · · · · · · · · · · · · · · · · · · 38

38. Pedicelled spikelet long-awned (1–12 cm) from the inferior glume, rarely awnless; racemes obliquely jointed and callus without a central peg · · · · · · · · · · · · · · · · 34. **Urelytrum**

− Pedicelled spikelet awnless or with an awnlet up to 5 mm long · · · · · · · · · · · · · · · · 39

39. Racemes numerous on a short common axis; callus with central peg · · · · 37. **Phacelurus**

− Racemes single · 40

40. Jointing of racemes oblique · 41

− Jointing of racemes transverse · 42

41. Callus inserted in the crateriform internode, the node fringed with a ring of hairs; inferior glume of sessile spikelet usually longitudinally ridged, entire · · · · · · · · · · · 35. **Loxodera**

− Callus not inserted in the internode; inferior glume of sessile spikelet not ridged, often bifid · 36. **Elionurus**

42. Inferior glume of sessile spikelet not (or rarely very obscurely) winged · · · 40. **Rhytachne**

− Inferior glume of sessile spikelet conspicuously winged · · · · · · · · · · · · · 41. **Coelorachis**

43. Inferior glume of sessile spikelet smooth · 44

− Inferior glume of sessile spikelet rough · 46

44. Raceme dorsally compressed, tough · 39. **Hemarthria**

− Raceme cylindrical, fragile · 45

45. Pedicelled spikelet present · 42. **Rottboellia**

− Pedicelled spikelet absent · 45. **Oxyrhachis**

46. Sessile spikelet broadly elliptic · 43. **Heteropholis**

− Sessile spikelet globose · 44. **Hackelochloa**

1. SACCHARUM L.

By W.D. Clayton

Saccharum L., Sp. Pl. **1**: 54 (1753).

Inflorescence a panicle, often large and plumose, bearing numerous racemes on its branches; racemes loose, fragile, with slender internodes. Spikelets alike, paired, one sessile, one pedicelled, lanceolate, enveloped in long silky hairs from the callus. Inferior glume membranous or sometimes subcoriaceous below, flat or rounded on the back. Inferior floret represented by a short hyaline lemma. Superior lemma entire or bidentate, awned or awnless, sometimes almost suppressed; stamens 2–3.

A genus of 35–40 species. Throughout the tropics and subtropics.

Saccharum officinarum L. (sugar cane) is widely cultivated on moist soils. It can be distinguished by its glabrous or pubescent panicle axis, and leaf laminas up to 4 cm wide. Its origins are discussed by Stevenson, Genetics and Breeding of Sugar Cane (1965).

Saccharum spontaneum L., Mant. Pl. **2**: 183 (1771). —Panje in Indian J. Agric. Sci. **3**: 1013 (1933). —Artschwager, Techn. Bull. U.S.D.A. No. 811 (1942). —Bor, Grasses Burma Ceyl. Ind. Pak.: 214 (1960). Type from India.

Rhizomatous perennial. Culms 200–400 cm high or more. Leaf laminas 5–15(40) mm wide. Panicle 25–60 cm long, the axis hirsute; racemes 3–15 cm long, usually much longer than supporting branches, the internodes and pedicels hirsute. Spikelets 3.5–7 mm long, the callus barbate with silky white hairs 2–3 times as long as the spikelet. Glumes subcoriaceous in the lower third, glabrous on the back, ciliate on the margins above. Superior lemma subulate and up to 3 mm long, or suppressed.

Subsp. **aegyptiacum** (Willd.) Hack. in A. & C. de Candolle, Monogr. Phan. **6**: 115 (1889). — Clayton & Renvoize in F.T.E.A., Gramineae: 704 (1982). TAB. **1**. Type from Egypt.
   Saccharum biflorum Forssk., Fl. Aegypt.-Arab.: 16 (1775). —Jackson & Wiehe, Annot. Check List Nyasal. Grass.: 56 (1958). Type from Egypt.
   Saccharum aegyptiacum Willd., Enum. Pl. **1**: 82 (1809). Type as for subsp. *aegyptiacum*.
   Saccharum spontaneum var. *aegyptiacum* (Willd.) Hack. in A. & C. de Candolle, Monogr. Phan. **6**: 115 (1889). —Stapf in Prain, F.T.A. **9**: 95 (1917). —Clayton in F.W.T.A., ed. 2, **3**: 466 (1972).
   Saccharum spontaneum subsp. *biflorum* (Forssk.) Pilg. in Fries, Wiss. Ergebn. Schwed. Rhod.-Kongo-Exped. **1**: 191 (1916). Type as for *S. biflorum*.

Lamina extending to the base of the leaf midrib; ligule crescent-shaped.

Malawi. N: Karonga Distr., Rukuru River, 8.vi.1954, *Jackson* 1332 (K).
Northwards to the Mediterranean region and Syria. River banks.
Subsp. *spontaneum*, from tropical and warm temperate Asia, has the lamina more or less suppressed towards the base of the leaf midrib, and the ligule is triangular.

2. ERIOCHRYSIS P. Beauv.

By W.D. Clayton

Eriochrysis P. Beauv., Ess. Agrostogr.: 8, fig. 4/11 (1812).

Inflorescence a panicle, bearing short racemes appressed to its central axis; racemes dense, fragile, with clavate internodes. Spikelets alike (though pedicelled female and slightly smaller), paired, one sessile, one pedicelled, elliptic, barbate with rufous or tawny hairs from the callus. Inferior glume chartaceous to coriaceous, broadly convex, copiously ciliate on the margins. Inferior floret represented by a hyaline lemma. Superior lemma entire, awnless; stamens 3.

A genus of 7 species. Africa, tropical America and India.

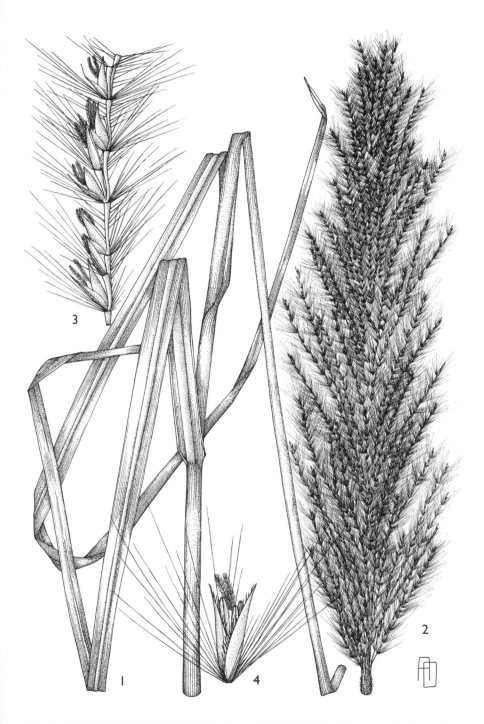

Tab. 1. SACCHARUM SPONTANEUM subsp. AEGYPTIACUM. 1, leaves (× ¹/₂), from *Grassl* 46–47; 2, inflorescence (× ¹/₂); 3, portion of raceme (× 3); 4, spikelet (× 5), 2–4 from *Greenway* 8458. Drawn by Ann Davies. From F.T.E.A.

1. Callus hairs exceeding the spikelet; inferior glume thinly coriaceous · · · · · · · · 1. *pallida*
 - Callus hairs shorter than the spikelet ·2
2. Inferior glume coriaceous, the nerves obscure · 2. *brachypogon*
 - Inferior glume thinly chartaceous, the nerves conspicuous · · · · · · · · · · · · · 3. *purpurata*

1. **Eriochrysis pallida** Munro in Harvey, Gen. S. Afr. Pl., ed. 2: 440 (1868). —Stapf in Prain, F.T.A. **9**: 93 (1917). —Stent & Rattray in Proc. & Trans. Rhodesia Sci. Assoc. **32**: 7 (1933), in part. —Sturgeon in Rhodesia Agric. J. **51**: 14 (1954). —Chippindall in Meredith, Grasses & Pastures S. Africa: 475 (1955). —Jackson & Wiehe, Annot. Check List Nyasal. Grass.: 41 (1958). —Vesey-FitzGerald in Kirkia **3**: 102 (1963). —Simon in Kirkia **8**: 20, 54 (1971). — Clayton in F.W.T.A., ed. 2, **3**: 466 (1972). —Clayton & Renvoize in F.T.E.A., Gramineae: 706 (1982). Type from South Africa.
   Saccharum pallidum (Munro) Benth. in J. Linn. Soc., Bot. **19**: 66 (1881).
   Saccharum munroanum Hack. in A. & C. de Candolle, Monogr. Phan. **6**: 124 (1889) *nom. superfl.*, based on *Eriochrysis pallida*.
   Eriochrysis munroana (Hack.) Pilg. in Notizbl. Bot. Gart. Berlin-Dahlem **11**: 648 (1932) *nom. superfl.*, based on *E. pallida*.

Caespitose perennial. Culms 30–90 cm high. Inflorescence 3–14 cm long, linear to oblong, of 3–14 racemes each 1–4 cm long. Sessile spikelet 3.5–6 mm long, its callus barbate with tawny hairs longer than the spikelet. Inferior glume thinly coriaceous, shining, its nerves indistinct.

Botswana. N: Moanachira River, 18.vi.1979, *P.A. Smith* 2792 (K; PRE). **Zambia.** B: Mongu Distr., Namushakende, 22.v.1964, *Verboom* 1197 (K). N: Mbala Distr., Ndundu, 1700 m, 13.ii.1959, *McCallum Webster* A18 (K). W: Mwinilunga Distr., SW of Dobeka Bridge, 13.x.1937, *Milne-Redhead* 2747 (K; PRE). C: Rufunsa to Lusaka, 7.ix.1947, *Greenway & Brenan* 8070 (K). **Zimbabwe.** N: Guruve Distr., Nyamunyeche Estate, 18.x.1978, *Nyariri* 428 (K; PRE). W: Matobo Distr., Besna Kobila Farm, 1460 m, iv.1957, *Miller* 4340 (K; PRE). C: Chirumanzu Distr., Mvuma (Umvuma), 15.ii.1971, *Chiparawasha* 347 (K). E: Mutasa Distr., Penhalonga, 1220 m, 25.i.1970, *Crook* 907 (K; PRE). S: Masvingo Distr., Great Zimbabwe National Park, 28.iii.1973, *Chiparawasha* 658 (K; PRE). **Malawi.** N: Mzimba Distr., Mzuzu, Katoto, 1370 m, 24.x.1969, *Pawek* 2915 (K). C: Mchinji Distr., 5 km NE of Mchinji on road to Kasungu, 1215 m, 27.iv.1970, *Brummitt* 10212 (K; MAL).
Also in Tanzania, Angola and South Africa. Swampy places; 1000–1700 m.

2. **Eriochrysis brachypogon** (Stapf) Stapf in Prain, F.T.A. **9**: 93 (1917). —Sturgeon in Rhodesia Agric. J. **51**: 14 (1954). —Vesey-FitzGerald in Kirkia **3**: 102 (1963). —Simon in Kirkia **8**: 20, 54 (1971). —Clayton in F.W.T.A., ed. 2, **3**: 466 (1972). —Clayton & Renvoize in F.T.E.A., Gramineae: 707, fig. 162 (1982). TAB. **2**. Types from Mali, Nigeria and Central African Republic.
   Saccharum brachypogon Stapf in Mém. Soc. Bot. France **8**: 97 (1908).
   Eriochrysis phaenostachys Pilg. in Notizbl. Bot. Gart. Berlin-Dahlem **11**: 648 (1932). Type from Tanzania.
   Eriochrysis brachypogon subsp. *australis* J.G. Anderson in Bothalia **8**: 170 (1964). Type from Swaziland.
   Eriochrysis pallida sensu Stent & Rattray in Proc. & Trans. Rhodesia Sci. Assoc. **32**: 7 (1933), in part, non Munro.

Like *E. pallida*, but callus barbate with rufous hairs up to half the length of the spikelet and inferior glume coriaceous, shining, the nerves obscure.

Zambia. B: Senanga Distr., Kataba Valley, 19.xii.1964, *Verboom* 1554 (K; SRGH). N: Kasama Distr., Chambeshi River near Mwuishi (Mwarushi), 1320 m, 27.ii.1960, *Richards* 12625 (K; SRGH). **Zimbabwe.** C: Harare (Salisbury), 1460 m, 27.ii.1927, *Eyles* 4706 (K). E: Mutasa Distr., Farm Charity, 29.vi.1934, *Gilliland* 536 (K; PRE). **Malawi.** N: Mzimba Distr., Mzuzu, Lunyangwa River waterworks, 1370 m, 30.i.1975, *Pawek* 9024 (K; PRE).
Widely distributed in tropical and South Africa. Swampy grassland; 1300–1700 m.

3. **Eriochrysis purpurata** (Rendle) Stapf in Prain, F.T.A. **9**: 92 (1917). —Jackson & Wiehe, Annot. Check List Nyasal. Grass.: 41 (1958). —Simon in Kirkia **8**: 54 (1971). —Clayton & Renvoize in F.T.E.A., Gramineae: 707 (1982). Types: Malawi, Mulanje, *Whyte* 8 (K, isosyntype) and without locality, *Buchanan* 977 (K, isosyntype).
   Saccharum purpuratum Rendle in Trans. Linn. Soc. London, Bot. **4**: 56 (1894).
   Eriochrysis brachypogon sensu Simon in Kirkia **8**: 20 (1971), non (Stapf) Stapf.

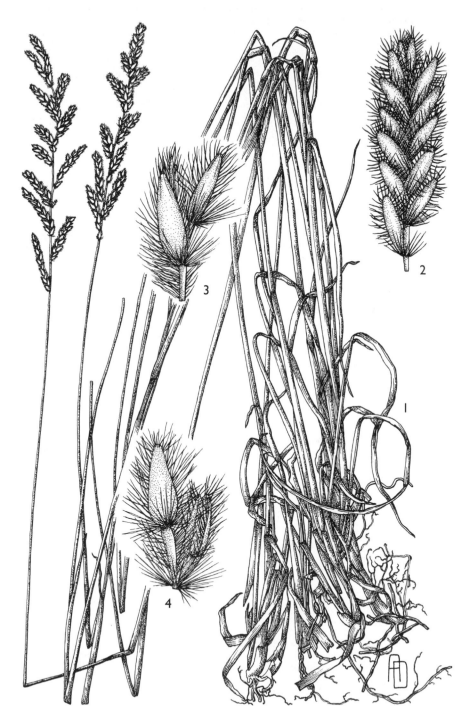

Tab. 2. ERIOCHRYSIS BRACHYPOGON. 1, habit (× ²⁄₃); 2, raceme (× 4); 3 & 4, spikelet pair, front and rear views (× 8), 1–4 from *Thomas* 4500. Drawn by Ann Davies. From F.T.E.A.

Like *E. pallida*, but spikelets 5–7 mm long, callus barbate with tawny hairs $\frac{1}{3}$–$\frac{1}{2}$ the length of the spikelet, and inferior glume thinly chartaceous, dull, the nerves visible.

Zambia. N: Kawambwa Distr., Chishinga Ranch, 13.ix.1961, *Astle* 915 (K). W: 16 km east of Mwinilunga Boma, 9.ix.1930, *Milne-Redhead* 1068 (K; PRE). C: Mkushi Distr., Mkushi River, 1430 m, 14.x.1967, *Simon & Williamson* 980 (K; PRE; SRGH). **Zimbabwe**. E: Nyanga Distr., Pungwe Gorge, 26.x.1946, *Rattray* 1033 (K). **Malawi**. N: Rumphi Distr., Nyika Plateau, Lake Kaulime, 2340 m, 16.v.1970, *Brummitt* 10804 (K; MAL; SRGH). S: Mulanje Distr., Mt. Mulanje (Mlanje), Lichenya Plateau, 1920 m, 16.x.1941, *Greenway* 6305 (K; PRE).

Also in Angola and Tanzania. Swampy places; 1400–2340 m.

3. MISCANTHUS Andersson

By W.D. Clayton

Miscanthus Andersson in Öfvers. Förh. Kongl. Svenska Vetensk.-Acad. **1855**: 165 (1856).
Miscanthidium Stapf in Prain, F.T.A. **9**: 89 (1917).

Inflorescence a panicle (but subdigitate in Asia), often large and plumose, bearing numerous racemes on its branches; raceme rhachis tough with slender internodes. Spikelets alike, paired, both pedicelled, lanceolate to narrowly oblong, barbate from the callus. Inferior glume thinly coriaceous, broadly convex. Inferior floret represented by a hyaline lemma. Superior lemma entire or bidentate, with or without a short awn; stamens 2–3.

A genus of c. 20 species. Africa and SE Asia.

1. Leaf lamina terete, circular in section · 1. *junceus*
 – Leaf lamina expanded, though narrowing to the midrib towards the base · · · · · · · · · · 2
2. Callus beard $\frac{1}{4}$–$\frac{1}{2}$ the length of the spikelet, less prominent than the pilose glumes · 2. *sorghum*
 – Callus beard $\frac{1}{2}$ as long to as long as the spikelet, hairier than the glabrous to thinly pilose glumes · 3. *violaceus*

1. **Miscanthus junceus** (Stapf) Pilg. in Engler & Prantl, Nat. Pflanzenfam., ed. 2, **14e**: 113 (1940). Types from South Africa and Lesotho.
　　Erianthus junceus Stapf in Dyer, F.C. **7**: 324 (1898).
　　Cleistachne teretifolia Hack. in Oesterr. Bot. Z. **51**: 153 (1901). Type from Angola.
　　Erianthus teretifolius Stapf in J. Linn. Soc., Bot. **37**: 478 (1906) non *Cleistachne teretifola* Hack. Type: Zimbabwe, Victoria Falls, *Gibbs* 141 (K, holotype).
　　Erianthus teretifolius Pilg. in Bot. Jahrb. Syst. **39**: 597 (1907) non Stapf (1906).
　　Miscanthidium teretifolium (Stapf) Stapf in Prain, F.T.A. **9**: 89 (1917). —Stent & Rattray in Proc. & Trans. Rhodesia Sci. Assoc. **32**: 7 (1933), in part. —Sturgeon in Rhodesia Agric. J. **51**: 12 (1954). —Chippindall in Meredith, Grasses & Pastures S. Africa: 480 (1955). Type as for *E. teretifolius*.
　　Miscanthidium gossweileri Stapf in Prain, F.T.A. **9**: 90 (1917). Type as for *Cleistachne teretifolia*.
　　Miscanthidium junceum (Stapf) Stapf in Hooker's Icon. Pl. **31**: t. 3084 (1922). —Sturgeon in Rhodesia Agric. J. **51**: 13 (1954). —Chippindall in Meredith, Grasses & Pastures S. Africa: 480 (1955). —Launert in Merxmüller, Prodr. Fl. SW. Afrika, fam. 160: 131 (1970). Type as for *Erianthus junceus*.
　　Miscanthus gossweileri (Stapf) Pilg. in Engler & Prantl, Nat. Pflanzenfam., ed. 2, **14e**: 113 (1940). Type as for *Cleistachne teretifolia*.
　　Miscanthus teretifolius (Stapf) Pilg., loc. cit. Type as for *Erianthus teretifolius*.

Caespitose perennial. Culms 100–300 cm high. Leaf laminas 1–3 mm in diameter, terete, the midrib represented by a slender yellow line on the adaxial surface. Panicle 20–50 cm long, linear to narrowly ovate, bearing racemes 2–8 cm long. Spikelets 3–5 mm long; callus beard $\frac{1}{4}$–$\frac{1}{3}$ the length of the spikelet. Inferior glume pilose to villous. Superior lemma bidentate, with a weakly geniculate awn 2–10 mm long.

Botswana. N: Dabonga Is., 29.vi.1973, *P.A. Smith* 656 (K; PRE). **Zambia**. B: Kaoma Distr., Mangango Mission, 17.iv.1964, *Verboom* 1186 (K). N: 3 km from Samfya, 1160 m, 17.iv.1963, *Symoens* 10299 (K). C: c. 64 km south of Mpika, Pagastomo Village, 1400 m, 4.v.1954, *Hinds* 232 (K). S: Kalomo Distr., Ibula, 1220 m, iii.1934, *Trapnell* 1493 (K). **Zimbabwe**. N: Gokwe Distr.,

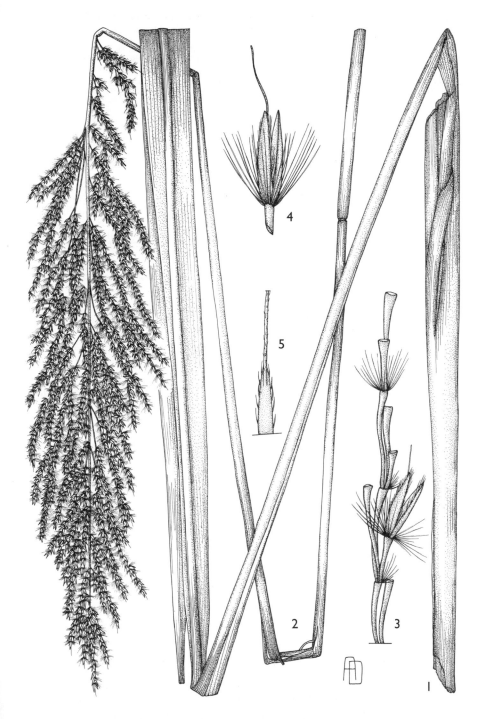

Tab. 3. MISCANTHUS VIOLACEUS. 1, leaf (× ¹⁄₂), from *Bogdan* 3451; 2, inflorescence (× ¹⁄₂), from *Haarer* 2467; 3, portion of raceme (× 5); 4, spikelet (× 7); 5, tip of superior lemma (× 8), 3–5 from *Johnston* s.n. Drawn by Ann Davies. From F.T.E.A.

Sengwa Research Station, 19.iii.1976, *P.R. Guy* 2410 (K; PRE). W: Hwange Distr., Kazuma Range, Kasetsheti (Katsatetsi) R., 1000 m, 10.v.1972, *Simon* 2193 (K; PRE).

Also in Dem. Rep. Congo (Shaba), Angola and South Africa. A waterside reed; 700–1400 m.

Specimens from the northern part of the range have been separated as *M. teretifolius*. They have hairier spikelets than those from South Africa, but variation appears to be continuous.

2. **Miscanthus sorghum** (Nees) Pilg. in Engler & Prantl, Nat. Pflanzenfam., ed. 2, **14e**: 113 (1940). Type from South Africa.

    Erianthus sorghum Nees, Fl. Afr. Austral. Ill.: 92 (1841).

    Miscanthidium sorghum (Nees) Stapf in Prain, F.T.A. **9**: 91 (1917). —Chippindall in Meredith, Grasses & Pastures S. Africa: 478 (1955).

    Miscanthidium teretifolium sensu Stent & Rattray in Proc. & Trans. Rhodesia Sci. Assoc. **32**: 4 (1933), in part, non (Stapf) Stapf.

    Miscanthidium junceum sensu Simon in Kirkia **8**: 20 (1971), non (Stapf) Pilg.

Caespitose perennial. Culms 150–400 cm high. Leaf laminas 4–15(25) mm wide, narrowing to the midrib near the base. Panicle 20–40 cm long, narrowly oblong to narrowly ovate, bearing racemes 1–7 cm long. Spikelets 3–6 mm long; callus beard $\frac{1}{4}$–$\frac{1}{3}$ the length of the spikelet, less prominent than the glume hairs. Inferior glume densely pilose. Superior lemma entire and awnless, or bidentate and bearing a straight or weakly geniculate awn up to 7 mm long.

    Zambia. S: Kalomo, 8.iv.1946, *Grassl* 46–64 (K). **Zimbabwe**. C: Shurugwi Distr., Gwenoro Dam, 19.i.1968, *Biegel* 2607 (K). E: Chimanimani Distr., 1310 m, 7.iii.1953, *Chase* 4825 (K; SRGH). S: Masvingo Distr., Great Zimbabwe, Oatlands Farm, 12.iv.1973, *Chiparawasha* 704 (K; PRE).

Also in Angola and South Africa. Streamsides; 1100–1350 m.

Perhaps not distinct from the South African *M. ecklonii* (Nees) Mabb. (= *M. capensis* (Nees) Andersson) which has the leaf lamina expanded down to the base.

3. **Miscanthus violaceus** (K. Schum.) Pilg. in Engler & Prantl, Nat. Pflanzenfam., ed. 2, **14e**: 113 (1940). —Clayton & Renvoize in F.T.E.A., Gramineae: 702, fig. 160 (1982). TAB. **3**. Type from Tanzania.

    Erianthus violaceus K. Schum. in Engler, Pflanzenw. Ost-Afrikas **C**: 96 (1895).

    Erianthus flavescens K. Schum. in Engler, Pflanzenw. Ost-Afrikas **C**: 96 (1895). Type from Tanzania.

    Miscanthidium violaceum (K. Schum.) Robyns, Fl. Agrost. Congo Belge **1**: 82 (1929). Type as for *E. violaceus*.

Like *M. sorghum* but the callus beard half as long to as long as the spikelet, hairier than the glumes which are glabrous to thinly pilose.

    Zambia. N: Mbala Distr., road to Kalambo Falls, 28.iii.1957, *Richards* 8910 (K).

Northwards to Uganda and Kenya. Stream and lake sides; 1000–1350 m.

Barely distinct from *M. sorghum*, with signs of introgression in Zambia.

4. IMPERATA Cirillo

By W.D. Clayton

Imperata Cirillo, Pl. Rar. Neapol. **2**: 26 (1792).

Inflorescence a silky spiciform panicle bearing numerous racemes, but these much abbreviated and not easily distinguished; raceme rhachis tough with slender internodes. Spikelets alike, paired, both pedicelled, terete, enveloped in long white hairs from the callus and glumes. Inferior floret represented by a hyaline lemma. Superior floret entire, awnless; lodicules 0; stamens 1–2.

A genus of 8 species. Tropical and warm temperate regions.

Imperata cylindrica (L.) Raeusch., Nomencl. Bot., ed. 3,: 10 (1797). —Stapf in Prain, F.T.A. **9**: 87 (1917). —C.E. Hubbard in Joint Publ. Imp. Agric. Bur. **7**: 5–13 (1944). —Launert in Merxmüller, Prodr. Fl. SW. Afrika, fam. 160: 122 (1970). —Simon in Kirkia **8**: 20, 54 (1971). —Clayton & Renvoize in F.T.E.A., Gramineae: 700, fig. 159 (1982). TAB. **4**. Type from Europe.

Tab. 4. IMPERATA CYLINDRICA.   1, habit (× ⅔), from *Richards* 26175; 2, portion of inflorescence axis (× 10); 3, spikelet (× 6), 2 & 3 from *Bruce* 5550. Drawn by Ann Davies. From F.T.E.A.

Lagurus cylindricus L., Syst. Nat., ed. 10, **2**: 878 (1759).
Saccharum koenigii Retz., Observ. Bot. **5**: 16 (1789). Type from Japan.
Saccharum thunbergii Retz., Observ. Bot. **5**: 17 (1789). Type from East Indies.
Imperata arundinacea Cirillo, Pl. Rar. Neapol. **2**: 27 (1792). Type from Italy.
Imperata koenigii var. *major* Nees, Fl. Afr. Austral. Ill.: 90 (1841). Type from South Africa.
Imperata arundinacea var. *africana* Andersson in Öfvers. Förh. Kongl. Svenska Vetensk.-Akad. **12**: 159 (1855). Type as for *Saccharum thunbergii*.
Imperata cylindrica var. *thunbergii* (Retz.) T. Durand & Schinz, Consp. Fl. Afric. **5**: 693 (1894). —Stapf in Prain, F.T.A. **9**: 88 (1917).
Imperata arundinacea var. *thunbergii* (Retz.) Stapf in Dyer, F.C. **7**: 320 (1898). —Stent & Rattray in Proc. & Trans. Rhodesia Sci. Assoc. **32**: 6 (1933).
Imperata angolensis Fritsch in Bull. Herb. Boissier, sér. 2, **1**: 1096 (1901). Type from Angola.
Imperata cylindrica var. *koenigii* (Retz.) Pilg. in Perkins, Fragm. Fl. Philipp.: 137 (1904). —Stapf in Prain, F.T.A. **9**: 88 (1917). —Stent & Rattray in Proc. & Trans. Rhodesia Sci. Assoc. **32**: 7 (1933). —Jackson & Wiehe, Annot. Check List Nyasal. Grass.: 46 (1958). Type as for *Saccharum koenigii* Retz.
Imperata dinteri Pilg. in Bot. Jahrb. Syst. **48**: 342 (1912). Type from Namibia.
Imperata cylindrica var. *major* (Nees) C.E. Hubb., Grasses Maur. Rodrig.: 96 (1940). —Chippindall in Meredith, Grasses & Pastures S. Africa: 477 (1955). —Bor, Grasses Burma Ceyl. Ind. Pak.: 169 (1960). Type as for *I. koenigii* var. *major*.
Imperata cylindrica var. *africana* (Andersson) C.E. Hubb. in Joint Publ. Imp. Agric. Bur. **7**: 10 (1944). —Sturgeon in Rhodesia Agric. J. **51**: 12 (1954). —Clayton in F.W.T.A., ed. 2, **3**: 464 (1972). Type as for *Saccharum thunbergii*.

Perennial, forming tufts of stiffly erect leaves from a scaly rhizome. Culms 10–120 cm high. Panicle 3–22 cm long, 1–3 cm wide, spiciform but occasionally the lowest branches loose. Spikelets 2.2–6 mm long, obscured in copious silky white hairs; stamens 2.

Botswana. N: Okavango Delta, Mboma Camp, 11.i.1973, *P.A. Smith* 339 (K; PRE). **Zambia**. B: Mongu, 29.iv.1964, *Verboom* 1187 (K; PRE). N: Kasama Distr., Chishimba Falls, 15.x.1960, *E.A. Robinson* 3991 (K). W: 64 km south of Mwinilunga, 13.viii.1930, *Milne-Redhead* 892 (K; PRE). C: 8 km west of Serenje, 14.x.1967, *Simon & Williamson* 990 (K; PRE; SRGH). S: Namwala, 16.xii.1962, *van Rensburg* 1102 (K). **Zimbabwe**. N: Mount Darwin Distr., Kandeya C.L. (Nat. Res.), 17.i.1960, *Phipps* 2286 (K; PRE). W: Hwange Distr., Kazuma Range, 1000 m, 11.v.1972, *Simon* 2212 (K; PRE). C: Harare Distr., Umwindsi River bridge, Borrowdale, 1400 m, 30.x.1965, *Simon* 484 (K; PRE; SRGH). E: Mutare Distr., Old Umtali Mission, 1400 m, 20.x.1968, *Crook* 829 (K; PRE). **Malawi**. N: Rumphi Distr., Mbuzinandi (Businande), 1830 m, 28.xii.1975, *E. Phillips* 769 (K). C: 3 km north of Nkhotakota (Nkhota Kota), 480 m, 16.vi.1970, *Brummitt* 11430 (K; MAL; SRGH). S: Zomba Mt., 21.viii.1950, *Jackson* 137 (K). **Mozambique**. N: Macomia Distr., Messalo (Msalu) River, 20.iii.1912, *C.E.F. Allen* 124 (K). Z: Mopeia, 20.x.1974, *Barnes* J380 (K). T: Angónia Distr., Ulónguè, 2.xii.1980, *Macuácua* 1373 (K; LISC; PRE). MS: Beira, 25.xii.1911, *Rogers* 4547 (K). GI: Xai-Xai Distr., Inhamissa, 1.xii.1957, *Macêdo* 48 (K; LISC). M: Marracuene, 2.x.1957, *Barbosa & Lemos* 7934 (K; LISC).

Extends through the Old World tropics and northwards to the Mediterranean region and SW Asia; also in Chile. Locally frequent in seasonally wet places (vleis) and streamsides. Aggressive weed of disturbed and cultivated places; 0–2000 m.

A number of geographical races can be discerned, but they overlap too much to justify formal taxonomic recognition.

5. EULALIA Kunth

By W.D. Clayton

Eulalia Kunth, Révis. Gramin. **1**: 160 (1829).

Inflorescence of 1–many subdigitate racemes; racemes fragile, conspicuously pilose, with slender internodes. Spikelets alike, paired, one sessile, one pedicelled, dorsally compressed, the callus obtuse, not long-barbate. Inferior glume cartilaginous to subcoriaceous, the back flat and usually nerveless, sharply inflexed on the flanks or becoming 2-keeled above, acute to obtuse or truncate, rarely biaristulate. Superior glume usually awnless. Inferior floret represented by a lemma or sometimes suppressed. Superior lemma linear to cordate, bidentate to bifid, usually with a glabrous awn.

A genus of c. 30 species, occurring in the Old World tropics.

Pogonatherum paniceum (Lam.) Hack., a delicate trailing perennial from tropical Asia allied to *Eulalia*, is sometimes grown as an ornamental (dwarf bamboo). It has single racemes of laterally compressed spikelets which are slenderly awned from the superior glume as well as the superior lemma.

1. Racemes with rusty-brown to golden hairs; spikelets 3.5–4.5 mm long · · · · · · · · · 1. *aurea*
 – Racemes with white or violet hairs; spikelets 5–8 mm long · 2
2. Inferior glume concave on the back, with 2 intercarinal nerves · · · · · · · · · · · · · 2. *villosa*
 – Inferior glume flat or subconvex on the back, with 5–7 intercarinal nerves · · 3. *polyneura*

1. **Eulalia aurea** (Bory) Kunth, Révis. Gramin. **1**: 359 (1830). —Clayton & Renvoize in F.T.E.A., Gramineae: 713 (1982). Type from Réunion.
 Andropogon aureus Bory, Voy. Îles Afrique **1**: 367 (1804).
 Eulalia ferruginea Stapf in Prain, F.T.A. **9**: 100 (1917). Type from Tanzania.
 Eulalia geniculata Stapf in Prain, F.T.A. **9**: 101 (1917). —Stent & Rattray in Proc. & Trans. Rhodesia Sci. Assoc. **32**: 7 (1933). —Sturgeon in Rhodesia Agric. J. **51**: 14 (1954). — Chippindall in Meredith, Grasses & Pastures S. Africa: 485 (1955). —Jackson & Wiehe, Annot. Check List Nyasal. Grass.: 41 (1958). —Launert in Merxmüller, Prodr. Fl. SW. Afrika, fam. 160: 116 (1970). —Simon in Kirkia **8**: 20, 54 (1971). Types: Zimbabwe, Harare, *Craster* 18 (K, syntype), *Rogers* 4088 (K, syntype); Bulawayo, *Eyles* 1137 (SRGH, syntype); also from Dem. Rep. Congo.
 Pollinia homblei De Wild., Notes Fl. Katanga **5**: 10 (1920). Types from Dem. Rep. Congo.
 Eulalia elata Peter, Fl. Deutsch Ost-Afrika **1**, Anh.: 117 (1936). Type from Tanzania.
 Eulalia elata var. *hirsuta* Peter, loc. cit. Type from Tanzania.
 Pogonatherum aureum (Bory) Roberty in Boissiera **9**: 391 (1960). Type as for *Andropogon aureus*.

Perennial. Culms 50–150 cm high, ascending from a tuft or geniculately decumbent and mat-forming. Leaf laminas 3–6 mm wide, glabrous, acuminate to a filiform apex. Inflorescence of 2–4(15) racemes each 3–14 cm long, villous with rusty-brown to golden hairs. Spikelets 3.5–4.5 mm long, narrowly elliptic-oblong. Inferior glume convex on the back, with 2–3 faint intercarinal nerves, brown, fulvously pilose, truncate. Inferior lemma reduced to a little scale or absent. Superior lemma bifid for $\frac{1}{3}$ its length, with or without a puberulous awn up to 20 mm long.

Botswana. N: Okavango Swamp, NE of Gwetshaa, 22.ii.1973, *P.A. Smith* 412 (K; PRE). **Zambia**. B: Mongu, 6.i.1966, *E.A. Robinson* 6772 (K; SRGH). N: Mpika Distr., Chibwa Swamp, 11.xi.1969, *Verboom* 2537 (K). C: Lusaka Distr., Chongwe River, west of Kasisi Mission, 1150 m, 3.xii.1972, *Strid* 2638 (K). E: Chipata (Fort Jameson) to Katete, 900 m, 9.i.1959, *Robson* 1119 (K; PRE). S: Choma Distr., Simansunda, 3 km east of Mapanza, 1070 m, 3.i.1954, *E.A. Robinson* 422 (K). **Zimbabwe**. N: 30 km west of Gokwe, 28.i.1964, *Bingham* 1119 (K; PRE). W: Matobo Distr., Two Tree Kop, 12.iv.1951, *West* 3229 (K). C: Harare (Salisbury), c. 1460 m, 25.i.1931, *Brain* 4941 (K; PRE). **Malawi**. N: Mzimba Distr., Mbawa Experiment Station, 17.x.1952, *Jackson* 982 (K; PRE). C: Kasungu National Park, 1000 m, 23.xii.1970, *Hall-Martin* 1354 (K). **Mozambique**. MS: Sussundenga Distr., entre Dombe e Matarara do Lucite, 21.x.1953, *Pedro* 4356 (K; PRE).
 Extends from Uganda and Kenya to South Africa; also in Réunion and Australia. Seasonal swamps; 150–1500 m.

2. **Eulalia villosa** (Thunb.) Nees, Fl. Afr. Austral. Ill.: 91 (1841). —Stapf in Prain, F.T.A. **9**: 99 (1917). —Sturgeon in Rhodesia Agric. J. **51**: 14 (1954). —Chippindall in Meredith, Grasses & Pastures S. Africa: 485 (1955). —Jackson & Wiehe, Annot. Check List Nyasal. Grass.: 41 (1958). —Simon in Kirkia **8**: 20, 54 (1971). —Clayton & Renvoize in F.T.E.A., Gramineae: 713, fig. 165 (1982). TAB. **5**. Type from South Africa.
 Andropogon villosus Thunb., Prodr. Pl. Cap. **1**: 20 (1794).
 Pogonatherum villosum (Thunb.) Roberty in Boissiera **9**: 383 (1960).

Caespitose perennial. Culms 40–120 cm high, erect. Leaf laminas 3–8 mm wide, appressedly pubescent, acute. Inflorescence of 3–12 racemes each 5–12 cm long, silky with white or violet hairs. Spikelets 5–7 mm long, lanceolate. Inferior glume shallowly concave on the back with 2 intercarinal nerves, light brown, glabrous or sparsely villous, ciliate on the keels above, minutely truncate. Inferior lemma 4–7 mm long, membranous below, hyaline above. Superior lemma bifid for $\frac{1}{3}$–$\frac{1}{2}$ its length, with an awn 10–20 mm long.

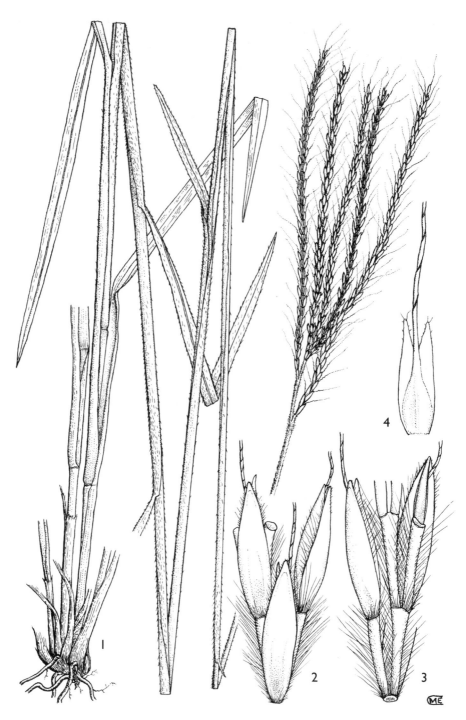

Tab. 5. EULALIA VILLOSA. 1, habit (× ²/₃); 2, portion of raceme, front view (× 6); 3, portion of raceme, rear view (× 6); 4, superior lemma (× 10), 1–4 from *Milne-Redhead & Taylor* 8948. Drawn by M.E. Church. From F.T.E.A.

Zambia. N: Mbala Distr., Ndundu, 1770 m, 23.ii.1959, *McCallum Webster* A134 (K). W: Mwinilunga Distr., River Lwau (Luao), 28.xii.1937, *Milne-Redhead* 3852 (K; PRE). **Zimbabwe**. E: Nyanga Distr., Pungwe Gorge, 1860 m, 6.ii.1966, *Crook* 778 (K; PRE). **Malawi**. N: Mzimba Distr., Mzuzu, 2 km north of Marymount, 1370 m, 28.iii.1971, *Pawek* 4545 (K). S: Mwanza Distr., Neno Hills, 1520 m, 1.vii.1949, *Wiehe* 158 (K). **Mozambique**. Z: Gurué, c. 5 km from the Falls, c. 1300 m, 21.ii.1966, Torre & Correia 14754 (LISC). MS: Gorongosa Distr., Gorongosa Mt., Pico Gogogo, 21.iv.1971, *Tinley* 2086 (K).
Extends from Ethiopia to South Africa; also in Madagascar and India. Wet grassland; 1300–1900 m.
E. villosa is barely distinct from the widespread Asiatic species *E. quadrinervis* (Hack.) Kuntze, which has narrower leaf laminas with filiform apex.

3. **Eulalia polyneura** (Pilg.) Stapf in Prain, F.T.A. **9**: 99 (1917). —Clayton & Renvoize in F.T.E.A., Gramineae: 715 (1982). Type from Kenya.
    Pollinia polyneura Pilg. in Bot. Jahrb. Syst. **39**: 597 (1907).

Caespitose perennial. Culms 50–100 cm high, erect. Leaf laminas 1–4 mm wide, often involute, glabrous or pubescent, acuminate. Inflorescence of 1–5 racemes each 7–12 cm long, villous with white or violet hairs. Spikelets 6–8 mm long, narrowly lanceolate. Inferior glume flat or broadly convex on the back with 5–7 intercarinal nerves, brown, sparsely villous, minutely truncate. Inferior lemma 5–6 mm long, hyaline. Superior lemma bifid for $^1/_3$–$^1/_2$ its length, with an awn 10–15 mm long.

Zambia. N: Chinsali Distr., Shiwa Ngandu, 1550 m, 19.ix.1938, *Greenway & Trapnell* 5716A (K). **Mozambique**. T: Angónia Distr., near Vila Coutinho, 9.v.1948, *Mendonça* 4154 (COI; LISC). Northwards to Ethiopia. Swampy places; 1300–1550 m.

6. HOMOZEUGOS Stapf

By W.D. Clayton

Homozeugos Stapf in Hooker's Icon. Pl. **31**: t. 3033 (1915). —Clayton in Garcia de Orta, Sér. Bot. **1**: 11–12 (1973).

Inflorescence of 1–several digitate racemes; racemes fragile, villous, with linear internodes. Spikelets alike, paired, one sessile, one pedicelled, terete, the callus acutely conical to pungent and obliquely attached to the internode. Inferior glume coriaceous, convex, obtuse at the apex. Inferior floret reduced to a hyaline lemma. Superior lemma linear, bidentate, with a puberulous or hirtellous awn.

A genus of 5 species. Central Africa.

Awn 1.5–3 cm long; callus 1.5–2 mm long · 1. *eylesii*
Awn 5.5–6 cm long; callus 3–4 mm long · 2. *katakton*

1. **Homozeugos eylesii** C.E. Hubb. in Bull. Misc. Inform., Kew **1936**: 295 (1936). —Jackson & Wiehe, Annot. Check List Nyasal. Grass.: 43 (1958). —Vesey-FitzGerald in Kirkia **3**: 108 (1963). —Simon in Kirkia **8**: 54 (1971). —Clayton & Renvoize in F.T.E.A., Gramineae: 711 (1982). TAB. **6**. Type: Zambia, Mufulira, *Eyles* 8369 (K, holotype).

Caespitose perennial. Culms 100–200 cm high; leaf sheaths with auricles 4–20 mm long. Inflorescence of 1–4 racemes, each 8–25 cm long, white-villous. Spikelet callus 1.5–2 mm long, conical, sharply acute. Inferior glume 6.5–8 mm long, villous. Superior lemma with a puberulous awn 1.5–3 cm long.

Zambia. N: Mbala Distr., Ndundu, 1710 m, 3.iv.1959, *McCallum Webster* A264 (K). W: Kitwe Distr., Mindolo, 1220 m, 28.iv.1953, *Hinds* 18 (K). **Malawi**. N: Mzimba, 14.iii.1953, *Jackson* 1148 (K; PRE).
Also in Dem. Rep. Congo and Tanzania. Wooded grassland; 1300–1750 m.

Tab. 6. HOMOZEUGOS EYLESII. 1, habit (× ½); 2, portion of raceme (× 2); 3, detail of raceme (× 3); 4, spikelet pair (× 6); 5, tip of superior lemma (× 5), 1–5 from *McCallum Webster* T237. Drawn by Ann Davies. From F.T.E.A.

2. **Homozeugos katakton** Clayton in Garcia de Orta, Sér. Bot. **1**: 11 (1973). Type from Angola.

Caespitose perennial. Culms 120–180 cm high; leaf sheaths with auricles 10–25 mm long. Inflorescence of 2–4 racemes, each 16–20 cm long, white villous. Spikelet callus 3–4 mm long, linear, pungent. Inferior glume 7–10 mm long, pilose. Superior lemma with a hirtellous awn 3.5–6 cm long

Zambia. B: Kaoma Distr., Mangango Mission, 17.iv.1964, *Verboom* 1383 (K). Also in Angola. Savanna woodland; 1000 m.

7. MICROSTEGIUM Nees

By W.D. Clayton

Microstegium Nees in Lindley, Nat. Syst. Bot., ed. 2: 447 (1836). —Bor in Kew Bull. 7: 209–223 (1952).

Inflorescence of 1–many subdigitate racemes; racemes fragile, sparsely pilose, with filiform to clavate internodes. Spikelets alike (though the pedicelled sometimes a little smaller), paired, one sessile, one pedicelled, dorsally compressed, the callus obtuse. Inferior glume herbaceous to cartilaginous, the back flat with a median groove, sharply inflexed on the flanks, acute to bidentate. Superior glume often shortly awned. Inferior floret very variable, male, reduced to a hyaline scale or suppressed. Superior lemma linear to cordate, entire to bifid, with a glabrous awn.

A genus of c. 15 species. Mainly tropical Asia.

Microstegium nudum (Trin.) A. Camus in Ann. Soc. Linn. Lyon, n.s. **68**: 201 (1921). —Bor, Grasses Burma Ceyl. Ind. Pak.: 194 (1960). —Clayton & Renvoize in F.T.E.A., Gramineae: 717 (1982). TAB. **7**. Type from Nepal.
    Pollinia nuda Trin. in Mém. Acad. Imp. Sci. St.-Pétersbourg, Sér. 6, Sci. Math. **2**: 307 (1832).
    Psilopogon capensis Hochst. in Flora **29**: 117 (1846). Type from South Africa.
    Eulalia capensis (Hochst.) Steud., Syn. Pl. Glumac. **1**: 412 (1854).
    Pollinia nuda var. *capensis* (Hochst.) Hack. in A. & C. de Candolle, Monogr. Phan. **6**: 179 (1889).
    Microstegium capense (Hochst.) A. Camus in Ann. Soc. Linn. Lyon, n.s. **68**: 201 (1921). —Chippindall in Meredith, Grasses & Pastures S. Africa: 484 (1955).

Rambling annual, rooting at the lower nodes. Culms 20–100 cm long. Leaf laminas 1–8 cm long, linear to lanceolate. Inflorescence of 2–10 slender racemes each 2–8 cm long; internodes filiform, glabrous, equalling or exceeding the loosely spaced spikelets. Spikelets 2.5–5 mm long, lanceolate-elliptic; callus shortly barbate. Inferior glume with a broad shallow median groove, glabrous, or finely ciliate near the apex; superior glume weakly keeled or rounded on the back, acute. Inferior floret represented by a lanceolate hyaline scale. Superior lemma 1.5–2 mm long, linear, entire or minutely bidentate, with a capillary awn 6–20 mm long; palea minute or absent; stamens 2.

Zimbabwe. E: Mutasa Distr., Mutare (Umtali), Mountain Home Farm, 13.iv.1969, *Crook* 867 (K; PRE). **Mozambique.** Z: Gurué, 28.vi.1943, *Torre* 5597 (LISC). MS: Manica Distr., Quinta da Fronteira (Border Farm), 32 km along Manica (Macequece) road, 20.vii.1947, *Fisher* 1340 (PRE).
    An Asiatic species, also occurring in Uganda, Dem. Rep. Congo, Tanzania and South Africa. Forest shade; 600–1700 m.

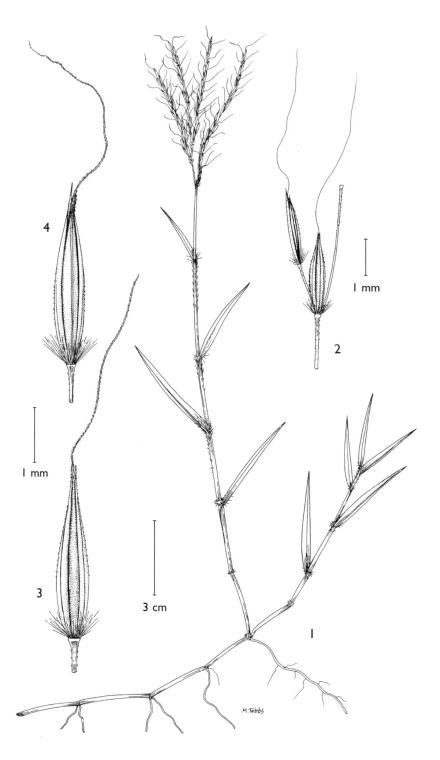

Tab. 7. MICROSTEGIUM NUDUM. 1, habit from *Schweicher* 6116; 2, spikelet pair; 3, sessile spikelet; 4, pedicelled spikelet, 2–4 from *Schweicher* 6117. Drawn by Margaret Tebbs.

8. TRACHYPOGON Nees

By W.D. Clayton

Trachypogon Nees, Agrost. Bras.: 341 (1829).

Inflorescence of single or digitate racemes, racemes tough, with linear internodes, bearing paired spikelets, one subsessile, one pedicelled. Subsessile spikelet male or barren, awnless (sometimes fertile and awned in *T. chevalieri*), persistent on the rhachis. Pedicelled spikelet bisexual, subterete, the callus pungent and obliquely attached to the internode. Inferior glume coriaceous, convex, obtuse at the apex. Inferior floret reduced to a hyaline lemma. Superior lemma linear, entire, with a pubescent or plumose awn.

A genus of 3 species. Tropical Africa and America.
This is one of the few genera of Andropogoneae in which the subsessile spikelet is the sterile and awnless member of the pair.

Plant annual; callus beard fulvous · 1. *chevalieri*
Plant perennial; callus beard white · 2. *spicatus*

1. **Trachypogon chevalieri** (Stapf) Jacq.-Fél. in J. Agric. Trop. Bot. Appl. **1**: 60 (1954). —Simon in Kirkia **8**: 52 (1971). —Clayton in F.W.T.A., ed. 2, **3**: 471 (1972). —Clayton & Renvoize in F.T.E.A., Gramineae: 709 (1982). Type from Central African Republic.
   Homopogon chevalieri Stapf in Mém. Soc. Bot. France **8**: 103 (1908). —Stapf in Prain, F.T.A. **9**: 409 (1919).

Annual. Culms 30–100 cm high. Leaf laminas flat. Racemes single, 2–8 cm long. Subsessile spikelet barren, smaller than the pedicelled and awnless (but sometimes fertile and resembling the pedicelled). Pedicelled spikelet 7–9 mm long, including the 2–3 mm fulvously barbate callus. Inferior glume glabrous or pubescent. Superior lemma with a stiffly geniculate, pubescent awn 3–7 cm long.

Zambia. N: 95 km east of Kasama, 3.iv.1961, *E.A. Robinson* 4572 (K). W: Mwinilunga Distr., 7 km north of Kalene Hill, 16.iv.1965, *E.A. Robinson* 6579 (K; SRGH).
Also in Tanzania, and extending northwestwards to Nigeria. Seasonally wet pans over rocks; 1400 m.

2. **Trachypogon spicatus** (L.f.) Kuntze, Revis. Gen. Pl. **2**: 749 (1891). —Sturgeon in Rhodesia Agric. J. **51**: 138 (1954). —Chippindall in Meredith, Grasses & Pastures S. Africa: 494 (1955). —Jackson & Wiehe, Annot. Check List Nyasal. Grass.: 63 (1958), in part, excl. *Jackson* 459 (= *Hyparrhenia nyassae*). —Launert in Merxmüller, Prodr. Fl. SW. Afrika, fam. 160: 204 (1970). —Simon in Kirkia **8**: 19, 52 (1971). —Clayton in F.W.T.A., ed. 2, **3**: 471 (1972). —Clayton & Renvoize in F.T.E.A., Gramineae: 709, fig. 163 (1982). TAB. **8**. Type from South Africa.
   Stipa spicata L.f., Suppl. Pl.: 111 (1781).
   Andropogon plumosus Willd., Sp. Pl. **4**: 918 (1806). Type from Venezuela.
   Trachypogon plumosus (Willd.) Nees, Agrost. Bras.: 344 (1829). —Stapf in Prain, F.T.A. **9**: 403 (1919). —Stent & Rattray in Proc. & Trans. Rhodesia Sci. Assoc. **32**: 15 (1933).
   Trachypogon capensis Trin. in Mém. Acad. Imp. Sci. St.-Pétersbourg, Sér. 6, Sci. Math. **2**: 257 (1832), nom. superfl., based on *Stipa spicata*.
   Heteropogon truncatus Nees, Fl. Afr. Austral. Ill.: 102 (1841). Type from South Africa.
   Andropogon spicatus (L.f.) Steud., Syn. Pl. Glumac. **1**: 368 (1854). Type as for *Stipa spicata*.
   Andropogon truncatus (Nees) Steud., Syn. Pl. Glumac. **1**: 368 (1854). Type as for *Heteropogon truncatus*.
   Trachypogon truncatus (Nees) Andersson in Öfvers. Förh. Kongl. Svenska Vetensk.-Acad. **14**: 49 (1857).
   Trachypogon polymorphus var. *truncatus* (Nees) Hack. in A. & C. de Candolle, Monogr. Phan. **6**: 327 (1889).
   Trachypogon involutus Pilg. in Fries, Wiss. Ergebn. Schwed. Rhod.-Kongo-Exped. **1**: 192 (1915). Type: Zambia, Kamindas, *Fries* 963 (UPS, holotype).
   Trachypogon durus Stapf in Prain, F.T.A. **9**: 405 (1919). Type from Angola.
   Trachypogon planifolius Stapf in Prain, F.T.A. **9**: 406 (1919). Type from Tanzania.
   Trachypogon glaucescens Pilg. in Notizbl. Bot. Gart. Berlin-Dahlem **11**: 803 (1933). Type from Tanzania.

Tab. 8. TRACHYPOGON SPICATUS. 1, habit (× ¹/₂), from *Chabwela* 5332; 2, raceme (× 2); 3, spikelet pair (× 3), 2 & 3 from *Mdehwa* 21; 4, detail of raceme (× 2); 5, tip of superior lemma (× 4), 4 & 5 from *Greenway & Kanuri* 14339. Drawn by Ann Davies. From F.T.E.A.

Perennial. Culms 30–200 cm high. Leaf laminas flat to acicular. Racemes 1(3), 4–30 cm long. Subsessile spikelet barren, as long as the pedicelled spikelet but awnless and inconspicuously winged on the margins above. Pedicelled spikelet 8–13 mm long, including the 1–3 mm white-barbate callus. Inferior glume glabrous to villous, usually pubescent. Superior lemma with a flexuous plumose awn 4–10 cm long.

Caprivi Strip. 32 km from Linyanti on road to Katima Mulilo, 910 m, 28.xii.1958, *Killick & Leistner* 3176 (K; PRE). **Botswana**. N: Ngamiland Distr., Shishikola Pan, 28.i.1978, *P.A. Smith* 2291 (K; PRE). **Zambia**. B: Mongu, 20.i.1966, *E.A. Robinson* 6807 (K). N: Kawambwa Distr., Ntumbachushi Falls, 1220 m, 22.xii.1967, *Simon & Williamson* 1508 (K). W: Mwinilunga Distr., on road to Kalene Hill, 21.xi.1972, *Strid* 2556 (K). C: Lusaka Distr., Chilanga, Mt. Makulu Agric. Research Station, 17.iv.1953, *Hinds* 85 (K). E: Nyika Plateau, 2130 m, 2.i.1959, *E.A. Robinson* 3010 (K; PRE). S: Choma Distr., Choma West Forest Reserve, 28.i.1960, *White* 6530 (K). **Zimbabwe**. N: Guruve Distr., Nyamunyeche Estate, 21.ii.1979, *Nyariri* 709 (K; PRE). W: Hwange Distr., Kazuma Depression, 1070 m, 12.ii.1985, *Gonde* 425 (K; PRE). C: Chikomba Distr., Chivhu (Enkeldoorn) area, 25.ii.1976, *Dye* 375 (K; PRE). E: Mutare (Umtali), 1600 m, 22.ii.1969, *Crook* 849 (K; PRE). S: Masvingo (Fort Victoria), 17.ii.1948, *D.A. Robinson* 243 (K; PRE). **Malawi**. N: Nyika Plateau, 29.vi.1952, *Jackson* 882 (K). C: Dedza Distr., Chongoni For. School, 8.vi.1967, *Salubeni* 739 (K). S: Blantyre Distr., Bvumbwe Agriculture Research Station (Tung Research Station), 6.iii.1951, *Jackson* 406 (K). **Mozambique**. N: Montepuez Distr., andados 14 km de Montepuez para Nantulo, 430 m, 8.iv.1964, *Torre & Paiva* 11744 (K; LISC; PRE). Z: Pebane, ao km 27, estrada para Naburi, c. 100 m, 16.i.1968, *Torre & Correia* 17181 (LISC). T: Angónia Distr., Posto Zootécnico, 12.v.1948, *Mendonça* 4182 (LISC). MS: Beira, ii.1972, *Tinley* 2394 (K). GI: between Inhambane and Inharrime, 9.xii.1944, *Mendonça* 3345 (LISC). M: Namaacha Mts., 22.xii.1944, *Torre* 6927 (K).

Throughout tropical and South Africa; also in South America. Wooded grassland and bushland, sometimes favouring the margins of flood plains and drainage tracts; 0–1700 m.

Variation in the vegetative characters is considerable but continuous, affording no justification for the recognition of segregate species.

9. SORGHUM Moench

By M.P. Setshogo

Sorghum Moench, Methodus: 207 (1794) *nom. conserv.* —C.E. Hubbard in Hooker's Icon. Pl. **34**: t. 3364 (1938). —Garber in Univ. Calif. Publ. Bot. **23**: 283–361 (1950). —Snowden in J. Linn. Soc., Bot. **55**: 191–260 (1955). —Celarier in Cytologia **23**: 395–418 (1959). —Ivanyukovich & Doronina in Trudy Prikl. Bot. **69**: 18–27 (1980). —Setshogo in Kirkia **17**: 138 (2001).

Blumenbachia Koeler, Descr. Gram.: 28 (1802), *nom. rejic.*, non Schrad. (1825).

Sarga Ewart in Proc. Roy. Soc. Victoria, n.s. **23**: 296 (1911).

Inflorescence a large panicle, its primary branches simple or subdivided, bearing short dense racemes with pilose internodes. Sessile spikelet dorsally compressed; callus obtuse or pungent; inferior glume cartilaginous, convex on the back, rounded on the flanks but becoming 2-keeled near the apex, usually pilose; superior lemma awned or awnless. Pedicelled spikelet male or barren, or reduced to the inferior glume.

A genus of c. 20 species, occurring in the Old World tropics and subtropics; 1 species endemic in Mexico.

Sorghum is the only genus in subtribe *Sorghinae* (genera 9–16 in this account) for which an extensive infrageneric classification exists. The most comprehensive subgeneric treatment of the genus is that of Garber, in Univ. Calif. Publ. Bot. 23: 283–361 (1950), who recognized six subgenera. One of these, *Sorghastrum*, has since been accorded generic status (genus 10 in this account); of the remainder, three (*Chaetosorghum*, *Heterosorghum* and *Stiposorghum*) are confined to Australia and two are represented in the Flora Zambesiaca area (*Sorghum* by S. × *drummondii*, S. × *almum*, S. *arundinaceum* and S. *halepense*; and *Parasorghum* by S. *versicolor*).

As is often the case with genera which include cultivated taxa, subgenus *Sorghum* contains a number of nomenclatural inconsistencies which have resulted in an extensive synonymy. The subgenus consists of cultivated grain sorghums, a complex of closely related annual taxa from Africa, and a complex of perennial taxa from southern Europe and Asia. The subgenus is sometimes divided into two groups, the *Halepensia* complex, and the *Arundinacea* complex (De Wet in Amer. J. Bot. **65**: 477–484 (1978)) and is considered by De Wet to consist of three species:

two rhizomatous taxa, *S. halepense* and *S. propinquum* (both in the *Halepensia*) and the large and complex *S. bicolor* which includes all annual wild, weedy and cultivated taxa in the *Arundinacea*. Snowden (The cultivated Races of Sorghum (1936)) recognized 24 'species' which are all now considered to be nothing more than races of *S. bicolor*.

The grain sorghums are artifacts of human selection and cultivation (De Wet & Huckaby in Evolution **21**: 787–802 (1967); De Wet, Harlan & Price in Amer. J. Bot. **57**: 704 (1970)) and the different cultivated kinds deserve at most racial status (Harlan & De Wet in Crop Science **12**: 172–176 (1972)). *S. bicolor* was divided by De Wet & Harlan (op. cit.) into three subspecies, *bicolor*, *arundinaceum* and *drummondii*, but in this account the first two are regarded as species in their own right and the last as a hybrid between them.

S. bicolor (L.) Moench, Methodus: 207 (1794) comprises all the grain sorghums, some of which are cultivated in the Flora Zambesiaca area, especially in Malawi, Zambia, Zimbabwe and Mozambique. Nine of Snowden's 'species' and many of his 'varieties' and 'forms' are grown in the Flora Zambesiaca area and his account should be consulted for further details.

S. × *drummondii* (Steud.) Millsp. & Chase in Publ. Field Mus. Nat. Hist., Bot. Ser. **3**: 21 (1903) is a tardily disarticulating element, often selected for cultivation as a fodder grass (and called 'Sudan Grass', *S. sudanense* (Piper) Stapf in Prain, F.T.A. **9**: 113 (1917)), derived from hybridization between *S. bicolor* and its presumed wild progenitor *S. arundinaceum* (see below). It is an annual, and therefore normally easily distinguished from *S. halepense*, but if the habit is not known the difficulties of identification are considerable. *S.* × *drummondii* has larger spikelets (6–7.5 mm compared with 4.5–5.5 mm) with 11–13 nerves on the inferior glume (compared with 7–9 of which only 2–4 are at all conspicuous in *S. halepense*) and a much less obviously 3-toothed tip to the glume. It is cultivated for fodder in Mozambique, Zambia and Zimbabwe and sometimes escapes into the wild; it also occurs as a spontaneous hybrid between the wild and the grain sorghums.

S. × *almum* Parodi in Revista Argent. Agron. **10**: 361 (1943) is a rhizomatous hybrid between *S. bicolor* and *S. halepense* and is occasionally cultivated as a fodder grass. If it backcrosses to *S. halepense* it can give rise to aggressive weeds, including the American 'Johnson Grass.' It has been cultivated in Zimbabwe in the past but it is not known whether it still is.

The numerous species and hybrids in *Sorghum* are extremely difficult to distinguish. It is essential, before any attempt is made, that material be complete, including any undergound parts, and mature to the point of spikelet disarticulation. Immature and incomplete specimens are almost invariably impossible to name. Hybrids and backcrosses can only be recognized if the user is fully familiar with the parents.

1. Nodes barbate · 3. *versicolor*
－ Nodes glabrous or pubescent · 2
2. Racemes tough or tardily disarticulating; plants cultivated or subspontaneous · · · · · · · 3
－ Racemes fragile; plants wild · 5
3. Grain large, commonly exposed by the gaping glumes; sessile spikelets persistent; cultivated · *bicolor* (see above)
－ Grain enclosed by the glumes; sessile spikelets persistent or tardily disarticulating; subspontaneous · 4
4. Annual, without rhizomes · × *drummondii* (see above)
－ Rhizomatous perennial · × *almum* (see above)
5. Rhizomatous perennial · 1. *halepense*
－ Annual or perennial without rhizomes · 2. *arundinaceum*

1. **Sorghum halepense** (L.) Pers., Syn. Pl. **1**: 101 (1805). —Stent & Rattray in Proc. & Trans. Rhodesia Sci. Assoc. **32**: 8 (1933). —Sturgeon in Rhodesia Agric. J. **51**: 15 (1954). — Chippindall in Meredith, Grasses & Pastures S. Africa: 460 (1955). —Clayton in F.W.T.A., ed. 2, **3**: 467 (1972). —Gibbs Russell et al., Grasses South. Africa [Mem. Bot. Surv. S. Africa No. 58]: 320 (1990). —Setshogo in Kirkia **17**: 138 (2001). Type locality in Syria.
    Holcus halepensis L., Sp. Pl. **2**: 1047 (1753).
    Andropogon arundinaceus Scop., Fl. Carniol., ed. 2: 274 (1772) non Berg. (1767). Type from Austria.
    Milium halepense (L.) Cav., Descr. Pl.: 306 (1802).
    Blumenbachia halepensis (L.) Koeler, Descr. Gram.: 29 (1802).
    Andropogon halepensis (L.) Brot., Fl. Lusit. **1**: 89 (1804).
    Andropogon miliaceus Roxb., Fl. Ind. **1**: 276 (1820). Type from India.
    Andropogon sorghum (L.) Brot. subsp. *halepensis* (L.) Hack. in A. & C. de Candolle, Monogr. Phan. **6**: 502 (1889).
    Sorghum miliaceum (Roxb.) Snowden in J. Linn. Soc., Bot. **55**: 205 (1955).

Sorghum miliaceum var. *parvispiculum* Snowden in J. Linn. Soc., Bot. **55**: 209 (1955). Type from India.

Perennial with abundant creeping rhizomes; culms 50–350 cm high, up to 2 cm wide, slender to robust, mostly simple but sometimes branched; nodes glabrous; leaf sheaths glabrous; ligule a glabrous membrane; leaf laminas 20–90 × 0.5–4 cm, broadly linear. Panicle 10–55 cm long, lanceolate to pyramidal; inferior primary branches naked for about 2–5 cm, glabrous; racemes 1–5-jointed; rhachis internodes and pedicels slender, densely ciliate with whitish hairs up to 1 mm long. Sessile spikelet 4.5–5.5 mm long, elliptic; callus short and blunt, shortly barbate; glumes coriaceous; inferior glume dorsally compressed, broadly lanceolate, 2-keeled on the margins, the keels winged and the wings broadening above and ending in prominent teeth forming, with the pointed apex, a ± equally 3-toothed tip, pilose on the back; superior glume sparsely pilose on the back, boat-shaped; inferior floret empty, its lemma lanceolate, ciliate; superior floret bisexual, ciliate, awned or awnless; awn (when present) up to 12 mm long, glabrous, untwisted; palea 0. Pedicelled spikelets male, c. 5.5 mm long, narrower than the sessile; glumes chartaceous; inferior glume lanceolate, ciliate on the lateral keels, glabrous on the back; superior glume sharply keeled, glabrous; inferior lemma c. 4 mm long, broadly lanceolate, ciliate on the margins, truncate at the apex; superior lemma c. 3 mm long ovate, ciliate on the margins; anthers c. 2.5 mm long.

Zimbabwe. N: Makonde (Lomagundi), 1070 m, 8.vii.1921, *Eyles* 3149 (K). W: Hwange Distr., Victoria Falls National Park, pathside in *Hyparrhenia* zone near Chimunzi, 850 m, 21.iv.1970, *Simon & Hill* 2129 (K). C: Harare, Agricultural Experimental Station (Salisbury Experiment Station), 5.vii.1921, *Mundy* 50b (K). E: Chipinge Distr., upper Rupembi R, east of the Save (Sabi) R, 22.i.1957, *Phipps* 86 (K; PRE). **Malawi**. S: Nsanje Distr., between Muona and Shire River, 80 m, 20.iii.1960, *Phipps* 2586 (K; PRE). **Mozambique**. N: Ribáuè, ix.1931, *Gomes e Sousa* 735 (LISC). T: Changara Distr., Boruma, rio Zambeze, v.1891, *Menyharth* 1046 (K). MS: Cheringoma Distr., Inhaminga, 10.ix.1942, *Mendonça* 178A (LISC). M: Maputo (Lourenço Marques), Costa do Sol, 4.xi.1963, *Balsinhas* 659 (LISC).

Throughout the Old and New World tropics. In moist areas on river banks, in clay soils and wet sandy soils. Rarely cultivated for fodder; 0–1500 m.

The distinguishing features of this species are the abundant slender wide-spreading rhizomes, the glabrous nodes, the narrow leaf laminas and the somewhat small contracted panicles. A frequent variation of the spikelets is the presence or absence of an awn to the superior lemma. The majority of plants have awned spikelets, but sometimes in specimens of the same gathering and number, the spikelets of one panicle may be awned and those of another awnless but accompanied by a few that are distinctly awned.

2. **Sorghum arundinaceum** (Desv.) Stapf in Prain, F.T.A. **9**: 114 (1917). —Stent & Rattray in Proc. & Trans. Rhodesia Sci. Assoc. **32**: 8 (1933). —Clayton in F.W.T.A., ed. 2, **3**: 467 (1972). — Clayton & Renvoize in F.T.E.A., Gramineae: 727, fig. 168 (1982). —Lowe, Fl. Nigeria, Grasses: 277 (1989). —Setshogo in Kirkia **17**: 139 (2001). TAB. **9**. Type from Ghana.
   Andropogon arundinaceus Willd, Sp. Pl. **4**: 906 (1805), non Berg. (1767).
   Rhaphis arundinacea Desv., Opusc. Sci. Phys. Nat.: 69 (1831).
   Andropogon verticilliflorum Steud., Syn. Pl. Glumac. **1**: 393 (1854). Type from Réunion.
   Andropogon sorghum var. *effusus* Hack. in A. & C. de Candolle, Monogr. Phan. **6**: 503 (1889), based on *Andropogon arundinaceus* Willd.
   Andropogon stapfii Hook.f., Fl. Brit. India **7**: 184 (1896). Type from India.
   Andropogon sorghum subsp. *effusus* (Hack.) Piper in Proc. Biol. Soc. Wash. **28**: 39 (1915).
   Sorghum verticilliflorum (Steud.) Stapf in Prain, F.T.A. **9**: 114 (1917). —Sturgeon in Rhodesia Agric. J. **51**: 15 (1954). —Jackson & Wiehe, Annot. Check List Nyasal. Grass.: 60 (1958). —Simon in Kirkia **8**: 20, 54 (1971).
   Sorghum stapfii (Hook.f.) C.E.C. Fisch. in Gamble, Fl. Madras **3**: 1735 (1934).
   Sorghum pugionifolium Snowden in J. Linn. Soc., Bot. **55**: 240 (1955). Type from India.
   Sorghum halepense sensu Simon in Kirkia **8**: 54 (1971), non (L.) Pers.
   Sorghum sudanense sensu Hall-Martin & Drummond in Kirkia **12**: 159 (1980), non (Piper) Stapf.

Annual, rarely short-lived perennial; culms 30–400 cm high, robust, branched; nodes mostly glabrous, sometimes pubescent; leaf sheaths glabrous; ligule a membrane, edged with a fringe of fine hairs and hairy on the back; leaf laminas variable, often large, 5–75 cm × 5–7 mm, broadly lanceolate, flat, glabrous on both surfaces, with a prominent whitish midrib. Panicle 10–60 cm long, broadly

Tab. 9. SORGHUM ARUNDINACEUM. 1, habit (× ½); 2, spikelet pair (× 3½); 3, sessile spikelet inferior glume (× 3½); 4, superior glume (× 3½); 5, caryopsis (× 3½). Drawn by W.E. Trevithick. From F.T.E.A.

spreading; main axis angular, glabrous; primary branches divided, pubescent at the nodes; racemes 2–7-jointed; rhachis internodes and pedicels pilose. Sessile spikelet (4)7(9) mm long, lanceolate to narrowly ovate; glumes coriaceous; inferior glume dorsally compressed, narrowly ovate, 2-keeled on the margins, the keels winged, the wings not or scarcely broadened above and ending in minute teeth well below the glume apex, the latter not equally 3-toothed, white-pubescent, sometimes tomentose or fulvously pubescent, slightly depressed longitudinally on the back; superior glume glabrescent or with sparse hairs on the back; inferior floret empty, its lemma c. 5.5 mm long, lanceolate, ciliate on the margins; superior floret bisexual, its lemma c. 3 mm long, deeply lobed, ciliate on the lobes and margins, awned; awn up to 20 mm long, glabrous; palea c. 2 mm long. Pedicelled spikelets neuter, c. 6.5 mm long, linear to lanceolate; glumes chartaceous; inferior glume glabrous; superior glume slightly shorter than the inferior, glabrous; inferior lemma glabrous, with a truncate apex.

Botswana. N: North East Distr., Tati River, SE of Francistown, 980 m, 4.iii.1985, *Long* 12242 (E). SE: Notwane Park at Notwane River, 24°38'S, 25°58'E, 1020 m, 18.iii.1978, *Hansen* 3387 (K). **Zambia**. B: Lukulu Distr., Kabompo River, near Lukulu, 1964, *Verboom* 1184 (K). C: South Luangwa National Park, Mfuwe, 26.iv.1965, *B.L. Mitchell* 2683 (BM; K; SRGH). E: Chipata (Fort Jameson), 11.v.1963, *van Rensburg* 2129 (K). S: Gwembe Distr., Zambezi R. c. 3 km upstream from Chirundu Bridge, 1.ii.1958, *Drummond* 5429 (K; PRE; SRGH). **Zimbabwe**. N: Hurungwe Distr., Mana Pools floodplain, 31.iii.1981, *Dunham* 73 (SRGH). W: Matobo Distr., Matopos Research Station, 1370 m, 26.ii.1954, *Rattray* 1672 (K). C: Mazowe Distr., Henderson Research Station, 10.iii.1964, *West* 4744 (K). E: Mudzi Distr., Lawleys Concession, 21.ii.1954, *West* 3413 (K). S: Beitbridge Distr., Sentinel Ranch, Limpopo River, 5 km east of Limpopo–Pazhi confluence, 25.iii.1959, *Drummond* 5992 (K; PRE). **Malawi**. N: Kondowe–Karonga, vii.1896, *Whyte* s.n. (K). C: Nkhotakota (Nkhota Kota), 490 m, 17.vi.1970, *Brummitt* 11519 (K; MAL; SRGH). S: Nsanje Distr., Mwanalundu Hills, 11.i.1990, *Salubeni & Nachamba* 5615 (MAL). **Mozambique**. N: Ribáuè, margens do rio Mepuipui (Meperipui) na base da serra, c. 480 m, 21.i.1964, *Torre & Paiva* 10078 (LISC). Z: Lugela Distr., Namagoa, *Faulkner* 37 (K; PRE). T: Magoe Distr., Msusa, Zambezi Valley, 25.vii.1950, *Chase* 2805 (BM; K). MS: Tambara Distr., 19 km from Tambara towards Lupata, c. 170 m, 16.v.1971, *Torre & Correia* 18495 (LISC). GI: Chibuto, Maniquenique, Estação Experimental do C.I.C.A., 13.vi.1960, *Lemos & Balsinhas* 98 (K; LISC; PRE). M: Umbelúzi State Farm near Boane, 19.ii.1985, *Timberlake* 3386 (SRGH).

Throughout Africa, extending eastwards to Australia. Swampy soils, streamsides, disturbed places and old farmland; 50–1400 m.

This species is the presumed wild progenitor of the grain sorghum, *S. bicolor*. The variability in the species-complex can be attributed to human selection of grain races and introgression with the wild species. The most obvious introgression products are generally referred to *S. × drummondii* (see above), but those that have progressed further from the grain sorghums by backcrossing are as likely to be assigned to *S. arundinaceum* as they are to the hybrid.

3. **Sorghum versicolor** Andersson in Peters, Naturw. Reise Mossambique, Bot.: 563 (1864). — Stapf in Prain, F.T.A. **9**: 138 (1917). —Stent & Rattray in Proc. & Trans. Rhodesia Sci. Assoc. **32**: 8 (1933). —Sturgeon in Rhodesia Agric. J. **51**: 15 (1954). —Chippindall in Meredith, Grasses & Pastures S. Africa: 459 (1955). —Bogdan, A Revised List of Kenya Grasses: 55 (1958). —Jackson & Wiehe, Annot. Check List Nyasal. Grass.: 60 (1958). — Napper, Grasses Tanganyika: 96 (1965). —Wild in Kirkia **5**: 55 (1965). —Hood, A Guide to the Grasses of Zambia: 58 (1967). —Simon in Kirkia **8**: 20, 54 (1971). —Hall-Martin & Drummond in Kirkia **12**: 159 (1980). —Clayton & Renvoize in F.T.E.A., Gramineae: 729 (1982). —Gibbs Russell et al., Grasses South. Africa [Mem. Bot. Surv. S. Africa No. 58]: 302 (1990). —Setshogo in Kirkia **17**: 139 (2001). TAB. **10**. Type: Mozambique Z: Boror, *Peters* (whereabouts uncertain).

Andropogon serratus Thunb. var. *versicolor* (Andersson) Hack. in A. & C. de Candolle, Monogr. Phan. **6**: 522 (1889).

Sorghum purpureo-sericeum (Hochst. ex A. Rich.) Asch. & Schweinf. var. *trinervatum* Chiov., Fl. Somala **2**: 439 (1932). Type from Somalia.

Annual; culms up to 250 cm high, slender; nodes with a conspicuous ring of long, spreading, silky white hairs; leaf sheaths glabrous except for a beard of long cilia at the mouth; ligule a glabrous membrane up to 3 mm long; leaf laminas 10–30 cm × up to 15 mm, hairy on both surfaces or occasionally glabrous on the superior. Panicle up to 25 cm long, open; branches simple; racemes 3–7-jointed; rhachis internodes and pedicels long-pilose. Sessile spikelet (5)5.5(7) mm long, elliptic-oblong; callus barbate with pallid to reddish hairs; glumes coriaceous, glossy;

Tab. 10. SORGHUM VERSICOLOR. 1, habit (× ¾); 2, ligule (× 1); 3, spikelet pair, a. joint,
b. sessile spikelet, c. pedicelled spikelet, d. pedicel; 4, joint of raceme; 5–9 sessile spikelet:
5, inferior glume; 6, superior glume; 7, inferior lemma; 8, superior lemma, showing base
of awn; 9, palea; 10 & 11 pedicelled spikelet: 10, inferior glume; 11, superior glume.
Drawn by W.E. Trevithick. From East Afr. Pasture Gr.

inferior glume reddish-brown to black, pilose on the back; superior glume pilose above the middle, glabrous and shiny below, tapering to a fine apex; inferior floret empty, its lemma c. 4 mm long and ciliate on the margins; superior floret bisexual; lemma c. 3 mm long, deeply lobed, ciliate on the lobes and margins, awned; awn c. (25)40 mm long, twisted, scabrid along the spiral; palea 0. Pedicelled spikelet neuter, (3)4(5) mm long, lanceolate, smaller than the sessile spikelet, greenish; inferior glume membranous, pilose on the back; superior glume membranous, glabrous to scabrid.

Botswana. N: Ngamiland Distr., 6 km SE of Tsao, 18.iii.1965, *Wild & Drummond* 7132 (BM; K; SRGH). SW: Kweneng Distr., c. 11 km NW of Lephepe Village, iii.1969, *Kelaole* 529 (K). SE: 101.3 km from Shoshong on Shoshong–Serowe road, 1140 m, 1.iv.1989, *Terry* et al. 146 (K). **Zambia.** C: Mumbwa Distr., Sala Reserve, Cheta River, 26.iii.1963, *Vesey-FitzGerald* 4019 (SRGH). E: Chipata Distr., Luangwa Valley, Nsefu, Lupande R., 10.iii.1968, *R. Phiri* 88 (K). S: Kalomo Distr., Siantambo, 7.ii.1963, *B.L. Mitchell* 17/63 (BM; PRE). **Zimbabwe**. N: Gokwe Distr., 30 km north of Gokwe on road to Chinyenyeti, 12.iii.1963, *Bingham* 512 (K; PRE). W: Nkayi Distr., Gwampa Forest Reserve, ii.1956, *Goldsmith* 6/56 (PRE). C: Gweru Distr., Gweru (Gwelo) Kopje, 1430 m, 5.iii.1967, *Biegel* 1959 (BM; K). E: Chimanimani Distr., Mutambara C.L. (Tribal Trust Land) on main road near Lisnacloon Farm, 1000 m, 26.iii.1969, *A.O. Crook* 850 (K; PRE). S: Masvingo (Fort Victoria), Victoria Reserve Zone, 19.iii.1956, *Cleghorn* 185 (BM; PRE). **Malawi**. N: Mzimba Distr., 40 km south of Rumphi on M1, 1220 m, 4.v.1974, *Pawek* 8571 (K; PRE; WAG). C: Salima Distr., Chitala, 8.iv.1952, *Jackson* 756 (K). S: Zomba Distr., near Macheleni Hill, Tembenu Village, 5.iii.1978, *Seyani & Patel* 800 (MAL). **Mozambique**. N: Malema Distr., Mutuáli, Estação Experimental do Instituto do Algodão, 5.iv.1962, *Lemos & Marrime* 318 (K; LISC; PRE). Z: between Águas Quentes and Mutarara, 6.v.1943, *Torre* 5303 (LISC). T: Tete, Cahora (Cabora) Bassa, 230–330 m, 2.v.1972, *Pereira & Correia* 2328 (WAG). MS: Chemba, Chiou, Estação Experimental do C.I.C.A., 14.iv.1960, *Lemos & Macuácua* 105 (K; LISC; PRE). GI: Massingir, 19.iv.1972, *Lousã & Rosa* 244 (K). M: 15 km from Macaene to Machatuíne, 22.i.1948, *Torre* 7179 (BM; LISC).

Eastern, central and southern Africa. In deciduous bushland or wooded grassland, commonly on waterlogged soils or black clays; 30–1450 m.

This species and *S. purpureo-sericeum* (from tropical Africa and Arabia) are so very similar that they may just represent variants of a single species. Differences between them mainly derive from their spikelet length (shorter in *S. versicolor*), their apparent inability to hybridize (Garber, op. cit.), and their different geographical distributions.

10. SORGHASTRUM Nash

By M.P. Setshogo

Sorghastrum Nash in Britton, Man. Fl. N. States: 71 (1901).
Porathera Raf. in Bull. Bot., Geneva **1**(8): 221 (1830), non Rudge (1811).
Dipogon Steud., Nomencl. Bot., ed. 2, **1**: 518 (1840), *nom. nud.*

Annuals or perennials. Inflorescence a panicle, the subdivided primary branches bearing short racemes, these sometimes reduced to triads. Sessile spikelet callus obtuse or pungent (with involucral hairs in *S. pogonostachyum*); inferior glume coriaceous, convex, keeled at the apex; superior glume broadly convex; superior lemma mostly entire, sometimes bilobed, awned. Pedicelled spikelets usually reduced to a barren pedicel, but present and bisexual in *S. pogonostachyum* and *S. fuscescens*.

A genus of c. 16 species, occurring in tropical Africa and tropical America.

1 Pedicelled spikelet bisexual; some pedicels barren · 2
– Pedicelled spikelets absent; all pedicels barren · 3
2. Culm nodes glabrous; racemes up to 7-jointed; callus with a beard of long hairs up to $^3/_4$ the length of the sessile spikelet; awn c. 4 mm long; superior lemma entire · · 1. *pogonostachyum*
– Culm nodes pilose; racemes 3–15-jointed; callus with a beard of short hairs only; awn c. 25 mm long; superior lemma bilobed · 2. *fuscescens*
3. Weak annual; racemes single-jointed; sessile spikelet 4–5.5 mm long; awn 25–40 mm long · 3. *incompletum* var. *bipennatum*
– Robust perennial; racemes 2–many-jointed; awns less than 20 mm long · · · · · · · · · · · · 4
4. Leaf-sheaths not produced into auricles; spikelets 5–7(8) mm long; awn 3–8(10) mm long, straight or geniculate and twisted · 4. *nudipes*

– Leaf-sheaths with auricles; spikelets 4–7(8) mm long; awns 8–16 mm long, never straight, always geniculate and twisted · 5. *stipoides*

1. **Sorghastrum pogonostachyum** (Stapf) Clayton in Kew Bull. **30**: 509 (1975). —Setshogo in Kirkia **17**: 136 (2001). Type from Angola.
   Sorghum pogonostachyum Stapf in Prain, F.T.A. **9**: 144 (1917).
   Miscanthidium gracilius Napper in Kirkia **3**: 120 (1963). Type from Tanzania.
   Sorghastrum sp. sensu Simon in Kirkia **8**: 54 (1971).

Caespitose perennial with creeping rhizomes; culms up to 100 cm high, glabrous at the nodes; leaf sheaths firm, tight, glabrous, smooth; ligule short, stout, pubescent on the back and margin; leaf laminas narrow, glabrous, smooth except towards the apex. Panicle up to 15 cm long, narrow; racemes loose, up to 7-jointed. Sessile spikelet lanceolate; callus short, rounded, pilose with hairs c. 4 mm long; inferior glume c. 5.5 mm long, chartaceous, sparsely pilose on the back; superior glume glabrous; inferior floret empty, the lemma oblong, ciliate on the margins; superior floret bisexual, epaleate, with lemma entire, 3-nerved, awned from the apex; awn c. 4 mm long, bristle-like; anthers c. 3 mm long. Pedicelled spikelet lanceolate with fertile superior floret; anthers c. 3 mm long; pedicels, some of which are barren, pilose.

Zambia. N: Kawambwa Distr., Chishinga Ranch, 1400 m, 28.x.1961, *Astle* 1012 (K). C: Mkushi Distr., Mkushi River dambo, 1430 m, 27.x.1967, *Simon & Williamson* 1222 (K; PRE). **Malawi**. C: Lilongwe Distr., Dzalanyama, Madzimaiera dambo, 4.xii.1951, *Jackson* 693 (K).
   Eastern and central Africa. Growing in wet places, especially along water courses.
   Many, if not most, of the pedicels bear perfect spikelets, a feature shared with *S. fuscescens*. In the remaining species the pedicels are, without exception, barren of any spikelet.

2. **Sorghastrum fuscescens** (Pilg.) Clayton in Kew Bull. **30**: 509 (1975). —Setshogo in Kirkia **17**: 136 (2001). Type from Tanzania.
   Miscanthidium fuscescens Pilg. in Notizbl. Bot. Gart. Berlin-Dahlem **11**: 806 (1933).

Caespitose perennial up to 200 cm high, glabrous at the nodes; leaf sheaths pubescent; ligule a short glabrous membrane; leaf laminas 10–60 cm × 2–7 mm, linear, densely and coarsely pubescent. Panicle 20–30 cm long, loose; racemes 3–10 cm long, 3–15-jointed. Sessile spikelet c. 6 mm long, lanceolate; callus c. 1 mm long, rounded, shortly pilose; inferior glume c. 5.5 mm long, lanceolate, coriaceous, pilose on the back, brown; superior glume slightly longer than the inferior, pilose along the keel, otherwise glabrous on the back, ciliate on the margins; inferior floret empty, the lemma ovate, ciliate on the margins; superior floret bisexual, epaleate, the lemma lobed, ciliate on the margins, awned; awn c. 25 mm long, bigeniculate, glabrous; anthers c. 4 mm long. Pedicelled spikelet similar to the sessile spikelet, also with a bisexual superior floret; glumes coriaceous; anthers c. 4 mm long.

Zambia. N: Kaputa Distr., Muzombwe, western side of Mweru Wantipa, 1070 m, 15.iv.1961, *Phipps & Vesey-FitzGerald* 3218 (BM). **Malawi**. C: Dedza Distr., Linthipe River, 10.iii.1951, *Jackson* 426 (K).
   Also in east and central Africa. Growing on grassy hillsides and on lake margins.
   Many, if not most, of the pedicels bear perfect spikelets, a feature shared with *S. pogonostachyum*. These two species differ from the rest of *Sorghastrum* in this respect, but otherwise conform in other characters.

3. **Sorghastrum incompletum** (J. Presl) Nash in N. Amer. Fl. **17**: 130 (1912). Type from Mexico.
   Andropogon incompletus J.Presl in Reliq. Haenk. **1**: 342 (1830).

Var. **bipennatum** (Hack.) Dávila in Ann. Missouri Bot. Gard. **76**: 1171 (1989). —Setshogo in Kirkia **17**: 137 (2001). TAB. **11**. Type from Sudan.
   Andropogon bipennatus Hack. in Flora **68**: 142 (1885).
   Sorghum bipennatum (Hack.) Kuntze, Revis. Gen. Pl. **2**: 791 (1891). —Stapf in Prain, F.T.A. **9**: 144 (1917).
   Sorghastrum bipennatum (Hack.) Pilg. in Notizbl. Bot. Gart. Berlin-Dahlem **14**: 96 (1938). —Jackson & Wiehe, Annot. Check List Nyasal. Grass.: 60 (1958). —Simon in Kirkia **8**: 20, 54 (1971). —Clayton in F.W.T.A., ed. 2, **3**: 468 (1972). —Clayton & Renvoize in F.T.E.A., Gramineae: 731 (1982).

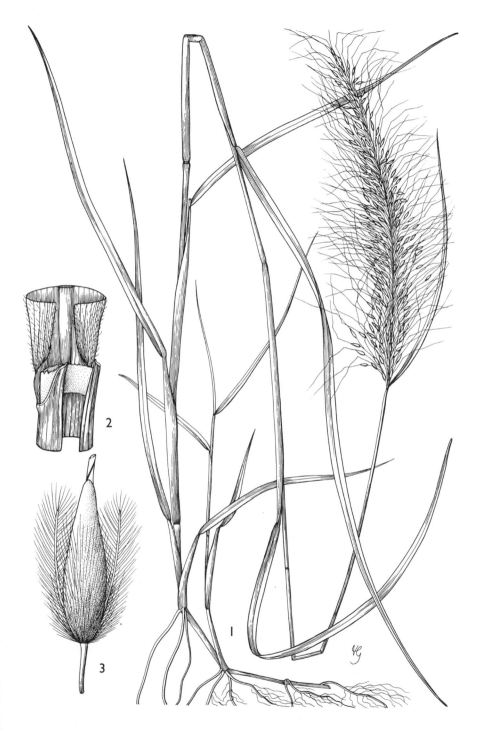

Tab. 11. SORGHASTRUM INCOMPLETUM var. BIPENNATUM. 1, habit (× ¹/₂); 2, ligule (× 4¹/₂); 3, spikelet and barren pedicels (× 12). Drawn by Victoria Goaman. From Ghana Gr.

Weak annual up to 150 cm high (rarely as much as 200 cm), often decumbent and rooting at the lower nodes and with short stilt roots; culms simple, the nodes shortly pilose; ligule a tough membrane; leaf laminas up to 30 cm × 3–10 mm, linear-lanceolate. Panicle loose, narrowly lanceolate; racemes reduced to a single sessile spikelet, always accompanied by 2 barren pilose pedicels c. 4 mm long. Spikelets 4–5.5 mm long, narrowly ovate, golden, paleate; callus rounded, without a noticeable fringe of hairs; inferior glume ovate, coriaceous, pilose on the back with white hairs; superior glume smooth on the back; inferior floret empty, the lemma hyaline, ciliate along the margins; superior floret bisexual, the lemma bilobed, sometimes inconspicuously so, awned; awn c. 40 mm long, bigeniculate, scabrid along the column, otherwise smooth; anthers not seen.

Zambia. W: Chizela Distr., Kafue National Park, Lufupa Camp, 6.iv.1963, *B.L. Mitchell* 19/56 (K; PRE). C: Luangwa Valley, South Luangwa National Park, Mfuwe Camp, 610 m, 1.iv.1966, *Astle* 4765 (K). E: Chipata Distr., without precise locality, 17.iii.1970, *Abel* 182 (SRGH). **Zimbabwe**. N: Gokwe Distr., without precise locality, 15.iii.1984, *Mahlangu* 975 (K; SRGH). **Malawi**. N: Chitipa Distr., Chisenga, 23 v 1962, *E.A. Robinson* 5227 (K). C: Dedza Distr., Mua-Livulezi forest, 9.iv.1964, *Adlard* 607 (SRGH). S: Zomba, Naisi road, 1070 m, 14.v.1949, *Wiehe* 106 (K). **Mozambique**. N: Malema Distr., 9.iv.1962, *Lemos & Marrime* 330 (K; PRE).
Tropical Africa. Wooded grasslands; 1000–1600 m.
Very similar to the tropical American var. *incompletum*, but this has a denser panicle, occasional racemes with more than one sessile spikelet, and generally thinner light brown glumes not more than 4 mm long.

4. **Sorghastrum nudipes** Nash, N. Amer. Fl. **17**: 129 (1912). Type from Mexico.
 Sorghum nutans (L.) Nash var. *angolense* Rendle in Hiern et al., Cat. Afr. Pl. Welw. **2**: 152 (1899). Type from Angola.
 Andropogon friesii Pilg. in R.E. Fries, Wiss. Ergebn. Schwed. Rhod.-Kongo-Exped. **1**: 195 (1916). Type: Zambia, Bangweulu, near Kamindas, *Fries* 965 and 974 (UPS, syntypes).
 Sorghum micratherum Stapf in Prain, F.T.A. **9**: 142 (1917). —Stent & Rattray in Proc. & Trans. Rhodesia Sci. Assoc. **32**: 8 (1933). Type as for *S. nutans* var. *angolense* Rendle.
 Sorghum friesii (Pilg.) C.E. Hubb. in Bull. Misc. Inform., Kew **1934**: 109 (1934). —Simon in Kirkia **8**: 20, 54 (1971).
 Sorghastrum friesii (Pilg.) Pilg. in Notizbl. Bot. Gart. Berlin-Dahlem **14**: 96 (1938). — Setshogo in Kirkia **17**: 137 (2001).

Caespitose, shortly rhizomatous perennial up to 120 cm high, often rooting from the lower nodes; culms with velvety-pubescent hairs; leaf sheaths without auricles, glabrous; ligule a glabrous membrane 0.5–1.5 mm long; leaf laminas up to 20 cm × 2–6 mm, linear to linear-lanceolate, usually reflexed, glabrous, expanded at the base. Panicle loose and open with divided primary branches; racemes (2)4–7-jointed, the internodes and pedicels pilose. Sessile spikelet 5–7(8) mm long, lanceolate; callus obtuse, barbate; inferior glume chartaceous, pilose on the back, slightly depressed longitudinally; superior glume chartaceous, glabrous; inferior floret empty, represented by a lemma only, this c. 5 mm long, ovate, hyaline, finely 2-nerved, ciliate on the margins above, emarginate at the apex; upper floret bisexual, epaleate, the lemma c. 4 mm long, bilobed at the apex, the lobes glabrous, awned between the lobes; awn 3–8(10) mm long, straight, glabrous; anthers 4–5 mm long. Pedicelled spikelet absent; pedicel c. 5 mm long, shorter than the sessile spikelet.

Caprivi Strip. Okavango R., 19 km north of Shakawe, 16.iii.1965, *Wild & Drummond* 7091 (BM; K). **Botswana**. N: Xudum drainage, Okavango, 1000 m, 15.iii.1961, *Vesey-FitzGerald* 3241 (BM). **Zambia**. B: Kabompo Distr., Lusongwa flood plain, 8 km south of Lusongwa School, 850 m, 26.xii.1969, *Simon & Williamson* 2054 (K). N: Mbala Distr., Saisi R. near Jericho Range, 27.iii.1958, *Vesey-FitzGerald* 1547 (K). C: Kabwe Distr., Mushwishi Agric. Station, 1160 m, 8.xii.1953, *Hinds* 175 (K). S: Mazabuka Distr., Mabwingombe Hills, 5.ii.1960, *White* 6837 (K). **Zimbabwe**. N: Gokwe Distr., c. 0.8 km north of Gokwe, 11.ii.1963, *Bingham* 484 (K; PRE; SRGH). W: Nkayi Distr., Gwampa vlei at Tunke Pan, Gwampa Forest Reserve, 910 m, ii.1956, *Goldsmith* 31/56 (K). C: Gweru Distr., Gwelo Teachers College, 10.ii.1967, *Biegel* 1989 (BM; K; PRE; SRGH). S: Masvingo (Victoria), 30.iii.1973, *Chiparawasha* 665 (K; PRE). **Malawi**. C: Lilongwe Distr., Naminyanga Dambo, Dzalanyama, 4.xii.1951, *Jackson* 699 (K). **Mozambique**. MS: Beira, ii.1912, *Rogers* 5939 (K).
Also in South Africa. Typically favours wet areas such as swamps, road drains and stream banks; often on poor sandy soils; 0–1400 m.

Tab. 12. **SORGHASTRUM STIPOIDES.** 1, habit (× ¹⁄₂); 2, raceme (× 3); 3, sessile spikelet and barren pedicels (× 5); 4, tip of superior lemma (× 7), 1–4 from *Bogdan* 5167. Drawn by Ann Davies. From F.T.E.A.

5. **Sorghastrum stipoides** (Kunth) Nash, N. Amer. Fl. **17**: 129 (1912). —Clayton & Renvoize in F.T.E.A., Gramineae: 732 (1982). —Setshogo in Kirkia **17**: 137 (2001). TAB. **12**. Type from Colombia.

  Andropogon stipoides Kunth in Humb., Bonpl. & Kunth, Nov. Gen. Sp. **1**: 189 (1816).
  Andropogon trichopus Stapf in Bull. Misc. Inform., Kew **1897**: 287 (1897). Type from Nigeria.
  Sorghum trichopus (Stapf) Stapf in Prain, F.T.A. **9**: 141 (1917).
  Sorghum rigidifolium Stapf in Prain, F.T.A. **9**: 143 (1917). Type from Kenya.
  Sorghastrum trollii Pilg. in Notizbl. Bot. Gart. Berlin-Dahlem **14**: 95 (1938). —Vesey-FitzGerald in Kirkia **3**: 104 (1963). Type from Tanzania.
  Sorghastrum trichopus (Stapf) Pilg. in Engler & Prantl, Nat. Pflanzenfam., ed. 2, **14e**: 142 (1940). —Clayton in F.W.T.A., ed. 2, **3**: 468 (1972).
  Sorghastrum rigidifolium (Stapf) Chippind. in Pole Evans, Mem. Bot. Surv. S. Africa **22**: 247 (1948). —Chippindall in Meredith, Grasses & Pastures S. Africa: 468 (1955). —Jackson & Wiehe, Annot. Check List Nyasal. Grass.: 60 (1958).

Robust, caespitose perennial up to 150 cm high, with hard creeping rhizomes; culms erect, usually unbranched, the nodes conspicuously pilose; leaf sheaths auriculate; ligule a glabrous membrane 1.5–4 mm long; leaf-laminas 15–45 cm × 3–6 mm, glabrous, usually inrolled, rigid and conspicuously narrowed towards the base. Panicle c. 20 cm long, linear-lanceolate; racemes 1–4-jointed; rhachis internodes and pedicels violet-hirsute. Sessile spikelets 4–7(8) mm long, lanceolate; callus rounded, pilose; inferior glume coriaceous, tawny with green nerves, pale violet-hirsute, broadly convex across the back, emarginate; superior glume as long as the inferior, paler, glabrous below, shortly pilose above, emarginate; inferior floret reduced to a hyaline lemma and minute palea; superior floret bisexual with linear bilobed lemma, this ciliate on the margins; awn 8–16 mm long, geniculate, glabrous; anthers not seen. Pedicelled spikelet represented by pedicel only (terminal sessile spikelet accompanied by 2 pedicels).

Zambia. N: Mbala Distr., Lunzua River Bridge, 32 km from Mbala (Abercorn), 1520 m, 5.iv.1959, *McCallum Webster* A283 (K). C: 25 km SE of Lusaka, Kanyanja, 1300 m, 25.ii.1996, *Bingham & Harder* 10940 (K). **Malawi**. N: Karonga, Mwentete dambo, 26.vi.1951, *Jackson* 552 (K). C: Nkhotakota Distr., Katimbira Village, 13.v.1986, *Patel & Kwatha* 3194 (MAL). S: Zomba, near Tongani, 20.iv.1950, *Wiehe* 492 (K; MAL). **Mozambique**. MS: Cheringoma Distr., Cheringoma Coastal Area, Zuni drainage, 5 km west of Nyamaruza Camp, v.1973, *Tinley* 2842 (K; PRE). M: Maputo, 20.ii.1952, *Myre & Carvalho* 1148 (K).

Tropical America and tropical Africa. Low lying areas subject to flooding; 0–550 m.

11. CLEISTACHNE Benth.

By M.P. Setshogo

Cleistachne Benth. in Hooker's Icon. Pl. **14**: t. 1379 (1882).

Inflorescence a panicle, its primary branches bearing racemes at regular intervals along their length, each raceme reduced to a single sessile spikelet (whose apparent pedicel is homologous with the raceme peduncle). Spikelets solitary, sessile, 2-flowered. Inferior floret barren. Superior floret bisexual, the lemma comprising the bilobed, hyaline base of the awn. Pedicelled spikelet and pedicel completely suppressed.

A genus of one species.

Cleistachne sorghoides Benth. in Hooker's Icon. Pl. **14**: t. 1379 (1882). —Stapf in Prain, F.T.A. **9**: 154 (1917). —Stent & Rattray in Proc. & Trans. Rhodesia Sci. Assoc. **32**: 8 (1933). —Sturgeon in Rhodesia Agric. J. **51**: 8 (1954). —Chippindall in Meredith, Grasses & Pastures S. Africa: 468 (1955). —Jackson & Wiehe, Annot. Check List Nyasal. Grass.: 33 (1958). —Bor, Grasses Burma Ceyl. Ind. Pak.: 119 (1960). —Napper, Grasses Tanganyika: 98 (1965). —Hood, A Guide to the Grasses of Zambia: 58 (1967). —Simon in Kirkia **8**: 20, 54 (1971). —Clayton & Renvoize in F.T.E.A., Gramineae: 734, fig. 170 (1982). —Gibbs Russell et al., Grasses South. Africa [Mem. Bot. Surv. S. Africa No. 58]: 87 (1990). —Setshogo in Kirkia **17**: 136 (2001). TAB. **13**. Type: Mozambique, Shupanga (Chupanga), *Kirk* s.n. (K, holotype).

Tab. 13. CLEISTACHNE SORGHOIDES. 1, habit (× ¹/₂); 2, inflorescence (× ¹/₂); 3, raceme (× 4); 4, spikelet (× 10); 5, tip of superior lamma (× 8), 1–5 from *Schlieben* 3653. Drawn by Ann Davies. From F.T.E.A.

Erect stilt-rooted annual up to 300 cm high, slender; nodes and internodes glabrous; leaf sheaths pilose or glabrous; ligule an entire membrane; leaf laminas 20–90 cm × 4–15 mm, linear-lanceolate, pilose on both surfaces but less so on the inferior, the midrib prominent. Panicle 8–40 cm long, compact; racemes 4–8 cm long, the peduncles pilose; spikelets varying from straw-coloured to chestnut-brown. Spikelets 4–5 mm long, oblong; callus obscure. Glumes similar in size, slightly coriaceous; inferior glume finely 7-nerved, narrowly truncate at the apex, pilose; superior glume pilose, especially at the apex and on the margins, often with a glabrous midline. Inferior floret barren, the lemma hyaline, oblong, ciliolate, 2-nerved. Superior floret with an awn 5–8 times the length of the lemma, geniculate with pilose column and scabrid limb; palea present.

Zambia. N: North Luangwa National Park, 690 m, 14.iv.1993, *P.P. Smith* 856 (K). E: Chipata (Fort Jameson) area, 1962, *Verboom* 585 (K). S: Kalomo Distr., Sichifula Controlled Hunting Area, 26.iii.1962, *B.L. Mitchell* 13/48 (K; PRE). **Zimbabwe**. N: Gokwe South Distr., Sengwa/Korwe confluence, 19.iii.1984, *Mahlangu* 981 (K). C: Mazowe Distr., Teviotdale, 20.iii.1965, *Bingham* 1429 (BM, K). E: Chimanimani Distr., Muchira River, Tarka Forest Reserve, 1100 m, iii.1971, *Goldsmith* 12/71 (BM; K; PRE). **Malawi**. N: Mzimba Distr., Kasitu Valley, 1040 m, 27.i.1938, *Fenner* 230 (K). C: Dedza Distr., Mua-Livulezi Forest, 9.iv.1964, *Adlard* 607A (MAL). S: Chiradzulu Distr., Chiradzulu Mt., 1250 m, 14.ix.1970, *Brummitt & Banda* 9831 (K). **Mozambique**. N: Ribáuè, 1000 m, 23.iii.1964, *Torre & Paiva* 11359 (K; LISC). Z: Chinde Distr., Luabo, 8.v.1966, *West* 7329 (K). T: Moatize (Muatize), c. 50 km from Zóbuè towards Tete, 350 m, 12.iii.1964, *Torre & Paiva* 11162 (K; LISC; PRE). MS: Manica Distr., Manica (Macequese), 17.vii.1948, *Fisher & Schweickerdt* 239 (K; PRE).

Tropical Africa, through Arabia to India. Grassland and savanna; 350–1500 m.

12. CHRYSOPOGON Trin.

By M.P. Setshogo

Chrysopogon Trin., Fund. Agrost.: 187 (1820), *nom. conserv.*
Rhaphis Lour., Fl. Cochinch.: 552 (1790).
Pollinia Spreng., Pl. Min. Cogn. Pug. **2**: 10 (1815).
Centrophorum Trin., Fund. Agrost.: 106 (1820).
Vetiveria Bory ex Lem.-Lisanc. in Bull. Sci. Soc. Philom. Paris **1822**: 42 (1822).
Trianthium Desv., Opusc. Sci. Phys. Nat.: 69 (1831).
Lenormandia Steud. in Flora **33**: 229 (1850).
Mandelorna Steud., Syn. Pl. Glumac. **1**: 359 (1854).
Chalcelytrum Lunell in Amer. Midl. Naturalist **4**: 212 (1915).

Caespitose perennials, sometimes rhizomatous; leaf sheaths sometimes keeled, the basal flabellate-imbricate; ligule a short membrane or a line of hairs; leaf laminas flat or conduplicate. Inflorescence a panicle, its primary branches whorled, simple; racemes 1–many-jointed; pedicels linear, filiform, never longitudinally grooved. Sessile spikelet laterally compressed, its callus elongated and obtuse to pungent, often large and conical; glumes subequal, unequally awned or awnless; inferior glume chartaceous to coriaceous, often spinulose on the keels and muricate or mammilate on the back; superior lemma entire or bilobed, awned or awnless; awn glabrous to pubescent and usually prominent. Pedicelled spikelet male or neuter.

A genus of c. 35 species, occurring in tropical and warm temperate regions of the Old World, principally in Asia and Australia; 1 species in tropical America.

There are too many intermediates between *Chrysopogon* and *Vetiveria* (particularly in Australia) to justify the recognition of more than one genus.

1. Raceme reduced to a triad of 1 sessile and 2 pedicelled spikelets · · · · · · · · · 1. *serrulatus*
– Raceme with several to many spikelet pairs · 2
2. Sessile spikelet mucronate or awnless · 2. *zizanioides*
– Sessile spikelet awned, the awn c. 5 mm long · 3. *nigritanus*

1. **Chrysopogon serrulatus** Trin. in Mém. Acad. Imp. Sci. St.-Pétersbourg, Sér. 6, Sci. Math. **2**: 318 (1832). —Bor, Grasses Burma Ceyl. Ind. Pak.: 118 (1960). —Clayton & Renvoize in

F.T.E.A., Gramineae: 736 (1982). —Gibbs Russell et al., Grasses South. Africa [Mem. Bot. Surv. S. Africa No. 58]: 85 (1990). —Setshogo in Kirkia **17**: 140 (2001). Type from Nepal.
Andropogon trinii Steud., Syn. Pl. Glumac. **1**: 395 (1854). Type from Nepal.
Andropogon ciliolatus Nees ex Steud., Syn. Pl. Glumac. **1**: 396 (1854). Type from East Indies.
Andropogon caeruleus Nees ex Steud., Syn. Pl. Glumac. **1**: 395 (1854). Type from East Indies.
Chrysopogon ciliolatus (Nees ex Steud.) Boiss., Fl. Orient. **5**: 458 (1884).
Andropogon trinii var. *increscens* Hack. in A. & C. de Candolle, Monogr. Phan. **6**: 558 (1889). Types from East Indies and Sri Lanka.
Andropogon monticola Schult. var. *trinii* (Steud.) Hook.f., Fl. Brit. India **7**: 193 (1896).
Andropogon tremulus Hack. in Bull. Herb. Boissier, sér. 2, **1**: 764 (1901). Type: Mozambique, Boroma, *Menyharth* 557 (W, holotype; K, photo).
Chrysopogon montanus Trin. var. *serrulatus* (Trin.) Stapf in Prain, F.T.A. **9**: 160 (1917).
Chrysopogon montanus var. *tremulus* (Hack.) Stapf in Prain, F.T.A. **9**: 160 (1917). —Stent & Rattray in Proc. & Trans. Rhodesia Sci. Assoc. **32**: 9 (1933). —Wild in Kirkia **5**: 54 (1965).
Chrysopogon fulvus (Spreng.) Chiov. var. *tremulus* (Hack.) Chiov., Racc. Bot.: 128 (1935).
Chrysopogon fulvus var. *serrulatus* (Trin.) R.B. Stewart in Brittonia **5**: 446 (1945).
Chrysopogon montanus sensu Simon in Kirkia **8**: 20, 54 (1971), non Trin.

Culms up to 150 cm high, sometimes robust; leaf-sheaths glabrous, rarely pilose above; ligule a ciliolate rim; leaf laminas 2–10 mm wide, linear-lanceolate, glabrous except near the ligule where there are sparse tubercle-based hairs on the margins. Panicle up to 15 cm long; peduncles fulvously bearded at the apex. Sessile spikelet c. 6 mm long, narrowly oblong; inferior glume awnless, slightly hispid at the apex; superior glume with a few white hairs on the back and with a glabrous awn c. 12 mm long; inferior lemma hyaline, 1-nerved and with sparsely ciliolate margins; superior floret bisexual, its lemma hyaline, bilobed and with a slightly puberulous awn c. 2.5 cm long; anthers c. 3 mm long. Pedicelled spikelet male or neuter, with awned glumes; inferior glume c. 6 mm long, with short white hairs on the back and with a hispidulous awn c. 5 mm long; superior glume c. 6.5 mm long, glabrous, with an obscurely hispidulous awn c. 2.5 mm long; superior lemma hyaline, sparsely ciliolate on the margins; anthers c. 3 mm long.

Botswana. N: 42 km west of Gweta on new road to Maun, 3.iii.1966, *Blair Rains* 60 (K). SE: Gaborone, Aedume Park, 1050 m, 9.xi.1977, *Hansen* 3276 (K; PRE). **Zambia**. S: Kalomo Distr., 8 km south of Kalomo, iii.1969, *G. Williamson* 1533 (K). **Zimbabwe**. N: Hurungwe Distr., 520 m, 16.x.1957, *Phipps* 797 (K; PRE). W: Nyamandhlovu, 21.ix.1956, *Plowes* 1880 (K; PRE). C: Shurugwi Distr., Good Hope Farm, 3.xii.1948, *Davies* in GHS 22541 (K). S: Chiredzi Distr., Sengwe C.L. (Tribal Trust Land), 15.xi.1973, *Cleghorn* 2910 (K; PRE). **Mozambique**. T: Cahora (Cabora) Bassa, River Mucangádzi, c. 230 m, i.1973, *Torre, Carvalho & Ladeira* 18855 (LISC). GI: Mabote Distr., 10 km west of Zinave Camp, 1973, *Tinley* 2945 (PRE).

Tropical Africa and NW India. Common in shallow sandy soils and in rocky places; 230–1400 m.
This species is very similar to *C. plumulosus* Hochst. (from NE Africa and Arabia) but differs by its longer callus (up to 1.5 mm long), longer superior glume of sessile spikelet (4–7 mm long, as compared with 4–6 mm) and glabrous or obscurely hispidulous (not plumose) awns of the glumes of both the sessile and the pedicelled spikelets. Also, the apex of the superior lemma is bilobed in *C. serrulatus* but entire in *C. plumulosus.*

2. **Chrysopogon zizanioides** (L.) Roberty in Bull. Inst. Franç. Afrique Noire, Ser. A, Sci. Nat. **22**: 106 (1960). Type from India.
Phalaris zizanioides L., Mant. Pl. **2**: 183 (1771).
Andropogon muricatus Retz., Observ. Bot. **3**: 43 (1783). Type from India.
Agrostis verticillata Lam., Encycl. **1**: 59 (1783). Type from India.
Anatherum muricatum (Retz.) P. Beauv., Ess. Agrostogr.: 150 (1812).
Vetiveria odoratissima Lem.-Lesanc. in Bull. Sci. Soc. Philom. Paris **1822**: 43 (1822). Types from Sri Lanka, Réunion and Mauritius.
Vetiveria odorata Virey in J. Pharm. Sci. Accessoires **13**: 501 (1827). Type from 'East Indies.'
Andropogon festucoides J. Presl ex C. Presl, Reliq. Haenk. **1**: 340 (1830). Type from Philippines.
Vetiveria muricata (Retz.) Griseb., Fl. Brit. West Indies: 560 (1864).
Vetiveria arundinacea Griseb., Fl. Brit. West Indies: 559 (1864). Types from West Indies.
Sorghum zizanioides (L.) Kuntze, Revis. Gen. Pl. **2**: 791 (1891).
Vetiveria zizanioides (L.) Nash in Small, Fl. S.E. U.S.: 67 (1903). —Stapf in Bull. Misc. Inform., Kew **1906**: 346-349, 362 (1906); in Prain, F.T.A. **9**: 157 (1917). —Sturgeon in Rhodesia Agric. J. **51**: 17 (1954). —Chippindall in Meredith, Grasses & Pastures S. Africa: 470 (1955). —Jackson & Wiehe, Annot. Check List Nyasal. Grass.: 70 (1958). —Bor, Grasses Burma Ceyl. Ind. Pak.: 238 (1960). —Napper, Grasses Tanganyika: 99 (1965).

—Clayton in F.W.T.A., ed. 2, **3**: 470 (1972). —Lowe, Flora of Nigeria, Grasses: 291 (1989).
—Setshogo in Kirkia **17**: 140 (2001).
  Andropogon zizanioides (L.) Urban, Symb. Antill. **4**: 79 (1903).
  Holcus zizanioides (L.) Kuntze ex Stuck. in Anales Mus. Nac. Hist. Nat. Buenos Aires **11**: 48 (1904).
  Anatherum zizanioides (L.) Hitchc. & Chase in Contr. U.S. Natl. Herb. **18**: 285 (1917).

Perennial up to 200 cm high or more; culms erect, the nodes all concealed within the leaf sheaths; ligule a scarious rim; leaf laminas up to 90 cm × 4–20 mm, linear. Panicle up to 30 cm long or more, oblong, contracted; main axis smooth and glabrous; whorls 6–10 each with up to 20 racemes; racemes very slender; rhachis internodes as long as or slightly longer than the spikelets, glabrous; pedicels shorter than the sessile spikelets, glabrous. Sessile spikelet linear-lanceolate, variable in colour (yellowish, olive, violet-brown, purplish); callus up to 1 mm long, obtuse, glabrous; inferior glume muricate on the back; superior glume spinulose on the keel; inferior lemma as long as the glumes, acute, retrorsely ciliate; superior floret bisexual, the lemma up to 3 mm long, narrow, oblong-lanceolate, mucronate, glabrous; anthers 2–3 mm long. Pedicelled spikelet sparingly aculeolate or almost smooth, male; inferior lemma as for the sessile spikelet; superior lemma acute, entire; anthers c. 2.5 mm long.

Zambia. B: Barotse Floodplain, 26.iii.1964, *Verboom* 1168 (K). **Zimbabwe**. N: Mazowe Distr., Makalanga, cultivated, 22.v.1950, *Pollitt* in *GHS* 27926 (K). E: Chipinge Distr., Newcastle Block, La Lucie Farm, cultivated, 19.iii.1981, *Nicoll* in *GHS* 267119 (K).
  Southeast Asia to tropical Africa; introduced to most parts of the world. Mostly in cultivated fields.
  This species is principally used for thatching and is not grazed except when young and tender. Its root is the scented *khas khas* which is used for *tattis* or grass screens which are kept wetted to cool rooms. Commercial oils are extracted from its roots and used in perfumery. It is also sometimes grown as edgings to roads.

3. **Chrysopogon nigritanus** (Benth.) Veldk. in Austrobaleya **5**: 526 (1999). TAB. **14**. Type from Nigeria.
  Andropogon nigritanus Benth. in Hooker, Niger Fl.: 573 (1849).
  Mandelorna insignis Steud., Syn. Pl. Glumac. **1**: 359 (1854). Type from Senegambia.
  Andropogon squarrosus L.f. var. *nigritanus* (Benth.) Hack. in A. & C. de Candolle, Monogr. Phan. **6**: 544 (1889).
  Vetiveria nigritana (Benth.) Stapf in Prain, F.T.A. **9**: 157 (1917). —Stent & Rattray in Proc. & Trans. Rhodesia Sci. Assoc. **32**: 8 (1933). —Sturgeon in Rhodesia Agric. J. **51**: 17 (1954). —Chippindall in Meredith, Grasses & Pastures S. Africa: 469 (1955). —Jackson & Wiehe, Annot. Check List Nyasal. Grass.: 65 (1958). —Napper, Grasses Tanganyika: 99 (1965). — Hood, A Guide to the Grasses of Zambia: 58 (1967). —Simon in Kirkia **8**: 20, 54 (1971). —Clayton in F.W.T.A., ed. 2, **3**: 470 (1972). —Clayton & Renvoize in F.T.E.A., Gramineae: 739, fig. 172 (1982). —Lowe, Fl. Nigeria, Grasses: 291 (1989). —Gibbs Russell et al., Grasses South. Africa [Mem. Bot. Surv. S. Africa No. 58]: 353 (1990). —Setshogo in Kirkia **17**: 141 (2001).
  Vetiveria zizanioides var. *nigritana* (Benth.) A. Camus in Bull. Mus. Hist. Nat. (Paris) **25**: 674 (1919).

Caespitose perennial; culms 150–300 cm high, unbranched, the uppermost nodes exposed, glabrous; ligule scarious with shortly ciliate margin or a line of hairs on an extremely short scarious rim; leaf laminas up to 90 cm × 7 mm, narrow. Panicle 15–40 cm long, lanceolate; main axis minutely ciliolate; whorls 8–10 each with up to 15 racemes; racemes very slender; rhachis internodes longer than the spikelets, glabrous; pedicels shorter than the sessile spikelets, glabrous. Sessile spikelet c. 7 mm long, narrowly linear-lanceolate; inferior glume coriaceous, spinulose on the back; superior glume coriaceous to chartaceous, sharply keeled in the middle and with inflexed margins, spinulose along the keel, drawn out into a shortly aristate tip; inferior floret with ovate hyaline lemma; superior floret bisexual, the lemma hyaline with a bilobed apex; awn c. 5 mm long, sometimes slightly exserted from the glumes, glabrous on the column, minutely scabrid on the bristle. Pedicelled spikelet neuter, shorter than the sessile; glumes similar to those of the sessile spikelet but less coriaceous and less spinulose; inferior glume c. 5 mm long, sparingly aculeolate along the keel towards the apex or almost smooth; superior glume smooth with flexible ciliate margins; inferior lemma hyaline with ciliate margins.

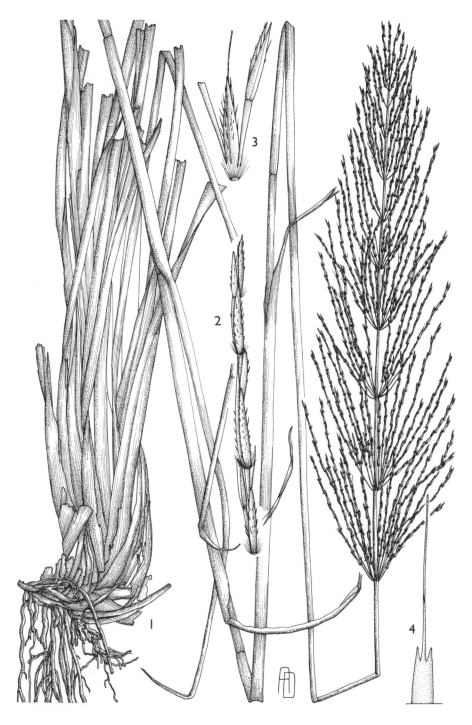

Tab. 14. CHRYSOPOGON NIGRITANUS. 1, habit (× ½); 2, raceme (× 3); 3, spikelet pair (× 4); 4, tip pf superior lemma (× 7), 1–4 from *Robinson* 414. Drawn by Ann Davies. From F.T.E.A.

Caprivi Strip. 7 km from Katima on road to Lisikili, 910 m, 24.xii.1958, *Killick & Leistner* 3084 (K; PRE). **Botswana**. N: Ngamiland Distr., Okavango River, Muhembo (Old Mohembo), 16 km north of Shakawe on Botswana border, 19.iii.1965, *Wild & Drummond* 7074 (BM; K; PRE). **Zambia**. B: Kaoma (Mankoya), near resthouse, 20.xi.1959, *Drummond & Cookson* 6644 (BM; E; PRE). W: Solwezi Distr., Mwombezhi River, west of Mutanda, 1250 m, 19.xii.1969, *Simon & Williamson* 1862 (K). C: Lusaka Distr., Chongwe River near Constantia, north of Kasisi (15°11'S, 28°27'E), 1120 m, 20.ii.1973, *Kornaś* 3338 (K). S: just south of Choma, 4.iv.1946, *Grassl* 46-62 (K). **Zimbabwe**. N: Gokwe North Distr., Copper Queen N.P.A., near Morowa River, 21.xii.1963, *Bingham* 901b (BM; K; PRE). W: Hwange Distr., Kazuma Depression, near Botswana border, 1070 m, 22.i.1974, *Gonde* 2/74 (K). **Malawi**. S: Nsanje (Port Herald), Tengani, 28.xi.1950, *Jackson* 310 (K; MAL). **Mozambique**. Z: Mopeia Distr., Mopeia, banks of the Zambezi River, 11.ix.1944, *Mendonça* 2037 (LISC). T: Sisitso Camp, Ulere Station, Zambezi River, 10.vii.1956, *Chase* 2671A (BM; K). MS: Marromeu Distr., between Lacerdónia and Marromeu, 9.v.1942, *Torre* 4123 (K; LISC).

Tropical Africa. Occurs mainly on river banks, occasionally on roadsides and in forested areas; 20–1300 m.

13. DICHANTHIUM Willemet

By M.P. Setshogo

Dichanthium Willemet in Usteri, Ann. Bot. **18**: 11 (1796). —De Wet & Harlan in Bol. Soc. Argent. Bot. **12**: 206–207 (1968).
Lepeocercis Trin., Fund. Agrost.: 203 (1820).
Diplasanthum Desv., Opusc. Sci. Phys. Nat.: 66 (1831).
Eremopogon Stapf in Prain, F.T.A. **9**: 182 (1917).

Annuals or perennials. Inflorescence of single, digitate or subdigitate racemes, these rarely branched below; spikelets conspicuously imbricate, those of a pair subequal to equal or unequal in size and shape, usually differing in sex except the lowermost 1–6 pairs which are, with rare exceptions, homogamous and male or neuter; rhachis internodes and pedicels solid. Sessile spikelet callus obtuse; inferior glume chartaceous to cartilaginous, broadly convex to slightly concave, sometimes pitted, acute to broadly obtuse at the apex; superior lemma forming the hyaline base to its awn, entire; awn glabrous. Pedicelled spikelet similar to the sessile spikelet, male or neuter, awnless, rarely bisexual and awned.

A genus of c. 20 species, occurring in the Old World tropics.

1. Inflorescence digitate with 1–15 sessile to subsessile racemes · · · · · · · · · · · · · · 3. *sericeum*
 – Inflorescence of 2–many digitate or subdigitate, shortly pedunculate racemes · · · · · · · 2
2. Culm nodes barbate; inferior glume of sessile spikelet pubescent in the lower third, with long tubercle-based cilia above the middle, along the margins and forming a subapical fringe · 1. *annulatum*
 – Culm nodes glabrous; inferior glume of sessile spikelet narrowly winged on the lateral keels, without tubercle-based cilia · 2. *aristatum*

1. **Dichanthium annulatum** (Forssk.) Stapf in Prain, F.T.A. **9**: 178 (1917). —Sturgeon in Rhodesia Agric. J. **51**: 19 (1954). —Bogdan, A Revised List of Kenya Grasses: 57 (1958). —Hood, A Guide to the Grasses of Zambia: 65 (1967). —Simon in Kirkia **8**: 17, 50 (1971). —Clayton in F.W.T.A., ed. 2, **3**: 471 (1972). —Clayton & Renvoize in F.T.E.A., Gramineae: 725 (1982). —Gibbs Russell et al., Grasses South. África [Mem. Bot. Surv. S. Africa No. 58]: 105 (1990). —Setshogo in Kirkia **17**: 142 (2001). TAB. **15**. Type from Egypt.
   Andropogon annulatus Forssk., Fl. Aegypt.-Arab.: 173 (1775).
   Andropogon scandens Roxb., Fl. Ind. **1**: 262 (1820). Type from India.
   Andropogon obtusus Nees in Hooker & Arnott, Bot. Beechey Voy.: 243 (1838) based on *A. bladhii* Roxb., Fl. Ind. **1**: 263 (1820), non Retz. (1781). Type from India.
   Lepeocercis annulata (Forssk.) Nees, Fl. Afr. Austral. Ill.: 98 (1841).
   Andropogon annulatus var. *decalvatus* Hack. in A. & C. de Candolle, Monogr. Phan. **6**: 572 (1889). Type from Egypt.
   Dichanthium annulatum (Forssk.) Stapf var. *decalvatum* (Hack.) Maire & Weiller, Fl. Afr. Nord **1**: 282 (1952).

Tab. 15. DICHANTHIUM ANNULATUM. 1, habit ($\times \frac{2}{3}$); 2, raceme ($\times 1\frac{1}{3}$); 3, ligule ($\times 4$); 4a & 4b, arrangement of spikelets ($\times 6$); 5, inferior glume of sessile spikelet ($\times 6$); 6, superior glume ($\times 6$); 7, inferior lemma ($\times 6$); 8, superior awned lemma ($\times 6$); 9, flower ($\times 6$); 10, caryopsis and cross section ($\times 10$), 1–10 from *Gillett & Rawi* 5915. Drawn by Derrick Erasmus. From Fl. Iraq.

Caespitose, decumbent or erect perennial with distinct short rhizomes; culms up to 200 cm high, woody and persistent, robust, geniculately ascending, simple or branched; nodes exposed and densely barbate; internodes glabrous; leaf sheaths glabrous or sparsely pilose; ligule a glabrous membrane c. 2 mm long; leaf laminas up to 30 cm × 4–4.5 mm, linear-lanceolate, glabrous for most of their length but long-pilose near the ligule. Racemes 2–many, subdigitate, shortly pedunculate, the peduncles glabrous; rhachis internodes and pedicels pilose, sometimes on one side only; the lowermost 1–6 spikelet pairs usually homogamous, male or neuter. Sessile spikelets 2–6 mm long, oblanceolate; inferior glume slightly concave, pubescent below the middle, with long tubercle-based cilia above; superior glume slightly shorter than the inferior, glabrous except for the often shortly ciliate keel and apex; inferior floret barren, the lemma c. 3.8 mm long, ovate, hyaline; superior floret stipitiform, the lemma c. 3 mm long; awn c. 25 mm long, minutely hispidulous on both column and bristle; anthers c. 1.5 mm long. Pedicelled spikelets similar to the sessile spikelets in size and shape, often slightly smaller, male or neuter.

Var. **annulatum**

Inferior glume of sessile spikelet pubescent to pilose below the middle, the bulbous-based hairs mainly confined to the margins above.

Zambia. C: Lusaka Distr., Kafue Bridge, near Kafue, 15°40'S, 28°11'E, 970 m, 9.xii.1971, *Kornaś* 628 (K). E: Chipata Distr., Luangwa Valley, Jumbe, 30.xii.1965, *Astle* 4228 (K). S: Mazabuka Distr., 40 km from Chirundu, 910 m, 3.i.1968, *Simon* 1630 (K; SRGH). **Zimbabwe**. N: Mudzi Distr., Mkota Reserve, Nyangombe R., 1.x.1948, *Wild* 2677 (K). W: Nkayi Distr., Gwampa Forest Land, ii.1955, *Goldsmith* 156/55 (K). C: Mazowe Distr., Passaford Road, off Sinoia road, 1460 m, 30.viii.1955, *Boughey* in *GHS* 55587 (K). **Malawi**. S: Lower Shire Valley, xii.1861–i.1862, *Kirk* s.n. (K). **Mozambique**. N: Pemba, 22.x.1993, *Bandeira & Boane* 407 (K). T: Changara Distr., Mazoe R., 8 km from Zimbabwe border, 300 m, 22.ix.1948, *Wild* 2586 (K). MS: Expedition Island, vii.1858, *Kirk* s.n. (K). GI: Massingir, 22.v.1972, *Myre, Lousã & Rosa* 5781 (K).
In the tropics and subtropics of Africa eastwards through Arabia and India to Southeast Asia; introduced in tropical America and Australia. Growing in sandy riverbeds and on river margins, lake shores, grassy floodplains, riverine forest margins and occasionally on roadsides; 100–1460 m.

Var. **papillosum** (Hochst. ex A. Rich.) de Wet & Harlan in Bol. Soc. Argent. Bot. **12**: 212 (1968). Type from Ethiopia.
 Andropogon papillosus Hochst. ex A. Rich., Tent. Fl. Abyss. **2**: 457 (1851).
 Dichanthium papillosum (Hochst. ex A. Rich.) Stapf in Prain, F.T.A. **9**: 179 (1917). —Jackson & Wiehe, Annot. Check List Nyasal. Grass.: 34 (1958). —Simon in Kirkia **8**: 17, 50 (1971).

Inferior glume of sessile spikelet pilose to villous, with a distinct subapical fringe of bulbous-based hairs.

Caprivi Strip. Mpalela (Mpilila) Island, 900 m, 15.i.1959, *Killick & Leistner* 3407 (K). **Botswana**. N: Ngamiland Distr., Qangwa R. (Quangwadum watercourse), 19°34'S, 21°01'E, 28.iv.1980, *P.A. Smith* 3498 (K; PRE). SE: Uitspan Farm, North Tuli Block, Majali River, 13.iii.1975, *Stephen* 1479 (K). **Zambia**. S: Livingstone Distr., Game Park, 14.ii.1961, *B.L. Mitchell* 5/78 (K). **Zimbabwe**. N: Gokwe South Distr., Sengwa Research Station, 10.i.1975, *P.R. Guy* 2256 (K; PRE). W: Hwange Distr., Victoria Falls National Park, 850 m, 24.iv.1970, *Simon & Hill* 2125 (K). C: Seke Distr., south of Beatrice, 13.iii.1975, *Cleghorn* 3030 (K; PRE). E: Chipinge Distr., east of Sabi R, between Musaswi (Musasve) R and Cilariati (Cikariati) R, 400 m, 22.i.1957, *Phipps* 104 (K; PRE). S: Mberengwa Distr., Ngezi R., Mt. Buhwa, 910 m, x.1973, *Gosden* 10 (K; PRE). **Malawi**. S: Zomba Distr., Chilwa (Shirwa) Lake, 15.xi.1950, *Wiehe* 677 (K). **Mozambique**. T: Mágoè Distr., c. 32 km from Chicoa towards Mágoè, c. 300 m, 18.ii.1970, *Torre & Correia* 18038 (LISC). M: Matutuíne (Bela Vista), 27.iv.1948, *Torre* 7726 (LISC).
Ethiopia to South Africa. Dry open places subject to overgrazing or disturbance; also in swampy depressions; 10–1300 m.
The species has two commonly recognized varieties, as above. *D. fecundum* S.T. Blake, from Australia and Papua New Guinea, in which some of the pedicelled spikelets in each raceme are fertile and awned, is probably best regarded as a third.

2. **Dichanthium aristatum** (Poir.) C.E. Hubb. in Bull. Misc. Inform., Kew **1939**: 654 (1939). —Sturgeon in Rhodesia Agric. J. **51**: 19 (1954). —Chippindall in Meredith, Grasses &

Pastures S. Africa 481 (1955). —Simon in Kirkia **8**: 17, 50 (1971). —Clayton & Renvoize in F.T.E.A., Gramineae: 723 (1982). —Gibbs Russell et al., Grasses South. Africa [Mem. Bot. Surv. S. Africa No. 58]: 105 (1990). —Setshogo in Kirkia **17**: 143 (2001). Type from Mauritius.

> *Dichanthium nodosum* Willemet in Usteri, Ann. Bot. **18**: 11 (1796), *nom. superfl.*, based on *Andropogon annulatus* Forssk. (but the description is that of *D. aristatum*).
> *Andropogon aristatus* Poir. in Lam., Encycl., Suppl. **1**: 585 (1810).
> *Andropogon mollicomus* Kunth, Révis. Gramin. **1**: 365 (1830). Type from Mauritius.
> *Diplasanthum lanosum* Desv., Opusc. Sci. Phys. Nat.: 67 (1831). Type from India.
> *Lepeocercis mollicoma* (Kunth) Nees in Edinburgh New Philos. J. **18**: 185 (1835).
> *Andropogon caricosus* L. subsp. *mollicomus* (Kunth) Hack. in A. & C. de Candolle, Monogr. Phan. **6**: 569 (1889).
> *Andropogon nodosus* (Willemet) Nash, N. Amer. Fl. **17**: 122 (1912).
> *Diplasanthum caricosum* (L.) A. Camus var. *mollicomus* (Kunth) Haines, Bot. Bihar Orissa **5**: 1039 (1924). —Stent & Rattray in Proc. & Trans. Rhodesia Sci. Assoc. **32**: 9 (1933).

Caespitose perennial with short stolons; culms up to 110 cm high, robust, erect or sometimes decumbent; nodes exposed and glabrous; internodes glabrous; leaf sheaths glabrous; ligule a scarious membrane; leaf laminas up to 25 cm × 2–5 mm, linear-lanceolate, glabrous for most of their length but long-pilose near the ligule. Racemes 1–5, subdigitate, shortly pedunculate, the peduncles and the culm below the inflorescence pubescent; rhachis internodes and pedicels usually pilose on one side, glabrous on the other; the lowermost 1–6 spikelet pairs usually homogamous, male or neuter. Sessile spikelets 2–5 mm long, obovate; inferior glume slightly concave, pilose below the middle, glabrous or shortly pilose above the middle towards the margins and apex, narrowly winged towards the apex; superior glume as long as the inferior, glabrous except for the often shortly ciliate keel and margins; inferior floret barren, the lemma hyaline; superior floret stipitiform, the lemma hyaline; awn c. 22 mm long, minutely pubescent on both column and bristle; anthers c. 2 mm long. Pedicelled spikelets similar to the sessile spikelets in size and shape, male or neuter.

Zambia. C: Luangwa Valley, South Luangwa National Park, near the Lundu Plains, 760 m, 4.v.1966, *Astle* 4865 (K; SRGH). **Zimbabwe**. C: Harare, university campus on southern slope, 14.iii.1985, *K.E. Bennett* in GHS 283137 (K; PRE). **Mozambique**. M: Matutuíne (Bela Vista), bank of Maputo River, 11.vii.1958, *Mogg* 27911A (K; PRE).
Native of India, introduced to Africa, Australia and America for its value as a fodder grass. Grows in damp places on disturbed ground; 10–1600 m.
The species is characterized by its robust, suberect habit, by having the peduncles to the racemes villous and by having the culm below the inflorescence strongly pilose. It is related to, and sometimes regarded as conspecific with, *D. caricosum* (L.) A. Camus (Tanzania and tropical Asia), but this has a glabrous culm below the inflorescence.

3. **Dichanthium sericeum** (R. Br.) A. Camus in Bull. Mus. Hist. Nat. (Paris) **27**: 549 (1921). —Burbidge, Australian Grasses **1**: 146 (1966). —Lazarides, The Grasses of Central Australia: 70 (1970). —Wheeler, Jacobs & Norton, Grasses of New South Wales: 163 (1982). —Simon, A Key to Australian Grasses, ed. 2: 98 (1993). —Simon & Latz, A Key to the Grasses of the Northern Territory, Australia: 26 (1994). Type from Australia.

> *Andropogon sericeus* R. Br., Prodr. Fl. Nov. Holl.: 201 (1810).
> *Andropogon affinis* R. Br., Prodr. Fl. Nov. Holl.: 201 (1810). Type from Australia.
> *Andropogon chrysantherus* F. Muell. in Linnaea **25**: 443 (1852). Type from Australia.
> *Andropogon jubatus* Balansa in Bull. Soc. Bot. France **19**: 322 (1872). Type from New Caledonia.
> *Andropogon acutiusculus* Hack. in A. & C. de Candolle, Monogr. Phan. **6**: 575 (1889). Type from Australia.
> *Andropogon sericeus* var. *mollis* F.M. Bailey in Queensland Agric. J. **30**: 316 (1913). Type from Australia.
> *Dichanthium acutiusculum* (Hack.) A. Camus in Bull. Mus. Hist. Nat. (Paris) **27**: 549 (1921).

Caespitose perennial with short rhizomes; culms up to 100 cm high, robust, erect or slightly decumbent; nodes exposed and bearded; internodes glabrous; leaf sheaths glabrous to pilose; ligule a scarious membrane up to 1.5 mm long; leaf laminas up to 15 cm long, linear-lanceolate, glabrous to pilose throughout. Racemes 1–15, digitate, sessile to subsessile, white-villous, the culm below the inflorescence glabrous; rhachis internodes and pedicels thinly pilose; the lowermost 1–6 spikelet pairs usually homogamous, male or neuter. Sessile spikelets elliptic-oblong; inferior glume pilose or rarely glabrous on the back below the middle, always with tubercle-based cilia (up to 6 mm long) along the margins and forming a transverse subapical fringe; superior glume slightly longer than the inferior, glabrous; inferior floret barren, the lemma hyaline; superior floret stipitiform, the lemma hyaline, without a palea; awn c. 30 mm long; anthers c. 2 mm long. Pedicelled spikelets c. 4 mm long, obovate-oblong.

Zimbabwe. C: Harare (Salisbury), cult., 1460 m, xii.1919, *Eyles* 1971 (K; PRE).
Native of Australia; cultivated in tropical Africa.
The racemes are sessile or very shortly pedunculate. The inferior glume of the sessile spikelet is covered in long white silky hairs and has an arch of extremely long cilia near the apex, many of which arise from tubercles.

14. CAPILLIPEDIUM Stapf

By M.P. Setshogo

Capillipedium Stapf in Prain, F.T.A. **9**: 169 (1917).
Filipedium Raiz. & Jain in J. Bombay Nat. Hist. Soc. **49**: 682 (1951).

Annuals or perennials; mostly rambling. Inflorescence a delicate loose panicle bearing short, 1–5(8)-jointed racemes at the ends of capillary branches and branchlets; rhachis internodes and pedicels filiform, with a translucent longitudinal groove. Sessile spikelet callus obtuse; inferior glume cartilaginous, broadly convex to slightly concave, acute or obtuse; superior lemma forming the hyaline base to its awn, entire; awn glabrous. Pedicelled spikelet male or neuter.

A genus of c. 14 species, occurring in eastern Africa and extending through tropical Asia to Australia.

Capillipedium parviflorum (R. Br.) Stapf in Prain, F.T.A. **9**: 169 (1917). —Stent & Rattray in Proc. & Trans. Rhodesia Sci. Assoc. **32**: 9 (1933). —Sturgeon in Rhodesia Agric. J. **51**: 17 (1954). —Bogdan, A Revised List of Kenya Grasses: 56 (1958). —Jackson & Wiehe, Annot. Check List Nyasal. Grass.: 32 (1958). —Napper, Grasses Tanganyika: 99 (1965). —Simon in Kirkia **8**: 17, 50 (1971). —Clayton & Renvoize in F.T.E.A., Gramineae: 718 (1982). —Setshogo in Kirkia **17**: 141 (2001). TAB. **16**. Type from Australia.
Holcus parviflorus R. Br., Prodr. Fl. Nov. Holl.: 199 (1810).
Sorghum parviflorum (R. Br.) P. Beauv., Ess. Agrostogr.: 132 (1812).
Anatherum parviflorum (R. Br.) Spreng., Syst. Veg. **1**: 290 (1824).
Holcus caerulescens Gaud. in Freycinet, Voy. Uranie: 411 (1829). Type from Australia.
Andropogon micranthus Kunth, Révis. Gramin. **1**: 165 (1829). Type from Australia.
Andropogon alternans J. Presl in C. Presl, Reliq. Haenk. **1**: 342 (1830). Type said to be from Peru, but this is thought to be an error.
Rhaphis caerulescens (Gaud.) Desv., Opusc. Sci. Phys. Nat.: 69 (1831).
Chrysopogon violascens Trin. in Mém. Acad. Imp. Sci. St.-Pétersbourg, Sér. 6, Sci. Math. **2**: 319 (1832). Type from Australia.
Andropogon quartinianus A. Rich., Tent. Fl. Abyss. **2**: 469 (1851). Type from Ethiopia.
Andropogon violascens (Trin.) Nees ex Steud., Syn. Pl. Glumac. **1**: 396 (1854).
Andropogon parvispicus Steud., Syn. Pl. Glumac. **1**: 397 (1854). Type from Nepal.
Andropogon capilliflorus Steud., Syn. Pl. Glumac. **1**: 397 (1854). Type from Java.
Andropogon villosulus Nees ex Steud., Syn. Pl. Glumac. **1**: 397 (1854). Type from Nepal.
Rhaphis microstachya Nees ex Steud., Syn. Pl. Glumac. **1**: 397 (1854), *nom. inval.*, in syn.
Andropogon parvispicus Steud.
Andropogon serratus Miq. in Ann. Mus. Bot. Lugduno-Batavum **2**: 290 (1866), non Thunb. (1784). Type from Japan.
Chrysopogon parviflorus (R. Br.) Benth., Fl. Austral. **7**: 537 (1878).
Chrysopogon parvispicus (Steud.) W. Watson in Atkinson, Gaz. NW. India: 392 (1882).

Caespitose perennial; culms mostly 75–100 cm high with barbate nodes; leaf sheaths tight or eventually slipping from the culms, the lowermost longer than and the remainder shorter than the internodes, ± covered in tubercle-based hairs and frequently villous on the collar; ligule a very short, truncate, ciliolate membrane; leaf laminas up to 30 cm × 2.5 mm, glaucous, erect and straight, linear from an often narrowed and slightly contracted base, flat, scabrid on the margins, finely pointed at the apex. Panicle 8–25 cm long; racemes reduced to 1 sessile and 2 pedicelled spikelets, rarely with 2 sessile and 3 pedicelled spikelets; rhachis internodes glabrous,

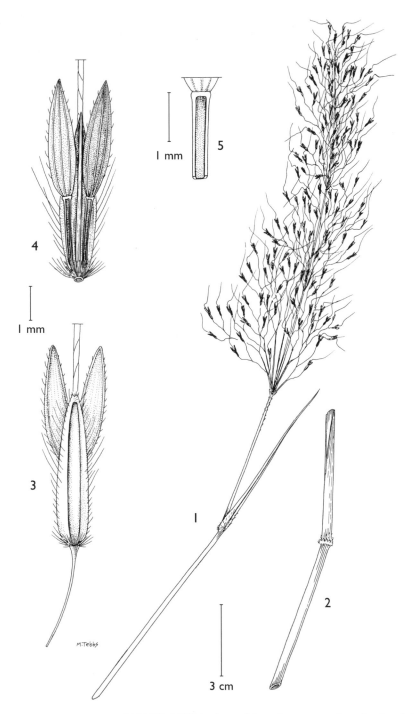

Tab. 16. **CAPILLIPEDIUM PARVIFLORUM.** 1, panicle and upper portion of culm; 2, leaf sheath showing villous node; 3, dorsal view of sessile spikelet; 4, dorsal view of pedicelled spikelets; 5, detail of pedicel, base cut away to show shape, 1–5 from *Kerr* 6. Drawn by Margaret Tebbs.

the pedicels hairy on the margins. Sessile spikelet 2.8–5 mm long, narrowly oblong; inferior glume with short hairs on the back, shallowly concave in the middle, with 2 intercarinal nerves, ciliate on the keels above, truncate at the apex; superior glume 1-keeled, scabrid on the keel, the median nerve excurrent as a short mucro; inferior lemma empty, 2–2.5 mm long, ovate; superior lemma stipitiform, linear, flattened, 1-nerved, entire; awn c. 20 mm long (including the lemma), minutely hispidulous; anthers c. 1.5 mm long. Pedicelled spikelets 2.5–3.5 mm long, lanceolate, male or neuter; inferior glume shortly hairy on the keels, with 3 intercarinal nerves, acute at the apex; superior glume glabrous, with 2 intercarinal nerves; inferior lemma oblong, truncate; anthers c. 1.5 mm long.

Zambia. E: Chipata (Fort Jameson), 25.v.1963, *Verboom* 761 (BM; K; SRGH). **Zimbabwe**. N: Zvimba Distr., Trelawney, 26.iv.1931, *Brain* 3757 (K). C: Gweru Distr., 10 km south from Gweru, 1400 m, 18.iii.1967, *Biegel* 2000 (BM; K). E: Mutare (Umtali), 1300 m, 3.ii.1974, *A.O. Crook* 2012 (K; PRE). **Malawi**. C: Dowa Distr., Uperere Mission, Chankalamu Dambo, 2.xi.1950, *Jackson* 249 (K; MAL). **Mozambique**. N: Lago Distr., serra Jeci, prox. de Malulo, a c. 60 km de Lichinga (Vila Cabral), 1700 m, 3.iii.1964, *Torre & Paiva* 11005 (K; LISC). MS: Parque Nacional da Gorongosa (Parque Nacional de Caça), andados 2 km do Acampamento de Chitengo para o batelao do rio Urema, c. 40 m, 2.v.1964, *Torre & Paiva* 12207 (LISC).

Widely distributed throughout the tropics of the Old World; also in China and Japan. Grassland; 40–1400 m.

Most specimens have the racemes reduced to a single sessile spikelet; specimens with more than one sessile spikelet include *Verboom* 761 (from Zambia E) and *Verboom* 978 (from Malawi C).

15. BOTHRIOCHLOA Kuntze

By M.P. Setshogo

Bothriochloa Kuntze, Revis. Gen. Pl. **2**: 762 (1891).
Andropogon subgen. *Gymnandropogon* Nees, Fl. Afr. Austral. Ill.: 103 (1841).
Gymnandropogon (Nees) Duthie in Atkinson, Gaz. NW. Prov. & Oude **10**: 638 (1882).
Amphilophis Nash in Britton, Man. Fl. N. States: 71 (1901).

Perennials. Inflorescence of few- to many-jointed pedunculate racemes, these digitate or subdigitate; racemes simple or the lower ones sparsely branched, without homogamous pairs; sessile and pedicelled spikelets similar in size and shape or the pedicelled reduced and smaller, always differing in sex; rhachis internodes and pedicels linear-filiform, with a translucent longitudinal groove. Sessile spikelet callus obtuse; inferior glume mostly cartilaginous, broadly convex to slightly concave, sometimes with 1–3 circular pits on the back, acute at the apex; superior lemma forming the hyaline base to its awn, entire (rarely bilobed; not in the Flora Zambesiaca area). Pedicelled spikelet male or neuter.

A genus of c. 35 species, occurring throughout the tropics.

1. Inflorescence paniculate, with a long central axis · 2. *bladhii*
– Inflorescence subdigitate · 2
2. Inferior glume of sessile spikelet pitted · 3. *insculpta*
– Inferior glume of sessile spikelet not pitted · 1. *radicans*

1. **Bothriochloa radicans** (Lehm.) A. Camus in Ann. Soc. Linn. Lyon, n.s. **76**: 164 (1931). —
   Sturgeon in Rhodesia Agric. J. **51**: 18 (1954). —Chippindall in Meredith, Grasses & Pastures S. Africa: 48 (1955). —Bogdan, A Revised List of Kenya Grasses: 56 (1958). — Napper, Grasses Tanganyika: 100 (1965). —Hood, A Guide to the Grasses of Zambia: 57 (1967). —Simon in Kirkia **8**: 17, 50 (1971). —Clayton & Renvoize in F.T.E.A., Gramineae: 721 (1982). —Müller, Grasses of South West Africa/Namibia: 86 (1984). —Gibbs Russell et al., Grasses South. Africa [Mem. Bot. Surv. S. Africa No. 58]: 63 (1990). —Setshogo in Kirkia **17**: 143 (2001). Type cultivated in Hamburg from seed from South Africa.
   Andropogon radicans Lehm., Sem. Hort. Bot. Hamburg. (1828), *fide* Kunth, Enum. Pl. **1**: 499 (1833).
   Andropogon ischaemum L. var. *radicans* (Lehm.) Hack. in A. & C. de Candolle, Monogr. Phan. **6**: 476 (1889).
   Amphilophis radicans (Lehm.) Stapf in Prain, F.T.A. **9**: 172 (1917).

Caespitose, often stoloniferous, perennial, shrubby or decumbent and rooting at the nodes; culms up to 70 cm high, branching from the inferior nodes to form cushions; culm nodes mostly exposed and barbate, the uppermost glabrous; leaf sheaths glabrous except at the mouth, those of the basal leaves loose and open, the remainder clasping the culms; ligule membranous; leaf laminas 6–20 cm × 2–6 mm, linear-lanceolate, gradually tapering to a fine point. Inflorescence of 5–16 subdigitate racemes; main axis slender, shorter than the lowest racemes; racemes up to 70 mm long, 10–20-jointed, shortly pedunculate; rhachis internodes densely long-pilose. Sessile spikelet 2.5–4 mm long, lanceolate; inferior glume firmly membranous, slightly concave on the back and pilose in the lower half; superior glume minutely hispid along the keel, acute at the apex; inferior floret empty, the lemma c. 3 mm long, oblong-linear, ciliolate at the apex; superior floret bisexual, epaleate, the lemma c. 1.5 mm long, stipitiform, glabrous, awned; awn c. 25 mm long, geniculate; anthers c. 1.5 mm long. Pedicelled spikelet neuter; inferior glume c. 3.5 mm long, oblong, with sparse long white hairs along the lateral keels and ciliolate towards the apex; superior glume c. 3 mm long, lanceolate, delicate, ciliate on the margins; inferior lemma c. 2 mm long, oblong, obtuse at the apex.

Caprivi Strip. Mpalela (Mpilila) Is., 900 m, 13.i.1959, *Killick & Leistner* 3350 (K). **Botswana**. N: Tsigara (Sigara) Pan, 48 km west of Nata River, 896 m, 26.iv.1957, *Drummond & Seagrief* 5240 (K; SRGH). SW: Kweneng Distr., c. 12 km NW of Lephepe Village, near Kalahari Sandveld Pasture Research Station, iii.1969, *Kelaole* 531 (K; PRE). SE: Seleka Ranch (23°00'S, 27°50'E), 790 m, 22.ii.1977, *Hansen* 3048 (K; PRE). **Zambia**. C/E: Luangwa Valley, c. 600 m, 11.ii.1966, *Astle* 4542 (K). **Zimbabwe**. W: Bulawayo Distr., Hlekwani, c. 10 km SW of Bulawayo, 1350 m, 19.xii.1990, *Laegaard* 15828 (K). S: Mwenezi Distr., Nuanetsi Ranch, Bubi Section, 23.ii.1967, *Cleghorn* 1421 (K). **Mozambique**. T: Cahora (Cabora) Bassa, 5 km from the Dam on the edge of the River Mucangádze, c. 300 m, i.1973, *Torre, Carvalho & Ladeira* 18795 (LISC). GI: Massingir, c. 26 km from Lagoa Nova, 14.iv.1972, *Myre, Lousã & Rosa* 5767 (K; PRE).

Tropical and South Africa; introduced into tropical America. Growing in open bushland; 140–1350 m.

2. **Bothriochloa bladhii** (Retz.) S.T. Blake in Proc. Roy. Soc. Queensland **80**: 62 (1969). —Clayton in F.W.T.A., ed. 2, **3**: 470 (1972). —Clayton & Renvoize in F.T.E.A., Gramineae: 719 (1982). —Lowe, Fl. Nigeria, Grasses: 225 (1989). —Gibbs Russell et al., Grasses South. Africa [Mem. Bot. Surv. S. Africa No. 58]: 62 (1990). —Setshogo in Kirkia **17**: 144 (2001). TAB. **17**. Type from China.

    Andropogon bladhii Retz., Observ. Bot. **2**: 27 (1781).
    Andropogon intermedius R. Br., Prodr. Fl. Nov. Holl.: 202 (1810). Type from Australia.
    Andropogon punctatus Roxb., Fl. Ind. **1**: 268 (1820). Type from India.
    Andropogon glaber Roxb., Fl. Ind. **1**: 271 (1820). Type from India.
    Andropogon haenkei J. Presl in C. Presl, Reliq. Haenk. **1**: 340 (1830). Type from Philippines.
    Andropogon caucasicus Trin. in Mém. Acad. Imp. Sci. St.-Pétersbourg, Sér. 6, Sci. Math. **2**: 286 (1832). Type from Caucasus.
    Rhaphis stricta Nees in Hooker's J. Bot. Kew Gard. Misc. **2**: 99 (1850). Type from Philippines.
    Andropogon inundatus F. Muell. in Linnaea **25**: 444 (1853). Type from Australia.
    Andropogon leptanthus Steud., Syn. Pl. Glumac. **1**: 391 (1854). Type from East Indies.
    Andropogon perfossus Nees & Meyen ex Steud., Syn. Pl. Glumac. **1**: 391 (1954), in syn. *A. punctatus* Roxb.
    Bothriochloa anamitica Kuntze, Rev. Gen. Pl. **2**: 762 (1891). Type from Vietnam.
    Sorghum intermedium (R. Br.) Kuntze, Rev. Gen. Pl. **2**: 792 (1891).
    Sorghum intermedium var. *haenkei* (J. Presl) Kuntze, Rev. Gen. Pl. **2**: 792 (1891).
    Amphilophis intermedia (R. Br.) Stapf in Agric. News (Barbados) **15**: 179 (1916).
    Amphilophis glabra (Roxb.) Stapf in Prain, F.T.A. **9**: 172 (1917). —Stent & Rattray in Proc. & Trans. Rhodesia Sci. Assoc. **32**: 9 (1933).
    Amphilophis glabra var. *haenkei* (J. Presl) E.G. & A. Camus in Lecomte, Fl. Indo-Chine **7**: 311 (1922).
    Amphilophis haenkei (J. Presl) Haines, Bot. Bihar Orissa: 1029 (1924).
    Bothriochloa intermedia (R. Br.) A. Camus in Ann. Soc. Linn. Lyon, n.s. **76**: 164 (1931).
    Bothriochloa glabra (Roxb.) A. Camus in Ann. Soc. Linn. Lyon, n.s. **76**: 164 (1931). —Jackson & Wiehe, Annot. Check List Nyasal. Grass.: 30 (1958). —Vesey-FitzGerald in Kirkia **3**: 103 (1963). —Simon in Kirkia **8**: 16, 49 (1971). —Jacobsen in Kirkia **9**: 147 (1973).
    Amphilophis insculpta sensu Stent & Rattray in Proc. & Trans. Rhodesia Sci. Assoc. **32**: 9 (1933), in part, non (Hochst. ex A. Rich.) Stapf.
    Bothriochloa inundata (F. Muell.) J.M. Black in Trans. & Proc. Roy. Soc. South Australia **60**: 163 (1936).

Tab. 17. BOTHRIOCHLOA BLADHII. 1, habit (× ½); 2, ligule (× 6); 3, portion of raceme (× 4½); 4, sessile spikelet (× 9); 5, superior lemma (× 12); 6, pedicelled spikelet (× 9). Drawn by Victoria Goaman. From Ghana Gr.

Bothriochloa caucasica (Trin.) C.E. Hubb. in Bull. Misc. Inform., Kew **1939**: 101 (1939).
Bothriochloa glabra subsp. *haenkei* (J. Presl) Henrard in Blumea **3**: 456 (1940).
Bothriochloa haenkei (J. Presl) Ohwi in Acta Phytotax. Geobot. **11**: 168 (1942).
Bothriochloa glabra (Roxb.) A. Camus var. *epunctata* Jackson, Annot. Check List Nyasal. Grass.: 30 (1958), *nom. nud.*
Dichanthium bladhii (Retz.) Clayton in Kew Bull. **32**: 3 (1977).

Caespitose perennial of variable habit; culms up to 100 cm high, rather straggling, the nodes glabrous, the internodes terete or channelled on one side; leaf sheaths barbate at the base; ligule membranous; leaf laminas up to 30 cm × 10 mm, almost glabrous, tapering gradually to a slender apex.   Inflorescence paniculate, with numerous slender branches 2–5 cm long on a common axis 3–10 cm long; lowermost branches sometimes themselves branched; spikelets usually dark purple or flushed with purple; rhachis internodes and pedicels pilose with hairs up to 2.5 mm long. Sessile spikelet 3–4 mm long, barbate at the base; inferior glume pilose below the middle on the back, rarely glabrous, rigidly ciliate on the margins above, depressed along the middle, with a deep or shallow pit in the upper part, the pit sometimes absent in some spikelets, rarely absent altogether; superior glume lanceolate, the margins sparsely ciliate above or glabrous, depressed along both sides of the keel, the keel scabrid above; inferior floret empty, the lemma oblong, obtuse, glabrous; superior floret epaleate, with stipitiform lemma; awn 10–18 mm long, geniculate, glabrous; anthers c. 1.5 mm long.   Pedicelled spikelet usually smaller than the sessile spikelet and reduced to 1 or 2 glumes, rarely well developed and with the inferior glume pitted.

Botswana. N: Tabu Island, east of Gumare, 15.i.1983, *Astle* 8228 (SRGH). SE: Central Distr., Madibeng, 980 m, 11.xii.1959, *de Beer* 870 (BM). **Zambia**. N: North Luangwa National Park (11°46'S, 32°03'E), 800 m, 25.ii.1994, *P.P. Smith* 938 (K). C: Lusaka Distr., Mt. Makulu, 30.i.1973, *Delmotte* 949 (SRGH). E: Chipata (Fort Jameson), iii.1962, *Verboom* 588 (K; WAG). S: Namwala Distr., Kafue National Park, Kafue River at Kalala, 8.xii.1963, *B.L. Mitchell* 24/29 (BM; PRE; SRGH). **Zimbabwe**. N: Hurungwe Distr., Karoi–Chinhoyi (Sinoia) road, 1070 m, ii.1972, *Davies* 3180 (SRGH). W: Nkayi Distr., Gwampa vlei, 910 m, i.1956, *Goldsmith* 45/56 (PRE). C: Harare (Salisbury), 14.iii.1945, *Weber* 143 (K). E: Lower Sabi area, c. 460 m, 27.i.–2.ii.1948, *Rattray* 1286 (K). S: Masvingo (Fort Victoria), 1070 m, 18.ii.1974, *Bezuidenhout* 59 (K). **Malawi**. N: Mzimba, Kasitu Valley, 1070 m, 26.ii.1938, *Fenner* 314 (K). C: Kasungu Game Reserve, 1000 m, 18.ii.1970, *Hall-Martin* 563 (PRE). S: Blantyre Distr., Limbe, Chichiri Campus, University of Malawi, 4.vii.1971, *Banda* 1160 (MAL). **Mozambique**. N: Montepuez Distr., andados 35 km de Montepuez para Namuno, 480 m, 31.xii.1963, *Torre & Paiva* 9791 (K). Z: Lugela Distr., Namagoa, 60–120 m, *Faulkner* 35 (K; PRE). MS: Gorongosa Distr., Parque Nacional da Gorongosa, Chitengo, 13 km from Acampamento, 2.v.1978, *Diniz* 164 (WAG); Parque Nacional da Gorongosa (Gorongosa National Park), iii.1971, *Tinley* 2055 (K; LISC). M: Manhiça Distr., Ilha Mariana, 7.iv.1954, *Myre & Carvalho* 1776 (PRE).

Africa to India, Australia and the Pacific; introduced in N America. Growing on streamsides, swamp margins and cracking clays; 40–1500 m.

The species is sometimes confused with *Capillipedium* (q.v.) on account of its paniculate inflorescence with capillary branches and branchlets.   It has 10–25-jointed panicle branches, compared with 1–10-jointed ones in *Capillipedium*.   Work on the genetic barriers around *B. bladhii* (under the name *B. intermedia*) has established that the species is capable of crossing not only with other species of *Bothriochloa* but also with species of the related genera *Dichanthium* and *Capillipedium* (De Wet & Harlan, 1966; Faruqi, 1969).   This led De Wet & Harlan to unite the three genera under *Dichanthium*, based on the biological species concept. *B. bladhii*, along with all of its hybrid progeny, was accommodated in a single taxon referred to as a 'compilospecies.'   Although we no longer accept the single genus concept, the compilospecies view of *B. bladhii* is still the most practical way of dealing with the problem of hybridization.   The species comprises all elements of *Bothriochloa* that have an elongated inflorescence axis.

3. **Bothriochloa insculpta** (Hochst. ex A. Rich.) A. Camus in Ann. Soc. Linn. Lyon, n.s. **76**: 165 (1931). —Sturgeon in Rhodesia Agric. J. **51**: 18 (1959). —Chippindall in Meredith Grasses & Pastures S. Africa: 483 (1955). —Bogdan, A Revised List of Kenya Grasses: 56 (1958). —Jackson & Wiehe, Annot. Check List Nyasal. Grass.: 31 (1958). —Napper, Grasses Tanganyika: 100 (1965). —Hood, A Guide to the Grasses of Zambia: 57 (1967). —Simon in Kirkia **8**: 17, 50 (1971). —Clayton & Renvoize in F.T.E.A., Gramineae: 720 (1982). —Gibbs Russell et al., Grasses South. Africa [Mem. Bot. Surv. S. Africa No. 58]: 63 (1990). —Setshogo in Kirkia **17**: 145 (2001). Type from Ethiopia.
Andropogon insculptus Hochst. ex A. Rich., Tent. Fl. Abyss. **2**: 458 (1851).

Andropogon pertusus (L.) Willd. var. *capensis* Hack. in A. & C. de Candolle, Monogr. Phan. **6**: 482 (1889). Type from South Africa.

Andropogon pertusus var. *insculptus* (Hochst. ex A. Rich.) Hack. A. & C. in de Candolle, Monogr. Phan. **6**: 482 (1889).

Andropogon pertusus subvar. *trifoveolatus* Hack. in A. & C. de Candolle, Monogr. Phan. **6**: 482 (1889). Type from Ethiopia.

Amphilophis insculpta (Hochst. ex A. Rich.) Stapf in Prain, F.T.A. **9**: 176 (1917). —Stent & Rattray in Proc. & Trans. Rhodesia Sci. Assoc. **32**: 9 (1933), in part.

Bothriochloa pertusa (L.) A. Camus var. —Jackson & Wiehe, Annot. Check List Nyasal. Grass.: 31 (1958).

Caespitose perennial; culms up to 200 cm high, either decumbent and rambling or developing into stout woody stolons, branching and rooting at the nodes; nodes barbate with white hairs; internodes channelled on one side; leaf sheaths glabrous, tightly clasping the internodes, barbate at the mouth; ligule membranous; leaf laminas 4–30 cm × 2–8 mm, glabrous on both sides, occasionally with tubercle-based hairs below, rough on the margins. Inflorescence of 3–20 subdigitate peduncled racemes; main axis up to 3 cm long, shorter than the lowermost racemes; raceme peduncles glabrous; racemes up to 10 cm long, but variable; rhachis internodes and pedicels pilose on the margins, the hairs increasing in length upwards to c. 3 mm. Sessile spikelet c. 4.5 mm long, narrowly elliptic; callus c. 0.5 mm long, barbate, obtuse; inferior glume with a single deep pit on the back just above the middle, glabrous or pubescent to shortly pilose below, shortly ciliate on the keels towards the apex; superior glume narrowly lanceolate, sharply carinate, glabrous, minutely ciliate on the keel towards the apex; inferior floret empty, its lemma c. 3.5 mm long, broadly ovate; superior floret epaleate, the lemma c. 2 mm long, stipitiform, awned from the apex; awn c. 25 mm long, glabrous; anthers c. 1.8 mm long. Pedicelled spikelet c. 5 mm long, neuter, glabrous; inferior glume oblong, with 0–4 shallow slit-like pits, irregularly toothed at the apex, ciliate along the keels; superior glume lanceolate, delicate, ciliate on the margins; inferior lemma c. 3 mm long, oblong, truncate at the apex.

Botswana. N: Ngamiland Distr., 25 km north of Tsao, 910 m, 31.iii.1987, *Long & Rae* 510 (E; K). SE: near Otse, north of Lobatse, 1000 m, 4.iii.1987, *Long & Rae* 34 (E; K). **Zambia**. N: Serenje Distr., Mote area, road to Kapalala from Kasanka, 1.iii.1996, *Renvoize* 5738 (K). C: Lusaka Distr., Kasoka Farm, 9.x.1929, *Sandwith* 64 (K). S: Mazabuka, Central Research Station, 1160 m, i.1932, *Trapnell* 906 (K). **Zimbabwe**. N: Gokwe Distr., 5 km north of Gokwe, 30.v.1963, *Bingham* 674 (SRGH). W: Bulilima Mangwe Distr., Dombodema Mission Farm, Tjonpani Store, 1300 m, 29.iv.1972, *Norrgrann* 145 (WAG). C: Chegutu (Hartley)–Norton, 45 km west of Harare (Salisbury), 10.ii.1946, *Weber* 201 (K). E: Bikita, Lower Sabi, 460 m, 27.i–2.ii.1948, *Rattray* 1262 (K; PRE; WAG). S: Masvingo Distr., Danson Claims, Brakfontein, 14.xii.1966, *Wild* 7582 (K; PRE). **Malawi**. N: Mzimba Distr., Kasitu River, near Ekwendeni, 19.i.1951, *Jackson* 367 (K). C: Ntcheu Distr., South Ntcheu, 20.iv.1956, *Jackson* 1851 (K; MAL). S: Zomba, 940 m, 5.vi.1949, *Wiehe* 123 (K). **Mozambique**. T: Moatize Distr., andados 50 km de Zóbuè para Tete, c. 350 m, 12.iii.1964, *Torre & Paiva* 11175 (LISC). GI: Chibuto Distr., Maniquenique, Estação Experimental do C.I.C.A., 13.vi.1960, *Lemos & Balsinhas* 100 (K; PRE; WAG). M: Namaacha Distr., Changalane, 22.viii.1980, *Zunguze & Mafumo* 269 (K).

Tropical Africa, Arabia and India; introduced to Australia. Occurring in overgrazed grasslands and weedy places; sometimes cultivated for fodder; 70–1300 m.

16. EUCLASTA Franch.

By M.P. Setshogo

Euclasta Franch. in Bull. Soc. Hist. Nat. Autun **8**: 335 (1895).
Indochloa Bor in Kew Bull. **9**: 75 (1954).

Inflorescences terminal and axillary, of delicate solitary or subdigitate pedunculate racemes; racemes with 1–3 large homogamous pairs at the base; rhachis internodes and pedicels with a translucent longitudinal groove. Sessile spikelet: inferior glume chartaceous, ± flat; inferior floret empty with a hyaline lemma; superior floret bisexual, the lemma entire, stipitiform. Pedicelled spikelet larger than the sessile, male or neuter.

A genus of 2 species, occurring in tropical Africa to India and tropical America.

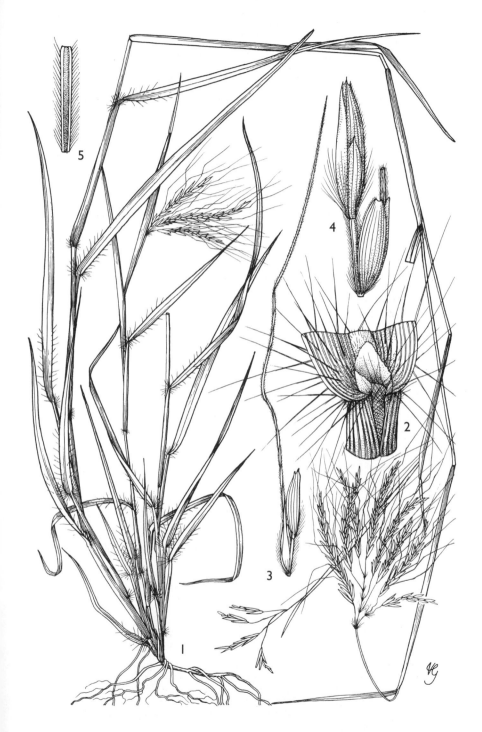

Tab. 18. EUCLASTA CONDYLOTRICHA. 1, habit ($\times \frac{2}{3}$); 2, ligule ($\times$ 8); 3, spikelet pair ($\times$ 4); 4, spikelet pair ($\times$ 8); 5, pedicel ($\times$ 12). Drawn by Victoria Goaman. From Ghana Gr.

Euclasta condylotricha (Hochst. ex Steud.) Stapf in Prain, F.T.A. **9**: 181 (1917). —Jackson &
Wiehe, Annot. Check List Nyasal. Grass.: 41 (1958). —Napper, Grasses Tanganyika: 100
(1965). —Hood, A Guide to the Grasses of Zambia: 65 (1967). —Clayton in F.W.T.A., ed. 2,
3: 471 (1972). —Clayton & Renvoize in F.T.E.A., Gramineae: 722 (1982). —Lowe, Fl. Nigeria,
Grasses: 237 (1989). —Setshogo in Kirkia **17**: 145 (2001). TAB. **18**. Type from Ethiopia.
   Andropogon condylotrichus Hochst. ex Steud., Syn. Pl. Glumac. **1**: 377 (1854).
   Andropogon piptatherus Hack. in Martius & Eichler, Fl. Bras. **2**(3): 293 (1883). Type from
Brazil.
   Sorghum piptatherum (Hack.) Kuntze, Revis. Gen. Pl. **2**: 792 (1891).
   Euclasta glumacea Franch. in Bull. Soc. Hist. Nat. Autun **8**: 336 (1895). Type from Congo
(Brazzaville).
   Euclasta graminea T. & H. Durand, Syll. Fl. Congol.: 649 (1909), *nom. superfl.* based on *E.
glumacea* Franch.
   Amphilophis piptatherus (Hack.) Nash, N. Amer. Fl. **17**: 127 (1912).

Annual; culms up to 150 cm high, geniculately ascending with stilt-roots, barbate
at the nodes; leaf sheaths with long tubercle-based hairs at the mouth; ligule
membranous with ciliolate to long-ciliate fringe; leaf laminas up to 25 cm × 2–10 mm,
linear-lanceolate, somewhat flaccid, contracted at the base and drawn out at the apex
into a setaceous point, glabrous on both sides or shortly pilose along the midrib and
adjacent nerves; midrib whitish, prominent. Inflorescence of subdigitate racemes.
Sessile spikelet: inferior glume c. 4 mm long, elliptic-oblong, with 5 nerves ending
below the hyaline truncate apex, pilose on the margins and on the back below,
scaberulous on the keels; superior glume c. 4 mm long, ovate, strongly carinate,
glabrous; inferior floret empty with ovate carinate lemma ciliate on the keel and
margins; superior floret bisexual; lemma hyaline, stipitiform; palea present; awn c. 37
mm long, minutely pilose; anthers c. 0.75 mm long. Pedicelled spikelet male;
inferior glume c. 5.5 mm long, oblanceolate, 7–10-nerved, pilose in the lower $^1/_3$;
superior glume 4–4.5 mm long; superior floret with lemma slightly shorter than the
glumes; anthers c. 1.5 mm long.

Zambia. N: Mbala Distr., Lake Tanganyika, Crocodile Is., 780 m, 9.ii.1964, *Richards* 18996
(K). C: Mazabuka Distr., Kafue Gorge, 980 m, 14.iv.1956, *E.A. Robinson* 1477 (K). E: Chipata
Distr., Jumbe area, 24.iii.1963, *Verboom* 941 (BM; PRE). **Zimbabwe**. N: Gokwe Distr., 16 km
from Gokwe in Ganye R. Gorge, 25.iii.1963, *Bingham* 571 (BM). C: Kadoma Distr., Sanyati
Reserve, 12.iii.1959, *Cleghorn* 451 (BM). **Malawi**. N: Mzimba Distr., Kasitu Valley, 1870 m,
31.i.1938, *Fenner* 247 (K). C: Salima Escarpment, 8.iv.1952, *Jackson* 757 (K; PRE). S: Mangochi
Distr., Jalasi, Misanje R., 27.iv.1955, *Jackson* 1648 (K). **Mozambique**. N: between Cuamba and
Mutuáli, near the bridge over the river Lúrio, 24.iv.1961, *Balsinhas & Marrime* 429 (K; LISC).
T: Cahora Bassa Distr., Serra do Songo, c. 900 m, 17.iii.1970, *Torre & Correia* 18298 (LISC).
   Tropical Africa. Growing mainly on rocky escarpments and along roadsides; 200–1900 m.

17. ISCHAEMUM L.

By T.A. Cope

Ischaemum L., Sp. Pl. **2**: 1049 (1753); Gen. Pl., ed. 5: 469 (1754).

Perennials, sometimes annuals, often decumbent. Leaf sheaths mostly with
auricles; ligule membranous; leaf laminas linear, sometimes sagittate or falsely
petiolate at the base. Inflorescence of paired, sometimes solitary or digitate, 1-sided
racemes, those of a pair often interlocked back to back and mimicking a solitary
raceme, terminal or axillary; rhachis internodes and pedicels stoutly linear to clavate
or inflated, often exposed on the back of the raceme as a U- or V-shaped segment,
sometimes the pedicels so short that the spikelets of a pair appear to be side by side.
Sessile spikelet dorsally compressed; callus obtuse and inserted in the concave top of
the internode; inferior glume chartaceous to coriaceous, convex to concave on the
back, laterally 2-keeled, often rugose, sometimes winged; superior glume awned or
awnless; inferior floret male, rarely barren, with a palea; superior lemma bifid,
passing between the teeth into a glabrous awn (rarely awnless). Caryopsis oblong to
lanceolate, dorsally compressed. Pedicelled spikelet as large as the sessile spikelet or
much smaller, often asymmetrical.

Species c. 65. Throughout the tropics but mainly in Asia.

1. Inferior glume of sessile spikelet concave on the back ·················1. *afrum*
– Inferior glume of sessile spikelet convex on the back ·······················2
2. Pedicelled spikelet as long as the sessile spikelet, geniculately awned ·····2. *fasciculatum*
– Pedicelled spikelet much reduced, awnless ····················3. *roseotomentosum*

1. **Ischaemum afrum** (J.F. Gmel.) Dandy in Andrews, Fl. Pl. Sudan **3**: 476 (1956). —Napper, Grasses Tanganyika: 95 (1965). —Simon in Kirkia **8**: 19, 52 (1971). —Clayton in F.W.T.A., ed. 2, **3**: 476 (1972). —Hall-Martin & Drummond in Kirkia **12**: 158 (1980). —Clayton & Renvoize in F.T.E.A., Gramineae: 747 (1982). —Gibbs Russell et al., Grasses South. Africa [Mem. Bot. Surv. S. Africa No. 58]: 191 (1990). TAB. **19**. Type from Ethiopia.
   Andropogon afer J.F. Gmel., Syst. Nat., ed. 13, **2**: 166 (1791).
   Andropogon brachyatherus Hochst. in Flora **27**: 241 (1844). Type from Sudan.
   Ischaemum brachyatherum (Hochst.) Fenzl ex Hack. in A. & C. de Candolle, Monogr. Phan. **6**: 239 (1889). —Stapf in Prain, F.T.A. **9**: 30 (1917). —Stent & Rattray in Proc. & Trans. Rhodesia Sci. Assoc. **32**: 4 (1933). —Sturgeon in Rhodesia Agric. J. **51**: 141 (1954). —Chippindall in Meredith, Grasses & Pastures S. Africa: 487 (1955). —Jackson & Wiehe, Annot. Check List Nyasal. Grass.: 46 (1958). —Wild in Kirkia **5**: 54 (1965).
   Ischaemum glaucostachyum Stapf in Dyer, F.C. **7**: 328 (1898); in Prain, F.T.A. **9**: 29 (1917). —Chippindall, op. cit.: 487. —Hood, A Guide to the Grasses of Zambia: 61 (1967). Type from South Africa.

Densely caespitose perennial with scaly rhizome; culms up to 200 cm high; leaf laminas 10–50 cm × 3–10 mm, glaucous, prominently nerved, tapering to a setaceous point. Inflorescence terminal, of (1)2–5 (rarely more) subdigitate racemes, each 6–20 cm long; rhachis internodes and pedicels glabrous to villous, yellowish to purple, subequal, the internodes clavate, the pedicels subinflated. Sessile spikelet 5–8 mm long, lanceolate; inferior glume chartaceous, 2-keeled along its whole length, concave between the keels, glabrous to villous on the back, wingless; superior lemma with an awn 5–20 mm long. Pedicelled spikelet 1–6 mm long, awnless or rarely awned from the superior lemma.

Botswana. N: Chobe Distr., floodplain of Sibuyu R. (18°31'S, 25°38'E), 12.iv.1983, *P.A. Smith* 4260 (K; PRE). SE: 60 km NW of Serowe, 25.iii.1965, *Wild & Drummond* 7308 (K). **Zambia.** S: Livingstone Distr., Maramba Quarantine Paddocks, 1005 m, 15.ii.1932, *Trapnell* 936 (K). **Zimbabwe.** N: Gokwe North Distr., Chirisa, Sengwa Research Area, 12.ii.1981, *Mahlangu* 433 (K; PRE). W: Umguza Distr., near Nyamandhlovu, cult. in Marondera (Marandellas), iii.1960, *West* in *GHS* 107814 (K; PRE). C: Gweru (Gwelo), 1400 m, v.1920, *Walters* 2237 (K). E: Lower Save (Sabi), 460 m, 27.i.–2.ii.1948, *Rattray* 1264 (K). S: 16 km south of Masvingo (Fort Victoria), 26.v.1973, *Vernon* 37 (K). **Malawi.** S: Nsanje Distr., Ngabu road, Chiromo, 40 m, 21.vii.1949, *Wiehe* 170 (K). **Mozambique.** N: Mecanhelas Distr., andados 22 km de Cuamba (Nova Freixo) para Mecanhelas, c. 600 m, 17.ii.1964, *Torre & Paiva* 10627 (LISC). Z: Morrumbala Distr., entre Águas Quentas e Mutarara, 4.v.1943, *Torre* 5290 (K; LISC). T: Cahora Bassa Distr., Zambezi Valley near Chicoa, 10.vi.1947, *R.M. Hornby* 2749 (K; PRE). MS: Gorongosa Distr., Parque Nacional da Gorongosa (Gorongosa National Park), iv.1973, *Tinley* 2792 (K). GI: Guija Distr., 4 km SW of Zulo, 6.xii.1981, *R.J. White* 34 (K). M: Umbelúzi, 14.i.1953, *Myre & Carvalho* 1448 (K).

Throughout tropical Africa from Nigeria eastwards to Ethiopia and southwards to South Africa; also in India. Growing in seasonally flooded grassland on heavy black clay soils; 15–1500 m.

2. **Ischaemum fasciculatum** Brongn. in Duperrey, Voy. Monde Phan.: 73 (1831). —Sturgeon in Rhodesia Agric. J. **51**: 141 (1954). —Clayton & Renvoize in F.T.E.A., Gramineae: 749 (1982). —Gibbs Russell et al., Grasses South. Africa [Mem. Bot. Surv. S. Africa No. 58]: 191 (1990). Type from Mauritius.
   Spodiopogon arcuatus Nees, Fl. Afr. Austral. Ill. **1**: 97 (1841). Type from South Africa.
   Andropogon arcuatus (Nees) Steud., Syn. Pl. Glumac. **1**: 374 (1854).
   Andropogon fasciculatus (Brongn.) Steud., Syn. Pl. Glumac. **1**: 382 (1854).
   Ischaemum fasciculatum var. *arcuatum* (Nees) Hack. in A. & C. de Candolle, Monogr. Phan. **6**: 235 (1889).
   Ischaemum junodii Hack. in Vierteljahrsschr. Naturf. Gas. Zürich **52**: 419 (1907). Type from South Africa.
   Ischaemum purpurascens Stapf in Prain, F.T.A. **9**: 32 (1917). —Stent & Rattray in Proc. & Trans. Rhodesia Sci. Assoc. **32**: 4 (1933). —Sturgeon in Rhodesia Agric. J. **51**: 141 (1954). —Jackson & Wiehe, Annot. Check List Nyasal. Grass.: 46 (1958). Syntypes: Zimbabwe, Victoria Falls, *Kolbe* 3143; and *Craster* 69 (K, isosyntypes) and several others.

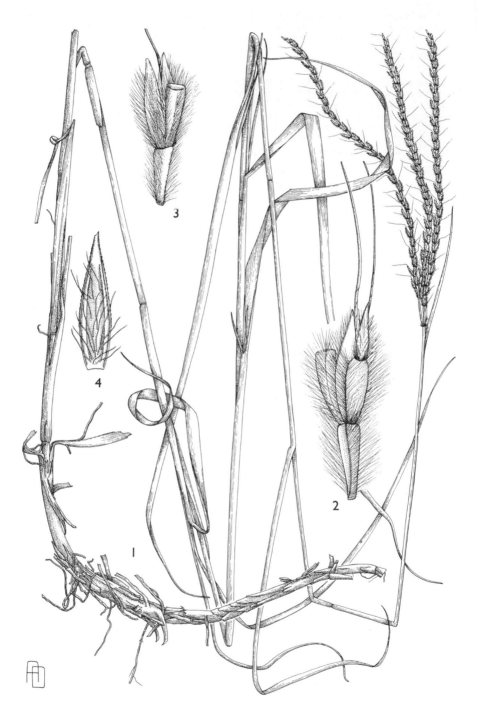

Tab. 19. ISCHAEMUM AFRUM. 1, habit (× ¹/₂); 2, portion of raceme, showing pedicelled
spikelet (× 4); 3, portion of raceme, showing sessile spikelet (× 5); 4, inferior glume of
sessile spikelet (× 5), 1–4 from *Welch* 527. Drawn by Ann Davies. From F.T.E.A.

Ischaemum arcuatum (Nees) Stapf, in Prain, F.T.A. **9**: 33 (1917). —Stent & Rattray in Proc.
& Trans. Rhodesia Sci. Assoc. **32**: 4 (1933). —Sturgeon in Rhodesia Agric. J. **51**: 141
(1954). —Chippindall in Meredith, Grasses & Pastures S. Africa: 487 (1955). —Hood, A
Guide to the Grasses of Zambia: 61 (1967). —Simon in Kirkia **8**: 19, 52 (1971).
Ischaemum melleri Stent in Bothalia **1**: 238, t. 1 (1924). Type from South Africa.

Rambling or prostrate rhizomatous perennial; culms up to 150 cm high; leaf
laminas 5–25 cm × 3–20 mm, glabrous or pubescent. Inflorescence mostly terminal,
of 2–5(8) subdigitate racemes each 3–15 cm long; rhachis internodes and pedicels
ciliate, yellowish to purple, subequal, clavate.   Sessile spikelet 4–7 mm long
(including the large oblong callus 1–1.5 mm long), narrowly elliptic; inferior glume
chartaceous or sometimes subcoriaceous in the inferior third, 2-keeled above but
merely incurved towards the base, convex, glabrescent to villous, the keels with or
without wings, these sometimes forming a prominent lateral tooth or lobe, usually
bidentate at the apex; superior lemma with an awn 5–15 mm long.   Pedicelled
spikelet as long as the sessile spikelet but laterally compressed, geniculately awned
from the superior lemma.

Zambia. B: Senanga Distr., Kataba Valley, 19.xii.1964, *Verboom* 1555 (K). N: Lufubu R.,
Chishinga Ranch, 13.ix.1961, *Astle* 898 (K). W: by R. Lunga, just east of Mwinilunga,
27.xi.1937, *Milne-Redhead* 3423 (K; PRE). C: Serenje Distr., upper Lukulu R., Kanona, 3.v.1970,
Verboom 2576 (K). S: Livingstone Distr., Victoria Falls, 29.xi.1964, *B.L. Mitchell* 25/80 (K).
Zimbabwe. N: Gokwe South Distr., Sengwe R. near eastern edge of Charama Plateau, 1220 m,
27.iv.1965, *Simon* 286 (K). W: Hwange Distr., Victoria Falls, c. 910 m, 31.vii.1941, *Greenway*
6246 (K; PRE). C: Goromonzi, 1590 m, 17.iv.1927, *Eyles* 4892 (K). E: Nyanga Distr., Rhodes
Inyanga National Park, 1.i.1965, *West* 6227 (K; PRE). S: Mberengwa Distr., southern slope of
Mt. Buhwa, c. 1200 m, 2.v.1973, *Simon, Pope & Biegel* 2425 (K). **Malawi**. N: Karonga Distr.,
foothills of the Misuku range, near Karonga, vi.1951, *Jackson* 568 (K). S: Zomba Mt.,
10.viii.1950, *Wiehe* 605 (K). **Mozambique**. T: Moatize Distr., outskirts of Zóbuè, 17.vi.1941,
Torre 2886 (LISC). MS: Gorongosa Mt., summit between Gogogo and Nhandore peaks
(18°26'S, 34°03'E), c. 1700 m, 10.xi.1971, *Ward* 7379 (K). GI: Xai-Xai Distr., Inhamissa,
16.xii.1957, *Macêdo* 39 (K; LISC). M: Inhaca Is., 37 km east of Maputo (Lourenço Marques),
0–150 m, 26.ix.1957, *Mogg* 27559 (K).
Throughout tropical Africa, extending eastwards through India to Indo-China. Growing in
rainforest, montane grassland and bogs, along streamsides and by waterfalls, and in low-lying
swamps on sandy soils; 0–1700 m.

3. **Ischaemum roseotomentosum** Phipps in Kirkia **3**: 30 (1963). —Simon in Kirkia **8**: 19 (1971).
Type: Zimbabwe W: Nyamandhlovu, *Vaughan-Evans* 27 (SRGH, holotype; BM; BR; EA; K;
LISC; LMA; PRE).

Caespitose perennial; culms up to 130 cm high, mostly erect; leaf laminas 4–15
cm × 5–8 mm, glaucous, tuberculately pilose near the base, prominently nerved,
tapering to an acute point. Inflorescence terminal, of single racemes each 5–7 cm
long and fragile even when young; rhachis internodes and pedicels densely villous,
the hairs pinkish or sometimes golden, the internodes and pedicels themselves
yellowish to purple, subequal, subinflated. Sessile spikelet c. 8 mm long, lanceolate
(not including the wings of the inferior glume); inferior glume coriaceous, 2-keeled
along its length, flat or convex between the keels, densely villous on the back in the
inferior $^2/_3$, pilose towards the margins above, winged on the keels, the wings
broadening above into lobes up to 1 mm wide, these, together with the emarginate
apex of the glume, giving the glume the appearance of being broadly 3-lobed, the
central lobe slightly the longest but exceeded by the superior glume; superior lemma
with a geniculate awn c. 30 mm long. Pedicelled spikelet 3–4 mm long, barren,
much reduced and represented only by its glumes, these similar to those of the sessile
spikelet but the inferior longer than the superior.

Zimbabwe. W: Umgusa Distr., Nyamandhlovu, Fountain Farm, 2.ii.1959, *Vaughan-Evans* 27
(K; SRGH). E: Lower Save (Sabi), 460 m, 27.i–2.ii.1948, *Rattray* 1253 (K). S: Gwanda Distr., 64
km from Beitbridge near main road to Bulawayo, 610 m, 4–5.i.1956, *Rattray* 1710 (K; PRE).
Not known elsewhere. Growing on black basaltic soils and in mopane woodland on alluvium;
270–1200 m.

18. THELEPOGON Roth ex Roem. & Schult.

By T.A. Cope

Thelepogon Roth ex Roem. & Schult., Syst. Veg. **2**: 46 (1817).

Annual; ligule shortly membranous; leaf laminas broad. Inflorescence a terminal cluster of shortly pedunculate racemes; rhachis internodes thickened, clavate, glabrous. Sessile spikelet dorsally compressed; callus obtuse and fitting the concave apex of the internode; inferior glume crustaceous, convex across the back, coarsely rugose, without keels or wings; superior glume broadly convex across the back, rugose, wingless; inferior floret male, with hyaline lemma and well developed palea; superior lemma bifid, passing between the teeth into a glabrous awn. Caryopsis ellipsoid. Pedicelled spikelet absent, represented only by a flattened linear pedicel.

A genus of one species. The genus differs from *Ischaemum* by little more than the barren pedicels.

Thelepogon elegans Roth ex Roem. & Schult., Syst. Veg. **2**: 788 (1817). —Stapf in Prain, F.T.A. **9**: 34 (1917). —Jackson & Wiehe, Annot. Check List Nyasal. Grass.: 62 (1958). —Hood, A Guide to the Grasses of Zambia: 72 (1967). —Simon in Kirkia **8**: 19, 52 (1971). —Clayton in F.W.T.A., ed. 2, **3**: 473 (1972). —Clayton & Renvoize in F.T.E.A., Gramineae: 744, fig. 174 (1982). —Gibbs Russell et al., Grasses South. Africa [Mem. Bot. Surv. S. Africa No. 58]: 334 (1990). TAB. **20**. Type from India.

Coarse annual; culms up to 150 cm high, erect, supported by stilt-roots; leaf laminas 4–20 cm × 5–30 mm, lanceolate, cordate at the base and often amplexicaul, glabrous or hispidulous, the margins finely pectinate-ciliate. Inflorescence composed of 2–17 racemes, the lowermost whorled, the uppermost borne on a short common axis 1–5 cm long; racemes 5–15 cm long, very fragile. Sessile spikelet 5–13 mm long; inferior glume narrowly ovate; superior lemma with an awn 1.5–2.5 cm long. Pedicel a little longer than the sessile spikelet, glabrous.

Zambia. C: Lusaka Distr., Chikupi Estate, 12.iv.1963, *van Rensburg* 1896 (K). E: Chipata Distr., Luangwa Valley, Chief Nsefu Area, 1.v.1963, *Verboom* 959 (K). S: Livingstone Distr., Livingstone to Kazungula (Kazangula), km 16, near Kapanda Bridge Camp, 11.iii.1960, *White* 7729 (K). **Zimbabwe**. N: Binga Distr., Lusulu Vet. Ranch, 24.ii.1965, *Bingham* 1408 (K; SRGH). W: Hwange Distr., Kazuma Range, c. 1000 m, 10.v.1972, *Simon* 2185 (K; PRE). **Malawi**. N: Rumphi Distr., Nyika Plateau, Mwenembwe (Mwanemba), 2290 m, ii–iii.1903, *McClounie* 18 (K). S: Mulanje Distr., Namphasa Dambo, road from Namasoko Village to Dzanje Road, Mulanje, 9.ii.1980, *Patel & Morris* 484 (K). **Mozambique**. N: east coast of Lake Malawi (Lake Nyasa), 22.ix.1900, *W.P. Johnson* 91 (K).
Throughout tropical Africa, extending through India to Indonesia. In grassland, on woodland margins and on lake shores in black clay soils; 550–2300 m.

19. SEHIMA Forssk.

By T.A. Cope

Sehima Forssk., Fl. Aegypt.-Arab.: 178 (1775).

Annual or perennial; ligule a line of hairs. Inflorescence a single terminal raceme, exserted from the uppermost leaf sheath; rhachis internodes and pedicels stoutly linear to subclavate, ciliate. Sessile spikelet laterally or dorsally compressed to ± square in section, fitting between the internode and pedicel; callus obtuse and inserted into the concave top of the internode; inferior glume coriaceous, concave or with a median groove, laterally 2-keeled or lyrate with the keels becoming dorsal towards the base, scarcely winged, bifid at the apex; superior glume awned; inferior floret male and with a palea; superior lemma bifid, awned from between the teeth with an awn puberulous to ciliate along the edges of the coils. Caryopsis lanceolate-oblong, dorsally compressed, concave on one side. Pedicelled spikelet large, lanceolate, strongly dorsally compressed, distinctively lyrate-nerved.

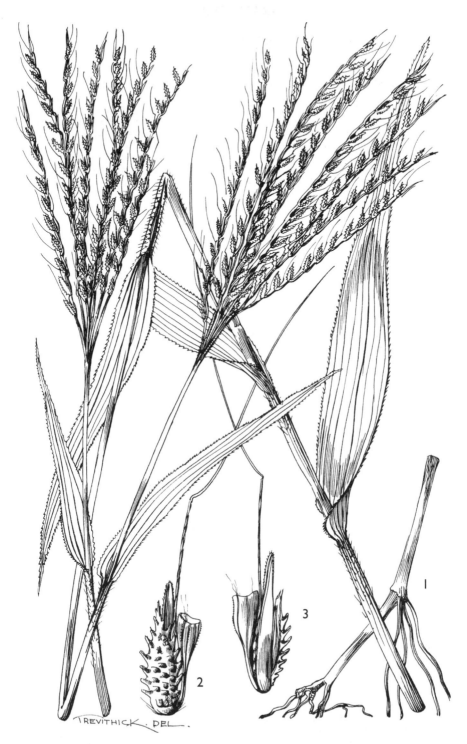

Tab. 20. **THELEPOGON ELEGANS.** 1, habit (× ⅔); 2, spikelet pair, showing sessile spikelet (× 4); 3, spikelet pair, rear view, showing pedicel (× 4). Drawn by W.E. Trevithick. From F.T.E.A.

A genus of 5 species; Old World tropics.
A distinctive genus, particularly in the lyrate venation of the pedicelled spikelet, that is
probably a segregate from *Ischaemum*.

1. Plant annual; inferior glume of sessile spikelet with a membranous apex as long as the body,
   deeply bifid; column of awn ciliate along the edges of the coils · · · · · · · · · 3. *ischaemoides*
 − Plant perennial; inferior glume of sessile spikelet with a membranous apex up to ¹/₃ the
   length of the body, entire or shallowly bifid; column of awn minutely ciliolate along the
   edges of the coils · 2
2. Robust plant with densely pilose basal sheaths; inferior glume of sessile spikelet ± flat on the
   back, polished-coriaceous below · 1. *galpinii*
 − Wiry plant with glabrous, or at most thinly pubescent, basal sheaths; inferior glume of
   sessile spikelet concave on the back, at least between the nerves, herbaceous throughout
   · 2. *nervosum*

1. **Sehima galpinii** Stent in Bothalia **1**: 239, t. 2 (1924). —Chippindall in Meredith, Grasses &
   Pastures S. Africa: 489 (1955). —Gibbs Russell et al., Grasses South. Africa [Mem. Bot.
   Surv. S. Africa No. 58]: 293 (1990). Type from South Africa (Transvaal).

Robust perennial up to 180 cm high; basal leaf sheaths densely pilose. Racemes
7–19 cm long, ± straight to gently curved. Sessile spikelet 12–15 mm long, dorsally
compressed; inferior glume oblong-elliptic, polished-coriaceous and faintly
nerved below, thinner and conspicuously nerved above, often with transverse
nerves connecting the longitudinal, with shallowly bifid to ± entire membranous
apex less than ¹/₅ the length of the body, 2-keeled, flat or shallowly convex between
the keels, inconspicuously lyrately-nerved with the nerves ± equally spaced;
superior lemma with an awn 2.5–4 cm long, the column minutely ciliolate along
the edges of the coils. Pedicelled spikelet 10–13 mm long, shortly ciliolate, the
inferior glume with a midnerve and 2–3 conspicuous lateral nerves adjacent to
each keel.

Mozambique. M: between Goba and Catuane, 27.x.1940, *Torre* 1931 (LISC).
Also in South Africa (Transvaal, KwaZulu-Natal) and Swaziland. Growing in open forest;
35–140 m.
The species is distinguished from the following by the pilose basal leaf sheaths and flat or
convex inferior glume of the sessile spikelet polished-coriaceous and inconspicuously nerved
below. It replaces *S. nervosum* in southern tropical Africa.

2. **Sehima nervosum** (Rottler) Stapf in Prain, F.T.A. **9**: 36 (1917). —Bogdan, A Revised List of
   Kenya Grasses: 62 (1958). —Napper, Grasses Tanganyika: 95 (1965). —Clayton & Renvoize
   in F.T.E.A., Gramineae: 750, fig. 176 (1982). TAB. **21**. Type from India.
      Andropogon nervosus Rottler in Ges. Naturf. Freunde Berlin, Neue Schriften **4**: 218 (1803).
      Ischaemum laxum R. Br., Prodr. Fl. Nov. Holl.: 205 (1810). Type from Australia.
      Ischaemum macrostachyum Hochst. ex A. Rich., Tent. Fl. Abyss. **2**: 472 (1851). Type from
   Ethiopia.

Caespitose wiry perennial up to 100 cm high; basal leaf sheaths glabrous or at
most thinly pubescent. Racemes 3–12 cm long, gently curved. Sessile spikelet 6–10
mm long, laterally compressed; inferior glume narrowly oblong-elliptic, thinly
coriaceous with a shallowly bifid membranous apex up to ¹/₃ the length of the body,
2-keeled, concave between the keels, lyrately nerved with 2–3 closely spaced nerves
adjacent to each keel; superior lemma with an awn 2–4 cm long, the column
minutely ciliolate along the edges of the coils. Pedicelled spikelet 6–10 mm long,
ciliate, the inferior glume with a midnerve and 2–3 conspicuous lateral nerves
adjacent to each keel.

Malawi. S: Nsanje Distr., between Thangadzi and Lalanje (Lilanje) Rivers, 90 m, 25.iii.1960,
Phipps 2683 (K). **Mozambique.** T: Changara Distr., km 48 on road from Ferrão to Tete,
18.v.1948, *Mendonça* 4324 (LISC).
East tropical Africa from Sudan to Mozambique, through southern Arabia to India, China
and Australia. Growing in mopane woodland.
The species is rare in southern tropical Africa where it is replaced by the more robust *S.
galpinii*.

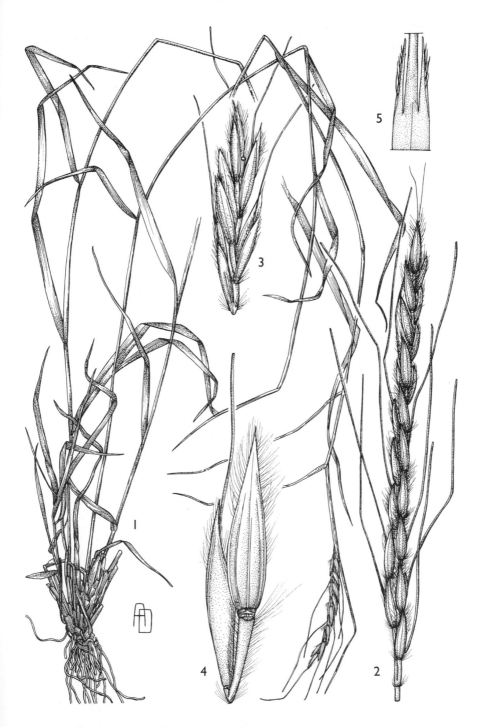

Tab. 21. SEHIMA NERVOSUM. 1, habit (× ½), from *Verdcourt, Hemming & Polhill* 2674; 2, raceme (× 2); 3, detail of raceme (× 3); 4, spikelet pair (× 5); 5, tip of superior lemma (× 10), 2–5 from *Greenway & Kanuri* 11384. Drawn by Ann Davies. From F.T.E.A.

3. **Sehima ischaemoides** Forssk., Fl. Aegypt.-Arab.: 178 (1775). —Stapf in Prain, F.T.A. **9**: 37 (1917). —Simon in Kirkia **8**: 19, 52 (1971). —Stent & Rattray in Proc. & Trans. Rhodesia Sci. Assoc. **32**: 5 (1933). —Sturgeon in Rhodesia Agric. J. **51**: 142 (1954). —Jackson & Wiehe, Annot. Check List Nyasal. Grass.: 58 (1958). —Simon in Kirkia **8**: 19, 52 (1971). —Clayton & Renvoize in F.T.E.A., Gramineae: 750 (1982). Type from Yemen.

  Andropogon sehima Steud., Syn. Pl. Glumac. **1**: 369 (1854), *nom. superfl.*, based on the preceding.

Annual up to 60 cm high. Raceme 3–15 cm long, gently curved. Sessile spikelet 9–15 mm long, laterally compressed; inferior glume linear, firmly coriaceous with a deeply bifid membranous apex ± as long as the body, 2-keeled above, the keels curving inwards towards the midline below, forming a deep slot between them and obscuring the intercarinal nerves; superior lemma with an awn 4–7 cm long, the column ciliate along the edges of the coils. Pedicelled spikelet 7–15 mm long, ciliate, the inferior glume with a midnerve and 2–3 conspicuous lateral nerves adjacent to each keel.

  Botswana. N: Ngwanalekau Hills, *Buerger* 1165 (K; PRE). **Zambia**. E: Chipata Distr., Machinje Hills, Jumbe, 760 m, 4.iii.1966, *Astle* 4624 (K; SRGH). S: Livingstone Distr., 4th and 5th gorges, Victoria Falls, 5–9.iii.1956, *Rattray* 1802 (K; PRE). **Zimbabwe**. N: Gokwe South Distr., Sengwa Research Station, 27.iii.1977, *P.R. Guy* 2491 (K; PRE). W: Hwange Distr., Matetsi Safari Area, 980 m, 4.ii.1980, *Gonde* 290 (K; PRE). E: Lower Save (Sabi), 460 m, 27.i.–2.ii.1948, *Rattray* 1258 (K; SRGH). S: Mwenezi Distr., Malumba Ranch, Mateke Hills, 23.iii.1966, *West* 7227 (K). **Malawi**. S: 63 km from Blantyre on Ntcheu road, 8.iii.1951, *Jackson* 410 (K). **Mozambique**. T: Magoé, 35 km from Mecumbura towards Chipembere (Chióco), c. 300 m, 11.iii.1970, *Torre & Correia* 18246 (LISC).

  Tropical Africa eastwards through Arabia to Pakistan. Growing in mopane woodland and shrubland on shallow soils, and in disturbed ground at roadsides; 270–1000 m.

20. ANDROPTERUM Stapf

By T.A. Cope

Andropterum Stapf in Prain, F.T.A. **9**: 38 (1917).

Perennial. Ligule a fringed membrane; leaf laminas broadly linear. Inflorescence of solitary racemes, these terminal and axillary; rhachis internodes and pedicels swollen, ± clavate, ciliate. Sessile spikelet laterally compressed; callus obtuse and fitting the concave top of the internode; inferior glume cartilaginous, 2-keeled, the keels dorsal, almost contiguous, separated by a narrow groove, not winged; superior glume strongly laterally compressed, conspicuously winged on the keel towards the apex; inferior floret male, with hyaline lemma and well developed palea; superior lemma bifid, passing between the teeth into a glabrous awn. Caryopsis narrowly ellipsoid. Pedicelled spikelet larger than the sessile, glabrous, asymmetrically winged, awnless, conspicuous.

  A genus of one species.

Andropterum stolzii (Pilg.) C.E. Hubb. in Mem. New York Bot. Gard. **9**: 112 (1954). —Jackson & Wiehe, Annot. Check List Nyasal. Grass.: 29 (1958). —Simon in Kirkia **8**: 19, 52 (1971). —Clayton & Renvoize in F.T.E.A., Gramineae: 752, fig. 177 (1982). TAB. **22**. Type from Tanzania.

  Ischaemum stolzii Pilg. in Bot. Jahrb. Syst. **54**: 280 (March 1917).

  Andropterum variegatum Stapf in Prain, F.T.A. **9**: 38 (July 1917); in Hooker's Icon. Pl. **31**: t. 3077 (1922). Syntypes: Malawi, Namadzi (Namadsi), *Cameron* 17; and Misuku Hills (Masuku Plateau), *Whyte* s.n. (K, syntypes).

  Sehima variegatum (Stapf) Roberty in Boissiera **9**: 381 (1960).

Rambling perennial, rooting at the nodes; leaf laminas 5–20 cm × 3–15 mm, flat, acuminate to a fine pungent apex. Racemes 4–11 cm long. Sessile spikelets 5–8 mm long; inferior glume narrowly lanceolate, bearing scattered setae on the keels, acute; superior glume narrowly oblong, truncate; superior lemma bifid for $\frac{3}{4}$ its length, with an inconspicuous awn 8–15 mm long. Pedicelled spikelets 6–10 mm long, narrowly oblong, purplish, truncate, the inferior glume broadly winged on one side, the superior glume winged from the keel near the apex.

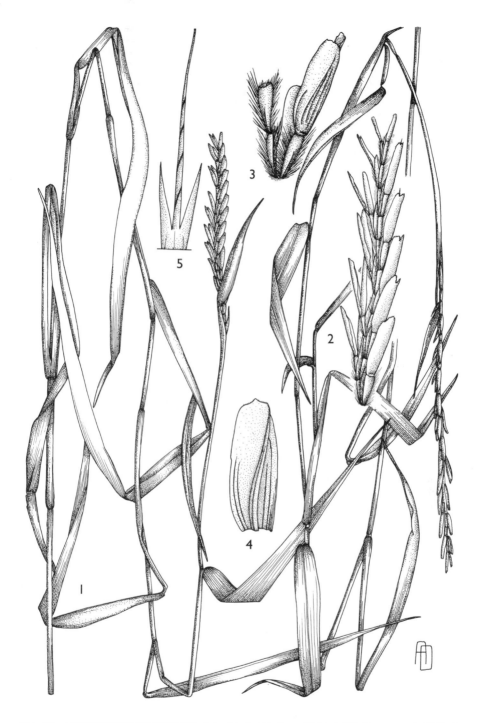

Tab. 22. ANDROPTERUM STOLZII. 1, habit (× ¹/₂), from *Richards* 195; 2, raceme (× 1¹/₂); 3, portion of raceme, showing pedicelled spikelet (× 4); 4, inferior glume of pedicelled spikelet (× 4); 5, tip of superior lemma (× 5), 2–5 from *Grassl* 46–97. Drawn by Ann Davies. From F.T.E.A.

Zambia. N: Mpika Distr., Kanchibya (Kanchibia) R., 48 km from Mpika on road to Kasama, 1520 m, 30.v.1964, *Vesey-FitzGerald* 4280 (K; PRE). E: Katete Distr., Mpangwe Hills, 16.vi.1963, *Verboom* 1101 (K). **Zimbabwe**. N: Guruve Distr., Mware protection block at foot of Great Dyke, 8.x.1965, *Cleghorn* 1151 (K). **Malawi**. N: Mzimba Distr., Katoto, 5 km west of Mzuzu, 1370 m, 19.v.1974, *Pawek* 8639 (K; PRE). C: Dedza Distr., slopes of Dedza Mt., 6.ix.1950, *Wiehe* 637 (K; PRE). S: Thyolo Distr., Bvumbwe Agriculture Research Station (Tung Experimental Station), 910–1070 m, 3.xi.1949, *Wiehe* 293 (K). **Mozambique**. T: Moatize Distr., Zóbuè, 27.viii.1943, *Torre* 5800 (K; LISC; PRE).

Also in Tanzania. Growing in dambos, along streamsides and pond margins and on the edges of riverine forest; 900–2000 m.

21. ANDROPOGON L.

By Fatíma Sales

Andropogon L., Sp. Pl.: 1045 (1753); Gen. Pl., ed. 5: 468 (1754). —Roberty in Boissiera **9**: 267–278 (1960). —J.G. Anderson in Bothalia **9**: 5 (1966). —Clayton in Hooker's Icon. Pl. **37**: t. 3644 (1967). —Heath in Calyx **4**(2): 52 (1994).

Racemes usually in pairs, sometimes solitary, arranged in a spatheate false panicle or in several subdigitate groups, usually exserted from the spatheoles; peduncles usually longer than the racemes, rarely the same length, erect at maturity; internodes and pedicels ± filiform or clavate, usually long-ciliate on both margins, occasionally inflated and canaliculate; lowermost pair of spikelets in the inferior raceme homogamous or not obviously different from the others. Sessile spikelet dorsally or laterally compressed, narrowly ovate; callus often rounded; inferior glume with a ± deep median groove or ± flat on the back, muticous, mucronate or awned; superior glume usually keeled and flat or ± strongly convex on the back; inferior floret reduced to a hyaline lemma; superior floret bisexual, its lemma ± deeply bifid or entire, usually awned. Pedicelled spikelet male or sterile, sometimes vestigial, very rarely absent.

A genus of c. 100 species, occurring throughout the tropics.

It is a polymorphic genus and was divided into 4 sections by Clayton, in Hooker's Icon. Pl. **37**: t. 3644 (1967). It merges with *Diheteropogon* and *Schizachyrium*, and is also related to *Cymbopogon* (see comments under those genera).

1. Racemes solitary, each subtended by a spatheole · 2
   – Racemes in pairs or in several subdigitate groups, each group subtended by a spatheole · · 4
2. Awns absent; ligule truncate; leaf sheaths distichously arranged, laterally compressed and keeled · 11. *festuciformis*
   – Awns present; ligule acute; leaf sheaths not distichously arranged nor laterally compressed and keeled · 3
3. Leaf laminas 1.5–4 mm wide; ligules 4–8 mm long; pedicelled spikelets sterile, chartaceous, twice the size of the sessile spikelets · 18. *fastigiatus*
   – Leaf laminas 5–6 mm wide; ligules 6–20 mm long; pedicelled spikelets usually male, about the same size as the sessile spikelets · 19. *textilis*
4. Racemes loose, zig-zag with divergent internodes and pedicels, often plumose · · · · · · · 5
   – Racemes ± compact and ± terete with appressed internodes and pedicels, glabrous or pubescent but not plumose · 7
5. Racemes usually in subdigitate groups; pedicelled spikelets sterile or absent · · 1. *eucomus*
   – Racemes in pairs; pedicelled spikelets male · 6
6. Culms not branched; ligule (1)1.5–4 mm long; spikelets 6–7 mm long · · · · · · 2. *ligulatus*
   – Culms usually branched; ligule c. 0.5 mm long; spikelets 4.5–5.5 mm long · · · 3. *africanus*
7. Culms not branched · 8
   – Culms branched · 12
8. Inferior glume with a conspicuous marginal wing; racemes exserted from the spatheole only at maturity · 9
   – Inferior glume wingless; racemes long-exserted from the spatheole · · · · · · · · · · · · · 11
9. Inferior glume with numerous conspicuous pits; wing very narrow · · · · · · · · 4. *lacunosus*
   – Inferior glume without pits or these present but inconspicuous; wing broad · · · · · · · · 10
10. Ligule fringed, c. 1 mm long; raceme internodes long-ciliate on one margin only and with a few shorter hairs on the other · 5. *distachyos*

- Ligule not fringed, c. 0.3 mm long; raceme internodes equally ciliate on the margins · · ·
 · 6. *amethystinus*
11. Racemes in pairs, switch-like; leaf laminas scabrid, reduced to a narrow wing on either side
 of the thickened midrib · 7. *lima*
- Racemes 1–5 in subdigitate groups, not switch-like; leaf laminas smooth, not reduced and
 midrib not conspicuously thickened · 8. *mannii*
12. Racemes numerous in subdigitate groups; basal leaf sheaths distichously arranged, laterally
 compressed and keeled · 13
- Racemes 2–3; basal leaf sheaths neither distichously arranged nor compressed and keeled
 · 14
13. Leaf margins scabrid; plant glaucous; awns 3.5–8 mm long · · · · · · · · · · · · · · 9. *brazzae*
- Leaf margins smooth; plant yellow tinged with red; awns 9–10 mm long · · 10. *appendiculatus*
14. Plant annual · 13. *pseudapricus*
- Plants perennial · 15
15. Inferior glume of sessile spikelet with shallow median groove; base of leaf lamina very
 narrow and petiole-like; abaxial second ligule usually present · · · · · · · · · · · · 12. *gayanus*
- Inferior glume of sessile spikelet with deep median groove; base of leaf lamina wide and not
 petiole-like; no abaxial second ligule present · 16
16. Racemes silvery with hairs up to c. 6 mm long; raceme internodes and pedicels clavate but
 not canaliculate, at most shallowly concave · 17
- Racemes with hairs 0.5–1 mm long, not silvery; raceme internodes and pedicels inflated-
 clavate and deeply canaliculate · 18
17. Pedicelled spikelet with an awn 3.5–7 mm long; callus of sessile spikelet short and shallowly
 inserted · 14. *chinensis*
- Pedicelled spikelet muticous or shortly mucronate (up to 2 mm); callus of sessile spikelet
 cuneate, 1–2 mm long and deeply inserted · 15. *schirensis*
18. Ligule (1.5)2.8–6 mm long; leaf lamina 1–3.2 mm wide; anthers 2–2.7 mm long · · · · · ·
 · 16. *perligulatus*
- Ligule c. 0.5 mm long; leaf lamina c. 4 mm wide; anthers 1–2 mm long · · 17. *canaliculatus*

1. **Andropogon eucomus** Nees, Fl. Afr. Austral. Ill.: 104 (1841). —Stapf in Prain, F.T.A. **9**: 230
 (1919). —Robyns, Fl. Agrost. Congo Belge **1**: 124 (1929). —Stent & Rattray in Proc. &
 Trans. Rhodesia Sci. Assoc. **32**: 10 (1933). —Sturgeon in Rhodesia Agric. J. **51**: 13 (1954).
 —Chippindall in Meredith, Grasses & Pastures S. Africa: 501 (1955). —Jackson & Wiehe,
 Annot. Check List Nyasal. Grass.: 28 (1958). —Napper, Grasses Tanganyika: 102 (1965).
 —J.G. Anderson in Bothalia **9**: 10 (1966). —Launert in Merxmüller, Prodr. Fl. SW. Afrika,
 fam. 160: 21 (1970). —Simon in Kirkia **8**: 16, 49 (1971). —Clayton in F.W.T.A., ed. 2, **3**: 482
 (1972). —Bennett in Kirkia **11**: 174 (1980). —Clayton & Renvoize in F.T.E.A., Gramineae:
 775 (1982). —Gibbs Russell et al., Grasses South. Africa [Mem. Bot. Surv. S. Africa No. 58]:
 40 (1990). Syntypes from South Africa.

Delicate or ± robust, caespitose perennial; culms up to 250 cm high, erect, much
branched above, tinged with red. Leaves glabrous except for the ligule area; sheath
shortly auriculate at the top; ligule a fimbriate membrane, 0.5–1.5 mm long; laminas
2–50 cm × 1.2–4 mm, flat or folded, abruptly acute at the apex and mucronate.
Racemes 2.5–4 cm long, in open subdigitate groups of (2)3–5, plumose, zig-zag,
exserted from the spatheole at maturity; peduncles longer than the racemes;
internodes and pedicels very delicate, filiform, ciliate with silky hairs 5–10(11) mm
long on both margins. Sessile spikelets 2–5.5 mm long, glabrous, often reddish;
inferior glume with a broad median groove, glabrous; superior glume ± flat on the
back; superior lemma entire or minutely bidentate; awn flexuous, 7–22 mm long,
often somewhat inconspicuous, often reddish; anthers 0.3–1.8 mm long, straw-
coloured to reddish. Pedicelled spikelets absent or reduced to a sterile glume up to
3 mm long.

A. eucomus, with its 2 subspecies, and *A. ligulatus* form a distinct species-group in the Flora
Zambesiaca area on account of their plumose, zig-zag racemes and similar ecology (damp soils).
Outside the Flora Zambesiaca area, *A. eucomus* subsp. *eucomus* and subsp. *huillensis* are fairly
distinct, the former being a delicate plant, the latter a robust one. But in the Flora Zambesiaca
area there is an almost continuous range of variation in inflorescence characters that also
includes *A. laxatus*, hitherto recognized as an independent species. Although the robust plants
have character-states associated with subsp. *huillensis*, smaller plants may have the same range in
their reproductive parts. There is not a clear-cut separation on awn length or pedicelled

spikelet length (the latter from absent or vestigial to 5.5 mm). Those differences in pedicel length cited by Anderson (loc. cit.) do not hold for the subspecies in the Flora Zambesiaca area. Also, no specimens of subsp. *huillensis* with fertile pedicelled spikelets were found in the Flora Zambesiaca area, in contradiction to the observations of Anderson and Chippindall (loc. cit.). In the Flora Zambesiaca area such specimens with pedicelled spikelets have other features which clearly assign them to the likewise plumose *A. ligulatus*. Plants with only 2 racemes named as *A. laxatus* do not show other distinct features and must be regarded as depauperate forms of subsp. *huillensis*. Among such is included the curious specimen 'Marandellas, 28.iii.1948, *Corby* 22551 (K)' with anomalously long ligules (2.2 mm); it had been named *A. laxatus* var. *ligulatus*.

This species-complex has apparently clear links with species in the New World and an overall revision is much needed.

Pedicelled spikelet totally suppressed or vestigial (careful observation with high magnification needed); sessile spikelet 2.5–3 mm long · subsp. *eucomus*
Pedicelled spikelet reduced to a sterile glume up to 3 mm long; sessile spikelet 3.5–5.5 mm long (see TAB 23, fig. 1 · subsp. *huillensis*

Subsp. eucomus

Botswana. N: Ngamiland Distr., Okavango Delta, Moremi Wildlife Reserve, near Mboma Camp, 8.i.1973, *P.A. Smith* 336 (K). SE: Gaborone Dam, Aedume Park, *O.J. Hansen* 3279 (K). **Zambia**. B: Mongu Distr., Mabumbu, 4.iii.1965, *van Rensburg* 3026 (K). N: Mbala Distr., Kawimbe Dambo, 26.iii.1959, *McCallum Webster* A129 (K). W: Ndola Distr., 2.xi.1957, *Fanshawe* 3830 (LISC). C: 22 km east of Lusaka, Great East Road, 5.xii.1971, *Kornaś* 593 (K). E: Katete, by Great East Road, 1070 m, 5.ii.1957, *Wright* 142 (K). S: near Kalomo, 8.iv.1946, *Grassl* 46-68 (K). **Zimbabwe**. N: Binga, Lake Kariba, 25.xii.1990, *Laegaard* 15865A (K). W: Nkayi Distr., Gwampa Forest Land (Reserve), c. 910 m, i.1956, *Goldsmith* 11/56 (K; PRE). C: Harare Distr., Cleveland Dam Park, c. 10 km west of Harare, 11.xii.1990, *Laegaard* 15755 (K). E: Mutare Distr., Zimunya C.L., SW slopes of Dangare Mt., 31.xii.1950, *Chase* 3543 (BM; LISC). S: Masvingo Distr., Makaholi Experimental Station, 10.iii.1978, *Senderayi* 181 (K; PRE). **Malawi**. N: Mzimba Distr., 5 km west of Mzuzu at Katoto, 15.vii.1973, *Pawek* 7194 (K). C: Ntchisi Distr., Ntchisi Forest Reserve near Chitembwene, 13.iv.1991, *Radcliffe-Smith* 5971 (K). S: Limbe Distr., Upper Hynde Dam, 2 km north of Limbe, *Brummitt* 8571 (K; LISC; MAL; PRE; SRGH; UPS). **Mozambique**. N: Montepuez Distr., andados 5 km de Montepuez para Namuno, 430 m, 26.xii.1963, *Torre & Paiva* 9694 (K; LISC). Z: Kongone, mouth of Zambezi, i.1861, *Kirk* s.n. (K). T: Angónia Distr., Vila Coutinho, 4.iii.1932, *Vincent* 65 (BM). MS: north of Beira, ii.1972, *Tinley* 2399 (K). GI: Inhassoro Distr., Bazaruto Is., 0–90 m, 20.x.–7.xi.1958, *Mogg* 29080 (K; LISC). M: Maputo Distr., Inhaca Is., 37 km east of Maputo (Lourenço Marques), 21.x.1964, *Mogg* 31915 (K).
Also in Uganda, Tanzania and Dem. Rep. Congo, southwards to South Africa and Angola; Madagascar. Growing in damp peaty, sandy or gravelly soils in dambos, marshes, meadows and open forest, along riverbanks and in coastal sand; 0–1500 m.

Subsp. **huillensis** (Rendle) Sales, comb. et stat. nov. TAB. **23**, fig. 1. Type from Angola.
    Andropogon huillensis Rendle in Hiern et al., Cat. Afr. Pl. Welw. **2**: 146 (1899). —Stapf in Prain, F.T.A. **9**: 231 (1919). —Robyns, Fl. Agrost. Congo Belge **1**: 125 (1929). —Stent & Rattray in Proc. & Trans. Rhodesia Sci. Assoc. **32**: 11 (1933). —Sturgeon in Rhodesia Agric. J. **51**: 14 (1954). —Chippindall in Meredith, Grasses & Pastures S. Africa: 500 (1955). —J.G. Anderson in Bothalia **9**: 12 (1966). —Launert in Merxmüller, Prodr. Fl. SW. Afrika, fam. 160: 21 (1970). —Simon in Kirkia **8**: 16, 49 (1971). —Bennett in Kirkia **11**: 174 (1980). —Gibbs Russell et al., Grasses South. Africa [Mem. Bot. Surv. S. Africa No. 58]: 41 (1990).
    Andropogon laxatus Stapf in Prain, F.T.A. **9**: 237 (1919). —Sturgeon in Rhodesia Agric. J. **51**: 16 (1954). —Chippindall in Meredith, Grasses & Pastures S. Africa: 501 (1955). —Napper, Grasses Tanganyika: 102 (1965). —J.G. Anderson in Bothalia **9**: 13 (1966). —Simon in Kirkia **8**: 16, 49 (1971). —Clayton & Renvoize in F.T.E.A., Gramineae: 775 (1982). —Gibbs Russell et al., Grasses South. Africa [Mem. Bot. Surv. S. Africa No. 58]: 41 (1990). —Zon, Gramin. Cameroun **2**: 430, t. 95/1–4 (1992). Lectotype: Zimbabwe, Harare (Salisbury) *Mundy* (K).

    Botswana. N: 22 km north of Nokaneng, 17.iii.1965, *Wild & Drummond* 7126 (BM; K). **Zambia**. B: Kabompo Distr., Kusokweji Dambo, 27 km north of Manyinga, 26.xii.1969, *Simon & Williamson* 2033 (BM; K; SRGH). N: Shiwa Ngandu, 19.ix.1938, *Greenway & Trapnell* 5706 (K). W: Mwinilunga Distr., Dobeka Bridge, 1937/8, *Milne-Redhead* 2700 (K). C: Serenje Distr., Kundalila Falls, xii.1968, *G. Williamson* 1344 (K). **Zimbabwe**. W: Matobo Distr., i.1957, *Miller* 4042 (LISC). C: Marondera (Marandellas), iii.1948, *Corby* in *GHS* 20990 (K; SRGH). E: Nyanga Distr., Rhodes Inyanga National Park, road to Nyamziwa Falls, 1980 m, 7.i.1970, *A.O. Crook* 904

(K; LISC; PRE). S: Masvingo (Victoria), 30.iii.1973, *Chiparawasha* 666 (K). **Malawi**. N: Nyika Plateau, Dam 1-11, 2255 m, 28.xi.1975, *E. Phillips* 380 (K). **Mozambique**. MS: Mossurize Distr., Espungabera (Spungabera), 12.xi.1943, *Torre* 6166 (LISC). GI: Inhassoro Distr., Bazaruto Is., Ponta Estone, x.1958, *Mogg* 28675 (K; LISC). M: Maputo Distr., Inhaca Is., 37 km east of Maputo (Lourenço Marques), 28.ix.1959, *Mogg* 31363 (K; PRE).

Also in South Africa (Transvaal, KwaZulu-Natal), Swaziland, Namibia and Angola. Growing in wet peaty and sandy soils in dambos, bogs, wet flushes and alluvial plains, and along roadsides; 0–2255 m.

2. **Andropogon ligulatus** (Stapf) Clayton in Kew Bull. **32**: 2 (1977). —Bennett in Kirkia **11**: 175 (1980). —Clayton & Renvoize in F.T.E.A., Gramineae: 774 (1982). Type: Zimbabwe, Charter Distr., *Mundy* s.n. (K, holotype).

    Andropogon laxatus Stapf var. *ligulatus* Stapf in Prain, F.T.A. **9**: 238 (1919). —Sturgeon in Rhodesia Agric. J. **51**: 14 (1954). —Simon in Kirkia **8**: 16, 49 (1971).

    Andropogon laxatus sensu Stent & Rattray in Proc. & Trans. Rhodesia Sci. Assoc. **32**: 11 (1933), non Stapf.

Delicate caespitose perennial; culms up to 120 cm high, erect, unbranched. Leaves glabrous except for scattered long hairs near the ligule; ligule (1)1.5–4 mm long, ± acute, shortly fimbriate; laminas 7–60 cm × 1–1.5 mm, linear. Racemes in terminal pairs, 4–7.5 cm long, ± plumose, long-exserted from the spatheoles; peduncles much longer than the racemes; internodes and pedicels filiform, ciliate with silky hairs up to 5 mm long on both margins. Spikelets similar, 6–7 mm long, narrowly lanceolate, reddish. Sessile spikelet: inferior glume with a broad median groove, glabrous; superior lemma bifid in the upper $^1/_4$–$^1/_3$; awn 9–20 mm long, geniculate; anthers 2.5–3.5 mm long, straw-coloured. Pedicelled spikelet male, 5–7 mm long, mucronate.

    Zambia. N: Mpika Distr., Danger Hill, 29 km north of Mpika, 1370 m, 18.xii.1967, *Simon & Williamson* 1436 (BM; K; PRE). W: Kitwe, 30.xi.1957, *Fanshawe* 4085 (LISC). C: Mkushi Distr., Mkushi River Dambo, 1430 m, 14.x.1967, *Simon & Williamson* 977 (BM; K; PRE; SRGH). **Zimbabwe**. C: Harare (Salisbury), i.1927, *Eyles* 2934 (K; PRE). E: Chimanimani (Melsetter), Pasture Res. sub-station, 1520 m, 21.xi.1949, *F.R. Williams* 30 (K). **Malawi**. C: Mchinji (Fort Manning), 11.x.1951, *Jackson* 618 (K).

Also in Tanzania. Growing in wet places in dambos and bogs; 1200–1550 m.

A member of the species-group that includes *A. eucomus* and *A. africanus*, with which it shares a number of inflorescence characters.

3. **Andropogon africanus** Franch. in Bull. Soc. Hist. Nat. Autun **8**: 325 (1895). —Stapf in Prain, F.T.A. **9**: 239 (1919). —Robyns, Fl. Agrost. Congo Belge **1**: 127 (1929). —Simon in Kirkia **8**: 49 (1971). —Clayton in F.W.T.A., ed. 2, **3**: 485 (1972). —Clayton & Renvoize in F.T.E.A., Gramineae: 775 (1982). —Zon, Gramin. Cameroun **2**: 432 (1992). Type from Congo-Brazzaville.

Caespitose perennial; culms up to 170 cm high, erect, usually branched. Leaves glabrous; ligule a very short (c. 0.5 mm) minutely fringed membrane; laminas 5–17 cm × 2.5–5 mm, flat or folded, abruptly acute at the apex. Racemes in pairs, 3–4.5 cm long, zig-zag, exserted from the spatheoles at maturity; peduncles longer than the racemes; internodes and pedicels filiform, divergent, clavate above, ciliate on both margins, the hairs much longer towards the apex. Spikelets similar, 4.5–5.5 mm long, reddish. Sessile spikelet: inferior glume with broad median groove, glabrous; superior lemma bifid to about half way; awn geniculate, 11–15.5 mm long; anthers 2–2.5 mm long, straw-coloured. Pedicelled spikelet male; inferior glume with an awn 1–2 mm long.

    Zambia. S: Choma Distr., Mochipapa, 15.iii.1962, *Astle* 1564 (BM; K). **Zimbabwe**. E: Nyanga Distr., Mare Dam, 8.iv.1966, *Simon* 781A (BM).

Throughout tropical Africa. Growing at the edges of dambos; 1230–1950 m.

Clearly related to *A. eucomus* and *A. ligulatus* in raceme structure, but in this species the racemes are not always so clearly plumose.

4. **Andropogon lacunosus** J.G. Anderson in Bothalia **8**: 113 (1962); & **9**: 20 (1966). —Simon in Kirkia **8**: 16, 49 (1971). —Clayton in F.W.T.A., ed. 2, **3**: 485 (1972). —Bennett in Kirkia **11**: 174 (1980). —Clayton & Renvoize in F.T.E.A., Gramineae: 770 (1982). —Gibbs Russell et al., Grasses South. Africa [Mem. Bot. Surv. S. Africa No. 58]: 41 (1990). —Zon, Gramin. Cameroun **2**: 425 (1992). Type from South Africa.

Small, delicate, decumbent straggling perennial; culms up to 40 cm high, unbranched; internodes compressed. Leaf sheaths hairy; ligule a very short (c. 1 mm) minutely fringed membrane; laminas 6–16 cm × 3–4 mm, spreading, flat, glabrous, tapering to a very fine point at the apex. Racemes in terminal pairs, 2.5–5 cm long, clearly exserted from the spatheoles only at maturity; peduncles much longer than the racemes; internodes and pedicels filiform, ciliate on both margins. Spikelets similar, 5–7.5 mm long (the pedicelled slightly the longer and tinged with purple); inferior glumes of both irregularly pitted between the nerves, narrowly winged, glabrous. Sessile spikelet: inferior glume with a deep median groove, convex on the back, bifid at the apex; superior lemma bifid to about the middle; awn geniculate, 14.5–16.5 mm long; anthers 3–3.5 mm long, straw-coloured. Pedicelled spikelet male.

Zimbabwe. E: Chimanimani Mts., 0.8 km north of Southern Lake, Bundi River, 26.ix.1966, *Simon* 843 (K).

Also in Tanzania, Cameroon and South Africa (Mpumalanga). Montane grassland, growing near water; c. 1460 m.

This is an interesting and distinctive species meriting anatomical research into the development of the pitted areas. It should be noted that inconspicuous pits are also present on the leaves. See comments under *A. distachyos*.

5. **Andropogon distachyos** L., Sp. Pl. **2**: 1046 (1753). —Stapf in Prain, F.T.A. **9**: 218 (1919). —Robyns, Fl. Agrost. Congo Belge **1**: 138 (1929). —Sturgeon in Rhodesia Agric. J. **51**: 13 (1954). —Chippindall in Meredith, Grasses & Pastures S. Africa: 496 (1955). —Napper, Grasses Tanganyika: 103 (1965). —J.G. Anderson in Bothalia **9**: 29 (1966). —Simon in Kirkia **8**: 16 (1971). —Clayton in F.W.T.A., ed. 2, **3**: 485 (1972). —Bennett in Kirkia **11**: 174 (1980). —Clayton & Renvoize in F.T.E.A., Gramineae: 770 (1982). —Gibbs Russell et al., Grasses South. Africa [Mem. Bot. Surv. S. Africa No. 58]: 40 (1990). —Zon, Gramin. Cameroun **2**: 426 (1992). TAB. **23**, fig. 2. Type from Europe.
   Andropogon flabellifer sensu Simon in Kirkia **8**: 49 (1971), non Pilg.

Slightly decumbent caespitose perennial; culms c. 90 cm high, unbranched. Leaf sheaths glabrous; ligule a very short (c. 1 mm) minutely fringed membrane; laminas 7–24 cm × 2–9 mm, loosely pilose, flat, tapering to a fine point at the apex. Racemes in terminal pairs, 6–6.5 cm long (up to 10.5 cm outside the Flora Zambesiaca area), cylindrical, clearly exserted from the spatheoles only at maturity; peduncles longer than the racemes; internodes and pedicels slightly clavate, ciliate with hairs up to 3.5 mm long on one margin and a few short hairs on the other (particularly noticeable on the internodes). Spikelets similar; inferior glumes asymmetrically beaked, with two broad (0.5 mm) marginal wings, pubescent; superior glumes awned. Sessile spikelet 11–12 mm long, obtuse at the base, inserted in the internode apex; inferior glume with a shallow median groove, asymmetrically winged, 2-toothed at the apex; superior glume ± flat on the back and with an awn 6–7 mm long; superior lemma bifid in the upper $^2/_3$; awn 23–26 mm long, geniculate; anthers 2.5–3 mm long, straw-coloured. Pedicelled spikelet male, 9–9.5 mm long; inferior glume with an awn c. 6 mm long; superior glume with an awn c. 2 mm long; superior lemma bidentate at the apex.

Zambia. N: Kasama Distr., Kayambi, Ntumba drainage, Chozi tributary, 8.v.1958, *Vesey-FitzGerald* 1774 (BM; SRGH). Zimbabwe. E: Nyanga Distr., Troutbeck, 20–21.iii.1948, *Rattray* 1425 (K; SRGH).

Throughout southern Europe, tropical and southern Africa, Arabia and ?Thailand. Growing in damp ground; 1400–2000 m.

Surprisingly rare or poorly collected in the Flora Zambesiaca area. The main distinguishing features are the cilia irregularly distributed on the raceme internodes and pedicels, and the broad asymmetrical brownish marginal wings on the inferior glumes. Although it does not have the well-marked pits of *A. lacunosus*, it does have some depressions on the inferior glume surface suggesting a relationship between the two species. The other species in the Flora Zambesiaca area that has broad marginal glume wings is *A. amethystinus*.

6. **Andropogon amethystinus** Steud., Syn. Pl. Glumac. **1**: 371 (1854). —Stapf in Prain, F.T.A. **9**: 216 (1919). —Robyns, Fl. Agrost. Congo Belge **1**: 129 (1929). —Jackson & Wiehe, Annot. Check List Nyasal. Grass.: 28 (1958). —Napper, Grasses Tanganyika: 103 (1965). —Clayton in F.W.T.A., ed. 2, **3**: 485 (1972). —Clayton & Renvoize in F.T.E.A., Gramineae: 772 (1982).

Tab. 23. Spikelet pairs, front and rear views of species of *Andropogon* (all × 5). 1, *A. eucomus* subsp. *huillensis*; 2, *A. distachyos*; 3, *A. gayanus*; 4, *A. chinensis*. Drawn by Ann Davies. From F.T.E.A.

—Gibbs Russell et al., Grasses South. Africa [Mem. Bot. Surv. S. Africa No. 58]: 40 (1990). —Zon, Gramin. Cameroun **2**: 427 (1992). Type from Ethiopia.

Andropogon pratensis Hack. in A. & C. de Candolle, Monogr. Phan. **6**: 463 (1889). —Napper, Grasses Tanganyika: 103 (1965). —Simon in Kirkia **8**: 49 (1971). Type from Ethiopia.

Creeping perennial; culms 50–80 cm high, branched below, reddish. Leaves often falcate, sparsely to distinctly pilose; ligule a very short (0.3–0.6 mm) membrane; laminas 4–13 cm × 1.8–3 mm, ± revolute, auriculate at the base, tapering to a fine point at the apex. Racemes in terminal pairs, 5–6.5 cm long, clearly exserted from the spatheoles only at maturity; peduncles much longer than the racemes; internodes and pedicels ± clavate, glabrous to ciliate on both margins. Spikelets similar; inferior glumes asymmetrical with 2 unequally broad hyaline marginal wings, 2-dentate or 2-mucronate at the apex; superior glumes awned. Sessile spikelet 6–12 mm long; inferior glume flat or with a shallow median groove, rarely slightly convex on the back, winged, minutely pilose or glabrous, often mucronate at the apex; superior glume convex on the back and with an awn 4–5 mm long; superior lemma bifid to about the middle; awn geniculate, 13–14.5 mm long. Pedicelled spikelet male, 6–8 mm long; inferior glume with an awnlet 2–3 mm long; superior glume with a mucro 0.7–1.4 mm long.

Zambia. W: Ndola Distr., Itawa Dambo, 25.v.1950, *Jackson* 28 (K). **Malawi**. N: Rumphi Distr., Nyika Plateau, Lake Kaulime, 2340 m, 16.v.1970, *Brummitt* 10789 (K; LISC; MAL; PRE; SRGH; UPS).

Nigeria and Ethiopia southwards to southern Africa; also in India (Nilgiris). Growing on stream banks and lake margins; c. 2340 m.

Rare or poorly collected in the Flora Zambesiaca area. Sometimes the culms are only divided at the base of the plant. See comments under *A. distachyos*.

7. **Andropogon lima** (Hack.) Stapf in Prain, F.T.A. **9**: 217 (1919). —Clayton in F.W.T.A., ed. 2, **3**: 485 (1972). —Clayton & Renvoize in F.T.E.A., Gramineae: 771 (1982). —Zon, Gramin. Cameroun **2**: 429 (1982). Type from Cameroon.

Andropogon amethystinus Steud. var. *lima* Hack. in A. & C. de Candolle, Monogr. Phan. **6**: 464 (1889).

Andropogon amethystinus sensu Jackson & Wiehe, Annot. Check List Nyasal. Grass.: 28 (1958) non Steud.

Caespitose perennial; culms up to 100 cm high, erect, unbranched; internodes terete. Leaves sparsely pilose, scabrid; sheaths auriculate; ligule a short (2–3 mm) membrane; laminas 6–47 cm × 2–4 mm, stiffly erect, almost reduced to just the keel, the narrow margins revolute, tapering to a very fine point at the apex. Racemes in terminal pairs (rarely solitary), switch-like, 4–6 cm long, long-exserted from the spatheoles; peduncles much longer than the racemes; internodes and pedicels filiform with scattered cilia up to 2 mm long on both margins. Sessile spikelet 8–9 mm long, longer than the internode, the callus conical and with a ring of long hairs; inferior glume slightly convex on the back, loosely pilose; superior glume ± flat on the back and with an awn 2–4 mm long; superior lemma bifid to the middle; awn geniculate, 9–15 mm long; anthers not seen. Pedicelled spikelet male, 5.5–8 mm long, with an awn 2–3 mm long.

Malawi. S: Mt. Mulanje, Chambe Plateau, 2130 m, 16.xi.1949, *Wiehe* 318 (K).

Also in Cameroon, Sudan and Ethiopia and southwards to Malawi. Occasional to rare in moist places in grassland; c. 2130 m.

The very narrow, almost glabrous racemes and the scabrid erect leaves are the most characteristic features of this distinct species. Apparently rare with few herbarium specimens from the Flora Zambesiaca area.

8. **Andropogon mannii** Hook.f. in J. Linn. Soc., Bot. **7**: 232 (1864). —Stapf in Prain, F.T.A. **9**: 226 (1919). —Napper, Grasses Tanganyika: 102 (1965). —Clayton in F.W.T.A., ed. 2, **3**: 485 (1972). —Bennett in Kirkia **11**: 175 (1980). —Clayton & Renvoize in F.T.E.A., Gramineae: 774 (1982). —Gibbs Russell et al., Grasses South. Africa [Mem. Bot. Surv. S. Africa No. 58]: 41 (1990). —Zon, Gramin. Cameroun **2**: 427 (1992). Syntypes from Bioko (Fernando Po).

Andropogon flabellifer Pilg. in Bot. Jahrb. Syst. **54**: 284 (1917). —Napper, Grasses Tanganyika: 102 (1965). —Simon in Kirkia **8**: 16, 49 (1971). Type from Tanzania.

Delicate caespitose perennial; culms 30–50 cm high, erect, unbranched, often reddish; internodes very delicate. Leaf sheaths distichous, compressed and ± keeled, glabrous; ligule a very short (c. 1 mm) minutely fringed membrane; laminas 2–17 cm × c. 0.2 mm wide, flat, loosely long-pilose, tapering to a fine point at the apex. Racemes 1–5 in terminal subdigitate groups, 3–7 cm long, slender and flexuous, long-exserted from the spatheoles, dark red; peduncles much longer than the racemes; internodes and pedicels filiform, almost glabrous or ciliate on both margins, the internodes shorter than the sessile spikelets. Spikelets similar, 4.5–8 mm long (the pedicelled slightly the longer), glabrous. Sessile spikelet 4.5–6 mm long, pilose at the base; inferior glume with a shallow median groove, glabrous, 1–3-mucronate at the apex; superior glume rounded on the back, usually with an awn 3–4.5 mm long; superior lemma bifid to the middle; awn geniculate, 6–16 mm long; anthers 2.5–3 mm long, straw-coloured. Pedicelled spikelet male, 5.5–8 mm long.

Zambia. N: Western Valleys, Nyika Plateau, 6.i.1959, *E.A. Robinson* 3092 (K). **Zimbabwe**. E: Chimanimani Distr., Bundi Valley between Skeleton Pass and Mountain Hut, c. 1540 m, 9.ix.1972, *Simon* 2281 (K; LISC; PRE). **Malawi**. S: Mt. Mulanje, Chambe Plateau, 23.iii.1958, *Jackson* 2184 (K). **Mozambique**. MS: Gorongosa Distr., Gogogo summit area, Gorongosa Mt., 1800 m, viii.1971, *Tinley* 2138 (K; LISC; PRE).
Also in tropical West Africa and Sudan, southwards to southern Africa. Growing in montane grassland and in shallow soil overlying rocks; 1540–2000 m.
A small plant with delicate awns and a tuft of white hairs at the raceme nodes. For affinities see comments under *A. festuciformis*.

9. **Andropogon brazzae** Franch. in Bull. Soc. Hist. Nat. Autun **8**: 326 (1895). —Stapf in Prain, F.T.A. **9**: 233 (1919). —Robyns, Fl. Agrost. Congo Belge **1**: 126 (1929). —J.G. Anderson in Bothalia **9**: 15 (1966). —Launert in Merxmüller, Prodr. Fl. SW. Afrika, fam. 160: 21 (1970). —Simon in Kirkia **8**: 16, 49 (1971). —Bennett in Kirkia **11**: 174 (1980). —Gibbs Russell et al., Grasses South. Africa [Mem. Bot. Surv. S. Africa No. 58]: 40 (1990). Type from Congo-Brazzaville.

Robust, caespitose, rhizomatous perennial; culms up to 230 cm high, erect, branched, with stout internodes, glaucous. Leaves glabrous, scabrid on the margins; basal sheaths distichous and keeled; ligule a very short (c. 1 mm) minutely fringed membrane; laminas 20–100 cm × 2–6 mm, keeled, often folded, ± revolute, tapering to a very fine point at the apex. Racemes up to 15 in subdigitate groups, 6–12 cm long, unequal, slender, flexuous, virtually glabrous, exserted from the spatheoles, dark red; peduncles longer than the racemes; internodes and pedicels filiform, scabrid, almost glabrous, the internodes slightly shorter than the sessile spikelets. Spikelets similar. Sessile spikelet 4.3–6.3 mm long; inferior glume with a shallow median groove, glabrous; superior glume ± flat on the back; superior lemma entire or bifid in the upper $^1/_5$–$^1/_2$, awnless or with a geniculate awn 3.5–8 mm long; anthers 2–3 mm long, straw-coloured. Pedicelled spikelet male, 4.5–7.5 mm long, awnless to shortly mucronate.

Botswana. N: Ngamiland Distr., Okavango River, 20.ii.1983, *P.A. Smith* 4119 (K; PRE). **Zambia**. B: Kabompo Distr., Mufuli River, 58 km west of Kabompo, 26.xii.1969, *Simon & Williamson* 2055 (BM; K). N: Mpika Distr., Chambeshi River Bridge, xii.1968, *G. Williamson* 1312 (K). W: Mwinilunga Distr., Luakela River, 25 km north of Mwinilunga, 10.vi.1963, *Drummond* 8274 (K). S: Kalomo Distr., Kazangula quarantine area, 1070 m, ii.1932, *Trapnell* 964 (K). **Zimbabwe**. W: Hwange Distr., Victoria Falls, Livingstone Island, 23.iii.1976, *Ellis* 2775 (K; PRE).
Also in Namibia, Angola and Dem. Rep. Congo. Growing in flood plains, along river banks and in sandy loams; 900–1370 m.
Two variants occur in the Flora Zambesiaca area but they do not merit formal recognition. One has awnless lemmas and the pedicelled spikelet longer than the sessile; the other has awned lemmas and the pedicelled spikelet shorter than the sessile. The only other species in the Flora Zambesiaca area with very distinctly scabrid leaves is *A. lima*.

10. **Andropogon appendiculatus** Nees, Fl. Afr. Austral. Ill.: 105 (1841). —Chippindall in Meredith, Grasses & Pastures S. Africa: 500 (1955). —Bennett in Kirkia **11**: 173 (1980). —Gibbs Russell et al., Grasses South. Africa [Mem. Bot. Surv. S. Africa No. 58]: 40 (1990). Types from South Africa.

Caespitose perennial; culms up to 110 cm high, erect, branched with slender internodes, yellowish with reddish tinge. Leaves glabrous except at the summit of the sheath; basal sheaths laterally compressed and keeled, distichous and shining; ligule a short (c. 1.5 mm) minutely fringed membrane; laminas 15–25 cm × c. 3 mm wide, folded and keeled, smooth on the margins, tapering to a very fine point at the apex. Racemes up to 10 in subdigitate groups, 9–10 cm long, slender, flexuous, sparsely pubescent, exserted from the spatheoles, dark red; peduncles longer than the racemes; internodes and pedicels filiform, ciliate on both margins, the internodes shorter than the sessile spikelets. Sessile spikelet 4.5–5 mm long; inferior glume with a shallow median groove or ± flat, with marginal hairs towards the apex; superior glume concave on the back; superior lemma bifid to the middle; awn geniculate, 9–10 mm long; anthers c. 1.5 mm long, straw-coloured. Pedicelled spikelet male, 3.5–4.5 mm long, shortly mucronate.

Zimbabwe. E: Mutasa Distr., Nuza Plateau, ii.1935, *Gilliland* 1424 (K).
Widespread in South Africa (all provinces except Northern Cape). Montane grassland and dambos; c. 2000 m.
The inflorescence of *A. appendiculatus* is similar to that of *A. brazzae*, but the taxa can be distinguished by the yellowish-red colour of the former, and by its smooth leaves and longer awns.

11. **Andropogon festuciformis** Rendle in Hiern et al., Cat. Afr. Pl. Welw. **2**: 145 (1899). —J.G. Anderson in Bothalia **9**: 14 (1966). —Simon in Kirkia **8**: 49 (1971). —Clayton in F.W.T.A., ed. 2, **3**: 485 (1972). —Gibbs Russell et al., Grasses South. Africa [Mem. Bot. Surv. S. Africa No. 58]: 41 (1990). Type from Angola.
   Andropogon schlechteri Hack. in Bull. Herb. Boissier, sér. 2, **6**: 703 (1906). —Chippindall in Meredith, Grasses & Pastures S. Africa: 516 (1955). Type from South Africa.
   Hypogynium schlechteri (Hack.) Pilg. in Engler & Prantl, Nat. Pflanzenfam., ed. 2, **14e**: 156 (1940).
   Hypogynium spathiflorum sensu Stapf in Prain, F.T.A. **9**: 168 (1917) non Nees (1829).

Delicate caespitose perennial; culms up to 140 cm high, erect, much-branched above, reddish throughout. Leaves glabrous; basal sheaths distichous, laterally compressed and keeled; ligule a very short (c. 0.5 mm) minutely fringed membrane; laminas 5–28 cm × 1.8–4 mm wide, keeled, tapering to a very fine point at the apex. Racemes solitary, gathered into fascicles of 1–6 at regular intervals along the culm, 3–4 cm long, slender, flexuous, virtually glabrous, short- or long-exserted from the spatheoles, dark red; peduncles about as long as the racemes; internodes and pedicels filiform, very short, glabrous, the internodes c. $\frac{1}{5}$ the length of the sessile spikelets. Spikelets similar, 4–5.3 mm long, awnless; inferior glumes flat on the back. Sessile spikelet with a small crown of hairs at the base; anthers 2–2.3 mm long, straw-coloured. Pedicelled spikelet male.

Zambia. B: Mongu Distr., 32 km east of Mongu, 10.xi.1959, *Drummond & Cookson* 6307 (BM). N: Kawambwa Distr., Chishinga Ranch, 13.ix.1961, *Astle* 894 (BM; K). W: Mwinilunga Distr., SW of Dobeka Bridge, 11.x.1937, *Milne-Redhead* 2705 (K). C: Serenje Distr., Kundalila Falls, 13 km SE of Kanona, 1400 m, 15.x.1967, *Simon & Williamson* 1009 (K; PRE; SRGH). **Malawi**. N: Mzimba Distr., 5 km west of Mzuzu at Katoto, 1370 m, 15.vii.1973, *Pawek* 7189 (K; PRE).
Also in South Africa (KwaZulu-Natal), Angola, Dem. Rep. Congo and Guinea. Growing in moist places and boggy grassland; 1000–1500 m.
A. festuciformis is related to the much more robust *A. brazzae*. Other characters that readily separate them are the arrangement of the racemes and the length of the raceme internodes; the internodes are long in *A. brazzae* but short in *A. festuciformis* resulting in open racemes in the former and condensed racemes in the latter.
A. festuciformis is also similar to *A. mannii* in that both (for the genus) are small plants with reddish racemes in clusters, and have similar spikelets; herbarium specimens of both often have burnt lower vegetative parts. However, in *A. mannii* the spikelets are awned, and the racemes are clearly exserted from the spatheoles.

12. **Andropogon gayanus** Kunth, Enum. Pl. **1**: 491 (1833). —Stapf in Prain, F.T.A. **9**: 261 (1919). —Robyns, Fl. Agrost. Congo Belge **1**: 140 (1929). —Bor, Grasses Burma Ceyl. Ind. Pak.: 90 (1960). —Launert in Merxmüller, Prodr. Fl. SW. Afrika, fam. 160: 22 (1970). —Simon in Kirkia **8**: 16, 49 (1971). —Bowden in Bot. J. Linn. Soc. **64**: 77 (1971). —Bowden, tom. cit. **69**: 77 (1974). —Clayton & Renvoize in F.T.E.A., Gramineae: 777 (1982). —Okoli &

Olorode in Bot. J. Linn. Soc. **87**: 263 (1983). —Müller, Grasses of South West Africa/Namibia: 50 (1984). —Gibbs Russell et al., Grasses South. Africa [Mem. Bot. Surv. S. Africa No. 58]: 40 (1990). —Zon, Gramin. Cameroun **2**: 443 (1992). TAB. **23**, fig. 3. Type from Senegal.

Andropogon squamulatus Hochst. in Schimper, Iter. Abyss. Sched. No. 715 (1842). Type from Ethiopia.

Andropogon appendiculatus Nees var. *polycladus* Hack. in Bull. Herb. Boissier **4**, app. 3: 11 (1896). Type from Namibia.

Andropogon gayanus var. *squamulatus* (Hochst.) Stapf in Prain, F. Trop. Afr. **9**: 263 (1919). —Stent & Rattray in Proc. & Trans. Rhodesia Sci. Assoc. **32**: 12 (1933). —Sturgeon in Rhodesia Agric. J. **51**: 15 (1954). —Chippindall in Meredith, Grasses & Pastures S. Africa: 499 (1955). —Jackson & Wiehe, Annot. Check List Nyasal. Grass.: 28 (1958). —Napper, Grasses Tanganyika: 105 (1965). —J.G. Anderson in Bothalia **9**: 28 (1966).

Andropogon gayanus var. *polycladus* (Hack.) Clayton in Kew Bull. **32**: 1 (1977). —Bennett in Kirkia **11**: 174 (1980).

Caespitose perennial; culms up to 350 cm high, erect, branched above, glaucous. Leaf sheaths occasionally woolly below; ligule a very short (c. 2 mm) minutely fringed membrane, the leaves often with small protuberances or a second ligule present abaxially to the ligule; laminas 20–50(80) cm × 4–22 mm, reduced to the midrib and petiole-like at the base, gradually much expanded above, folded or flat, tapering to a fine point at the apex, this often with minute dense white hairs. Racemes in pairs, 4–10 cm long, clearly exserted from the spatheole, silvery with white hairs; peduncles longer than the racemes; internodes and pedicels clavate, ciliate with hairs c. 4 mm long on both margins, bilobed at the apex, the lobes c. 2 mm long. Sessile spikelet 6.5–8.5 mm long; inferior glume with a shallow median groove, glabrous or with short hairs; superior glume convex on the back; superior lemma bifid in the upper $^1/_4$; awn geniculate, 13–32 mm long; anthers 3–5 mm long, straw-coloured. Pedicelled spikelet male, 5–7.5 mm long; inferior glume mucronate or with an awn up to 3.5 mm long.

Caprivi Strip. Caprivi side of the river at Andara Mission, 23.ii.1956, *de Winter & Marais* 4827 (K; PRE). **Botswana**. N: Ngamiland Distr., Tsodilo Hills, Female Hill, 'Rhino Valley', 1030 m, 3.iv.1987, *Long & Rae* 614 (E; K). **Zambia**. B: Kaoma (Mankoya), 21.iv.1964, *Verboom* 1358 (K). N: North Luangwa National Park, 750 m, 25.iii.1994, *P.P. Smith* 967 (K). W: Mwinilunga Distr., near Mavundu, ix.1934, *Trapnell* 1610 (K). C: Lusaka Distr., Chilanga Agric. Research Station, 15.iv.1953, *Hinds* 68 (K). E: Chipata Distr., Senegallia Farm, ii.1954, *Grout* 124 (K; PRE). S: Livingstone Distr., 16 km from Livingstone to Kazungula, 11.iii.1960, *White* 7745 (K). **Zimbabwe**. N: Binga, Lake Kariba, 500 m, 24–25.xii.1990, *Laegaard* 15867B (K). W: Matobo Distr., Besna Kobila Farm, iv.1951, *Miller* 4286 (LISC). C: Chegutu Distr., Poole Farm, 19.i.1946, *H.I.E. Hornby* 2273 (K). E: Chipinge Distr., east of Sabi R, between Musaswi R and Cilariati (Cikariati) R, 400 m, 22.i.1957, *Phipps* 101A (LISC). S: Chiredzi Distr., near Chiredzi African Township, 3.iii.1971, *Taylor* 139 (K). **Malawi**. N: Nkhata Bay Distr., Chikale Beach, 470 m, 15.v.1977, *Pawek* 12761 (K). C: Dowa Distr., 31 km west of Salima on road to Lilongwe, 760 m, 29.iv.1970, *Brummitt* 10258 (K). S: 29 km from Zomba on road to Blantyre, 975 m, 17.iii.1950, *Wiehe* 442 (K). **Mozambique**. N: Angoche Distr., andados 11 km de Angoche (António Enes) para Namaponda, c. 30 m, 31.iii.1964, *Torre & Paiva* 11525 (LISC). Z: Morrumbala Distr., M'bobo to Morrumbala, 28.iv.1943, *Torre* 5223 (LISC). T: between Chicoa and Magoé, 12.ii.1970, *Torre & Correia* 17917 (LISC). MS: Manica Distr., Nhaungue Mte. (Serra de Garuso), 2.v.1948, *Barbosa* 1345 (K; LISC). GI: Guijá Distr., Limpopo, entre Caniçado e o regulo Meginge, 18.v.1948, *Torre* 7850 (K; LISC). M: Matutuíne (Bela Vista) to Maputo, 20.xi.1940, *Torre* 2113 (K; LISC).

Throughout tropical and southern Africa. Growing in xerophytic grassland on doloritic, sandy or clayey soils, and in fallow fields and along roadsides; 400–1500 m.

Four varieties are recognized in West Africa (cf. Zon, loc. cit.). In the Flora Zambesiaca area all the material studied has internodes and pedicels ciliate all along both margins; about half the specimens, scattered throughout the Flora Zambesiaca area, have short scattered hairs on the pedicelled spikelets; all specimens are glaucous, sometimes reaching 3.5 m high. In fact, some of the features of our material relate to more than one of the West African varieties. The four varieties have not been described consistently and a deeper study of the whole range of variation in the species is much needed, especially of var. *tridentatus* (Hochst.) Hack., reputedly with a high proportion of diploids (2n = 20); cf. F.W.T.A.

13. **Andropogon pseudapricus** Stapf in Prain, F.T.A. **9**: 242 (1919). —Jackson & Wiehe, Annot. Check List Nyasal. Grass.: 28 (1958). —Clayton in F.W.T.A., ed. 2, **3**: 486 (1972). —Clayton & Renvoize in F.T.E.A., Gramineae: 778 (1982). —Zon, Gramin. Cameroun **2**: 434 (1992). Type from Nigeria.

Caespitose annual; culms up to 150 cm high, ascending, branched above. Leaf sheaths glabrous or rarely loosely pilose; ligule up to 2 mm long; leaf laminas 8–40 cm × 1–5 mm, glabrous. Inflorescence of paired racemes gathered into a copious leafy false panicle, the pairs embraced below by, or eventually exserted from, linear to narrowly lanceolate spatheoles c. 5 cm long; racemes 2–4 cm long; internodes and pedicels cuneate to narrowly ellipsoid, ciliate with hairs c. 2 mm long. Sessile spikelet 5–6 mm long, compressed between internode and pedicel; inferior glume linear, deeply depressed between the dorsal keels, glabrous; superior glume with an awn 8–16 mm long; superior lemma bidentate; awn geniculate, 30–50 mm long, rarely less. Pedicelled spikelet 4–5 mm long, narrowly elliptic, membranous, glabrous to villous, 2-awned, the longer awn 5–11 mm long, the other much shorter.

Malawi. S: Zomba Plateau, Mlumbe Mt. (Zomba Rock), 1896, *Whyte* s.n. (K).
West Africa from Senegal to Chad and Cameroon; also in Tanzania and probably introduced in Mexico and Brazil; c. 2085 m.
Collected only once in the Flora Zambesiaca area, quite a long way from its centre of distribution in West Africa.

14. **Andropogon chinensis** (Nees) Merr. in Philipp. J. Sci., Bot. **12**: 101 (1917). —Clayton & Renvoize in F.T.E.A., Gramineae: 779 (1982). —Müller, Grasses of South West Africa/Namibia: 52 (1984). —Gibbs Russell et al., Grasses South. Africa [Mem. Bot. Surv. S. Africa No. 58]: 40 (1990). —Zon, Gramin. Cameroun **2**: 435 (1992). TAB. **23**, fig. 4. Type from China.
   Homoeatherum chinense Nees in Lindley, Nat. Syst. Bot., ed. 2: 448 (1836).
   Andropogon ascinodis C.B. Clarke in J. Linn. Soc., Bot. **25**: 87 (1889). —Simon in Kirkia **8**: 16, 49 (1971). —Clayton in F.W.T.A., ed. 2, **3**: 486 (1972). —Bennett in Kirkia **11**: 173 (1980). Type from India.
   Andropogon schinzii Hack. in A. & C. de Candolle, Monogr. Phan. **6**: 458 (1889). —Stapf in Prain, F.T.A. **9**: 245 (1919). —Robyns, Fl. Agrost. Congo Belge **1**: 137 (1929). —Stent & Rattray in Proc. & Trans. Rhodesia Sci. Assoc. **32**: 11 (1933). —Sturgeon in Rhodesia Agric. J. **51**: 13 (1954). —Chippindall in Meredith, Grasses & Pastures S. Africa: 499 (1955). —Jackson & Wiehe, Annot. Check List Nyasal. Grass.: 29 (1958). —Vesey-FitzGerald in Kirkia **3**: 104 (1963). —Napper, Grasses Tanganyika: 104 (1965). —J.G. Anderson in Bothalia **9**: 22 (1966). —Launert in Merxmüller, Prodr. Fl. SW. Afrika, fam. 160: 22 (1970). —Simon in Kirkia **8**: 16, 49 (1971). Type from Namibia.
   Andropogon pseudoschinzii Stapf in Prain, F.T.A. **9**: 249 (1919). —Jackson & Wiehe, Annot. Check List Nyasal. Grass.: 29 (1958). Type: Malawi, Shire, *Buchanan* 67 (K, holotype).
   Andropogon patris Robyns, Fl. Agrost. Congo Belge **1**: 137 (1929). Type from Dem. Rep. Congo.
   Andropogon lindiensis Pilg. in Notizbl. Bot. Gart. Berlin-Dahlem **13**: 407 (1936). —Napper, Grasses Tanganyika: 104 (1965). —Simon in Kirkia **8**: 16, 49 (1971). Type from Tanzania.
   Andropogon sylvaticus C.E. Hubb. in Kew Bull. **4**: 371 (1949). —Jackson & Wiehe, Annot. Check List Nyasal. Grass.: 29 (1958). Type: Malawi, Zomba, *Brass* 16098 (K, holotype).

Caespitose perennial; culms up to 170 cm high, erect, branched, often glaucous. Leaf sheaths glabrous, with well developed auricles; ligule a short (1.5–4 mm) minutely fringed membrane; laminas 7–20 cm × 5–8 mm, very often involute, tapering to a very fine point. Racemes in pairs, 3.5–9.5 cm long, exserted from the spatheoles only at maturity, silvery with white hairs; peduncles much longer than the racemes at maturity; internodes and pedicels strongly clavate with cilia up to 6 mm long on both margins, the pedicel bilobed at the apex. Sessile spikelet 6–8 mm long, strongly compressed between internode and pedicel, the short callus shallowly inserted; inferior glume with a deep median groove, loosely pilose; superior glume convex on the back, 2-toothed at the tip and with an awn 5–9 mm long; superior lemma bifid to the middle; awn geniculate, 24–30 mm long; anthers 2.2–3.5 mm long, straw-coloured. Pedicelled spikelet male, 4.5–7 mm long; inferior glume loosely pilose, often 2-toothed at the apex, usually purplish-orange, with an awn 3.5–7 mm long; superior glume often with an awn 0.7–5 mm long.

Botswana. N: Ngamiland Distr., Aha Hills, 21.iv.1982, *Mavi* 1681 (K). SE: Mogobane, Lobatse, 3.iv.1957, *de Beer* 81895 (LISC). **Zambia**. N: Mbala Distr., 27.iii.1959, *McCallum Webster* A229a (LISC). W. Ndola Distr., 8.v.1959, *Vesey FitzGerald* 2592 (BM; SRGH). C: Mpika Distr., Kapamba River, Luangwa Valley Game Reserve South, 30.iii.1966, *Astle* 4711 (K). S: Choma Distr., Muzoka, iv.1946, *Grassl* 46-70 (K). **Zimbabwe**. N: Gokwe Distr., 11 km north of

Gokwe, 6.iii.1964, *Bingham* 1096 (K; LISC; PRE). W: Hwange Distr., Kazuma Range, 11.v.1971, *Simon* 2205 (LISC; PRE). C: Shurugwi Distr., above Ferny Creek, 1.iv.1967, *Biegel* 2031 (K). E: Mutare Distr., main road near Eastlands border, 28.iii.1970, *A.O. Crook* 919 (K; LISC). S: Masvingo Distr., Oatlands Farm, adjacent to Great Zimbabwe, near Chipopo River, *Chiparawasha* 718 (K). **Malawi**. N: Nkhata Bay Distr., above Chikale Beach, 1 km south of Nkhata Bay, 490–530 m, 10.v.1970, *Brummitt* 10586 (K; LISC; MAL; PRE; SRGH; UPS). C: Nkhotakota Distr., Chipata Hill, 4.v.1963, *Verboom* 983 (K). S: Zomba Distr., Zomba Plateau, 1430 m, 30.v.1946, *Brass* 16098 (K). **Mozambique**. N: Cuamba (Nova Freixo), 9.v.1971, *Bowbrick* JA148A (K; LISC). T: Macanga Distr., Serra da Pandalajala, Furancungo, 15.v.1948, *Mendonça* 4257 (LISC). MS: between Dondo and Inhaminga, 7.iv.1942, *Torre* 4013 (K; PRE). GI: Massingir, 20.iv.1972, *Lousã & Rosa* 252 (K).

Widespread in tropical and southern Africa; also in the Arabian Peninsula and from India to China. Growing in woodland and scrub and in dambos, usually in loamy or sandy soils; 150–1550 m.

This species can be confused with *A. schirensis* but readily distinguished by its larger ligules and awned pedicelled spikelets.

The awn of *A. chinensis* can be 2-geniculate as in *A. schirensis* and in genus *Diheteropogon*; it also has some short hairs at the base, a character unusual in subtribe *Andropogoninae*.

15. **Andropogon schirensis** Hochst. ex A. Rich., Tent. Fl. Abyss. **2**: 456 (1850). —Stapf in Prain, F.T.A. **9**: 246 (1919). —Robyns, Fl. Agrost. Congo Belge **1**: 134 (1929). —Stent & Rattray in Proc. & Trans. Rhodesia Sci. Assoc. **32**: 11 (1933). —Sturgeon in Rhodesia Agric. J. **51**: 15 (1954). —Chippindall in Meredith, Grasses & Pastures S. Africa: 497 (1955). —Jackson & Wiehe, Annot. Check List Nyasal. Grass.: 28 (1958). —Napper, Grasses Tanganyika: 104 (1965). —J.G. Anderson in Bothalia **9**: 24, 26 (1966). —Launert in Merxmüller, Prodr. Fl. SW. Afrika, fam. 160: 22 (1970). —Simon in Kirkia **8**: 16, 49 (1971). —Clayton in F.W.T.A., ed. 2, **3**: 486 (1972). —Bennett in Kirkia **11**: 175 (1980). —Clayton & Renvoize in F.T.E.A., Gramineae: 779, fig. 181 (1982). —Gibbs Russell et al., Grasses South. Africa [Mem. Bot. Surv. S. Africa No. 58]: 42 (1990). —Zon, Gramin. Cameroun **2**: 437, t. 96/1–5 (1992). TAB. **24**. Type from Ethiopia.

Andropogon schirensis var. *angustifolius* Stapf in Dyer, F.C. **7**: 340 (1898). —Chippindall in Meredith, Grasses & Pastures S. Africa: 497 (1955). —Jackson & Wiehe, Annot. Check List Nyasal. Grass.: 28 (1958). —J.G. Anderson in Bothalia **9**: 24, 26 (1966). Type from South Africa.

Andropogon schirensis var. *natalensis* Hack. in Mém. Herb. Boissier **20**: 9 (1900). Type from South Africa.

Andropogon dummeri Stapf in Prain, F.T.A. **9**: 248 (1919). Types from Uganda.

Andropogon ravus J.G. Anderson in Bothalia **7**: 417 (1960); & **9**: 24, 26 (1966). —Gibbs Russell et al., Grasses South. Africa [Mem. Bot. Surv. S. Africa No. 58]: 42 (1990). Type from South Africa.

Densely caespitose perennial; culms up to 200 cm high, erect, branched, sometimes glaucous, tinged with purple or dark red. Leaves glabrous, occasionally densely long-pilose; ligule a minute (0.5–1 mm) fimbriate membrane; laminas 10–26(30) cm × 2–(12) mm, flat, tapering at the apex, sometimes abruptly so, to a very fine point. Racemes in terminal pairs, rarely in threes, 5–20 cm long, clearly exserted from the spatheoles, dark red, the pedicelled spikelets clearly imbricate on one side; peduncles much longer than the racemes; internodes and pedicels clavate, ciliate on both margins, the internodes bilobed at the apex. Sessile spikelet 4.3–8.5 mm long, strongly compressed between internode and pedicel, the conical callus deeply inserted in the internode apex; inferior glume with a ± deep median groove, glabrous; superior glume convex on the back, awnless; superior lemma bifid to the middle; awn bigeniculate, 22–40 mm long; anthers 2.5–4.5 mm long (longer in the pedicelled spikelet), straw-coloured or dark red. Pedicelled spikelet male, 6–11 mm long, clearly longer than the sessile, papery, purplish to dark red, awnless or occasionally mucronate.

Botswana. N: Ngamiland Distr., near Movombe Village, 14.ii.1983, *P.A. Smith* 4040 (K). **Zambia**. B: Kabompo, 26.iii.1961, *Drummond & Rutherford-Smith* 7392 (BM; K; LISC; PRE). N: Mbala Distr., Lungu River, 1.iv.1958, *Vesey-FitzGerald* 1573 (BM). W: Ndola, 7.ii.1960, *E.A. Robinson* 3333 (K; PRE). C: Lusaka Distr., Chakwenga headwaters, 100–120 km east of Lusaka, 14.ii.1965, *E.A. Robinson* 6377 (K). E: Nyika Plateau, 18.ii.1961, *Vesey-FitzGerald* 2979 (BM). S: Livingstone Distr., Makoli, iii.1933, *Trapnell* 1673 (K). **Zimbabwe**. N: Gokwe South Distr., Sengwa Research Station, 8.iii.1975, *P.R. Guy* 2332 (K; PRE). W: Umguza Distr., Nyamandhlovu Pasture Research Station, iv.1956, *Plowes* 1942 (K; PRE). C: Makoni Distr., 15 km north of Rusape, 1520 m, 12.ii.1974, *Davidse, Simon & Pope* 6504 (K). E: Nyanga National Park, Mt.

Tab. 24. ANDROPOGON SCHIRENSIS. 1, habit (× ⅔); 2, ligule (× 1); 3, spikelet pair, front view, showing: a. internode, b. sessile spikelet, c. pedicel, d. pedicelled spikelet (× 4); 4, spikelet pair, rear view, showing: a. internode, b. sessile spikelet, c. pedicel, d. pedicelled spikelet (× 4). Drawn by W.E. Trevithick. From F.T.E.A.

Nyangani (Inyangani), 12.ii.1974, *Davidse, Simon & Pope* 6578 (K; PRE). S: Masvingo Distr., Zimbabwe National Park, 27.iii.1973, *Chiparawasha* 596 (K; PRE). **Malawi**. N: Chitipa Distr., Nyika Plateau, Rumphi Bridge, 2300 m, 16.iv.1975, *Pawek* 9258 (K). C: Lilongwe Distr., Dzalanyama Forest Reserve, near Chiunjika, 1550 m, 9.ii.1959, *Robson* 1532 (BM; K; LISC). S: Mt. Mulanje, Chambe Plateau, 2130 m, 16.xi.1949, *Wiehe* 328 (K; PRE). **Mozambique**. N: Nampula to Mecubúri (Mucuburi), 20.iii.1937, *Torre* 1298 (COI). Z: Pebane, near the lighthouse, 8.iii.1966, *Torre & Correia* 15091 (LISC). T: Angónia Distr., andados 11 km de Calobuè, 1500 m, 7.iii.1964, *Torre & Paiva* 11052 (LISC). MS: Gondola Distr., Tsetserra to Chimoio (Vila Pery), 3.iv.1966, *Torre & Correia* 15638 (LISC). GI: Inharrime Distr., between Inharrime and Chacane, 28.i.1941, *Torre* 2585 (K; LISC). M: Maputo (Lourenço Marques), *Torre* 7350 (LISC).

Throughout tropical and southern Africa. Woodland, grassland and savanna, in wet or dry soils, and in cultivated fields; 15–2600 m.

Closely related to *Diheteropogon amplectens* which has very similar sessile spikelets with bigeniculate awns, but it has rounded leaf lamina bases, branched culms, narrower pedicelled spikelets and pungent callus. Variation in the Flora Zambesiaca area includes plants with narrow leaves which have been named as var. *angustifolius*, but formal recognition is not merited.

The dorsiventral racemes are very distinct. On one side of the raceme only can be seen the imbricate backs of the pedicelled spikelets, whereas on the other side the sessile spikelets, internodes and pedicels are exposed.

See under *A. chinensis* for further comments.

16. **Andropogon perligulatus** Stapf in Bull. Misc. Inform., Kew **1908**: 410 (1908). —Stapf in Prain, F.T.A. **9**: 250 (1919). —Clayton & Renvoize in F.T.E.A., Gramineae: 782 (1982). Type from Togo.

Andropogon centralis Pilg. in Fries, Wiss. Ergebn. Schwed. Rhod.-Kongo-Exped. **1**: 193 (1916). Syntypes: Zambia, Kamindas, *Fries* 972 (UPS) and from Tanzania.

Andropogon tumidulus Stapf in Prain, F.T.A. **9**: 252 (1919). —Stent & Rattray in Proc. & Trans. Rhodesia Sci. Assoc. **32**: 11 (1933). —Sturgeon in Rhodesia Agric. J. **51**: 15 (1954). Types from Zimbabwe, Harare (Salisbury), *Craster* 61, and Angola (K, syntypes).

Andropogon canaliculatus sensu Simon in Kirkia **8**: 16 (1971), non Schumach.

Andropogon sp. sensu Simon in Kirkia **8**: 16 (1971).

Caespitose perennial; culms up to 120 cm high, erect, little branched above. Leaf sheaths with well developed auricles, glabrous; ligule (1.5)2.5–6 mm long, truncate; laminas 7.5–33 cm × 1–3.2 mm, revolute with a conspicuous keel, shortly and densely pilose, tapering to a very fine point at the apex. Racemes in pairs, 1.5–7.5 cm long, ± terete, clearly exserted from the spatheoles; peduncles much longer than the racemes; internodes and pedicels ± strongly canaliculate, very broad and thin, slightly clavate, ciliate on both margins with hairs 0.5–1 mm long. Spikelets dark red. Sessile spikelet 4.5–5 mm long, with rounded callus; inferior glume with a deep median groove, glabrous; superior glume strongly convex on the back, 2-toothed at the apex, with a mucro up to 2 mm long; superior lemma 2–3.5 mm long, bifid almost to the base; awn 1–2-geniculate, 18–29 mm long; anthers 2–2.7 mm long, bright red. Pedicelled spikelet male or sterile in the same plant, 4–6 mm long; inferior glume with scattered short hairs and an awnlet up to 1.5 mm long.

Zambia. W: Kabompo Distr., Lusongwa floodplain, 8 km south of Lusongwa School, 26.xii.1969, *Simon & Williamson* 2052 (BM; K). N: Mbala Distr., Lunzua River bridge, Kambole road, c. 32 km from Mbala (Abercorn), 5.iv.1959, *McCallum Webster* A284 (K). W: Mwinilunga Distr., Lisombo River, 10.vi.1963, *Loveridge* 923 (BM; K). C: near Mkushi River, east of Kabwe (Broken Hill), 26.iii.1946, *Grassl* 46-35 (K). S: 8 km east of Choma, 1310 m, 27.iii.1955, *E.A. Robinson* 1193 (K). **Zimbabwe**. N: Guruve Distr., Nyamunyeche Estate, Karoi Vlei, 7.ii.1979, *Nyariri* 672 (K). C: Hwedza Distr., Chingameri, *MacDonald* 128 (K). E: Chimanimani Distr., 22.5 km south of Chimanimani (Melsetter), 8.iii.1954, *Crook* 539 (K; PRE). **Malawi**. N: Chitipa Distr., 2.4 km north of Chisenga, 24.viii.1962, *Tyrer* 525 (BM). C: Lilongwe Distr., Dzalanyama, Chiputu Dambo, 4.xii.1951, *Jackson* 695 (K).

Also in Uganda, Kenya, Tanzania, Angola, Central African Republic and Ethiopia. Growing in wet acidic soils; 1230–1600 m.

Closely related to *A. canaliculatus*; see under that species for comments.

17. **Andropogon canaliculatus** Schumach., Beskr. Guin. Pl.: 52 (1827). —Stapf in Prain, F.T.A. **9**: 251 (1919). —Robyns, Fl. Agrost. Congo Belge **1**: 137 (1929). —Napper, Grasses Tanganyika: 104 (1965). —Bennett in Kirkia **11**: 174 (1980). —Zon, Gramin. Cameroun **2**: 438 (1992). Type from Guinea.

Andropogon canaliculatus var. *fastigiatus* Stapf in Prain, F.T.A. **9**: 252 (1919). —Simon in Kirkia **8**: 16, 49 (1971). Types from Guinea.

Very closely allied to *A. perligulatus* but differing in the following characters: ligule a fringed membrane c. 0.5 mm long; leaf laminas c. 4 mm wide with some long hairs at the base; pedicels more strongly inflated-clavate; sessile spikelets 4.5–6 mm long; lemma 3.2–4 mm long, bifid to the middle; pedicelled spikelets shorter than the sessile spikelet, 4.2–4.5 mm long; anthers 1–2 mm long, dark red.

Zimbabwe. E: Mutare Distr., Vumba Mts., 1710 m, 10.v.1956, *Chase* 6107 (K). **Malawi**. N: Nkhata Bay escarpment, 7.vii.1952, *Jackson* 929 (K). **Mozambique**. N: Nampula Distr., andados 21 km de Nampula para Nametil, 320 m, 1.iv.1964, *Torre & Paiva* 11557 (COI; K; LISC).

Also in Uganda, Kenya and Tanzania westwards to Mali. Growing in woodland and bogs; 320–1700 m.

Very closely related to *A. perligulatus*. The characters that clearly separate them are the shape and length of the ligules and the length of the lemmas. These seem to be reasonably significant differences and for this reason, and because of some additional if less distinct differences in their geographical distribution, they are maintained here as separate species. Nevertheless, the general distribution of character states is rather irregular. *A. canaliculatus* tends to have wider leaves, broader internodes and pedicels, shorter awns, pedicelled spikelets that are shorter than the sessile spikelet, and shorter anthers.

18. **Andropogon fastigiatus** Sw., Prodr.: 26 (1788). —Clayton in Kew Bull. **17**: 468 (1964). —Simon in Kirkia **8**: 16, 49 (1971). —Clayton in F.W.T.A., ed. 2, **3**: 485 (1972). —Bennett in Kirkia **11**: 174 (1980). —Clayton & Renvoize in F.T.E.A., Gramineae: 777 (1982). —Gibbs Russell et al., Grasses South. Africa [Mem. Bot. Surv. S. Africa No. 58]: 41 (1990). —Zon, Gramin. Cameroun **2**: 433 (1992). Type from Jamaica.

Diectomis fastigiata (Sw.) Kunth in Humboldt, Bonpland & Kunth, Nov. Gen. Sp. **1**: 193 (1816). —Stapf in Prain, F.T.A. **9**: 207 (1919). —Sturgeon in Rhodesia Agric. J. **51**: 12 (1954). —Chippindall in Meredith, Grasses & Pastures S. Africa: 504 (1955). —Vesey-FitzGerald in Kirkia **3**: 104 (1963).

Annual; culms up to 150 cm high, erect or sometimes geniculate at the base, much branched above, tinged with pink to dark red. Leaf sheaths glabrous; ligule 4–7 mm long, acute; laminas 8.5–34 cm × 1.5–4 mm, loosely and shortly pilose, tapering to a fine point at the apex. Racemes solitary, 3–5 cm long, shortly exserted from the spatheoles; peduncles longer than the racemes; internodes and pedicels clavate, ciliate on both margins with hairs c. 2 mm long. Sessile spikelet 3.5–4.5 mm long, strongly compressed between internode and pedicel; inferior glume with a deep median groove, glabrous; superior glume convex on the back, 2-toothed at the apex and with an awn 5.5–18 mm long; superior lemma bifid in the upper $^{1}/_{4}$; awn geniculate, 24–48 mm long; anthers 1–1.5 mm long, straw-coloured. Pedicelled spikelet sterile, 6.5–10 mm long; inferior glume broadly lanceolate, papery, glabrous, faintly to deeply flushed reddish-purple and with an awn 5.5–7(15) mm long; superior glume much smaller.

Botswana. N: Mpandama-tenga (Pandamatenga), Zimbabwe frontier, 28.iii.1961, *Vesey-FitzGerald* 3362 (BM). **Zambia**. N: Mbala Distr., Chisungu Estate, 20.iv.1959, *McCallum Webster* A342 (K). W: Mwinilunga Distr., 5.6 km north of Salujinga, 24.xii.1969, *Simon & Williamson* 1998 (BM; K). C: Lusaka Distr., Chakwenga Headwaters, c. 120 km east of Lusaka, 27.iii.1965, *E.A. Robinson* 6548 (K; PRE). S: Choma Distr., Mapanza Mission, 9.iii.1953, *E.A. Robinson* 114 (K). **Zimbabwe**. N: Gokwe Distr., c. 29 km north of Gokwe on road to Chinyenyeti, 12.iii.1963, *Bingham* 506 (K; LISC). W: Bubi Distr., Alfalfa Ranch, *Nyathi* in R.R.T. 76/93 (K; PRE). C: KweKwe Distr., Sebakwe River, 6.iii.1961, *Vesey-FitzGerald* 3101 (BM). S: Chiredzi Distr., west end of Chionja Hills, 30.iii.1961, *Phipps* 2899 (BM; K; PRE). **Malawi**. N: Mzimba Distr., 32 km west of Mzuzu, Lunyangwa River bridge, 1220 m, 18.iv.1974, *Pawek* 8349 (K). C: Dowa Distr., Salima escarpment, 8.iv.1952, *Jackson* 758 (K). S: Zomba Plateau, near Malumbe Peak, 4.iv.1985, *Long* 12425 (E). **Mozambique**. N: Malema Distr., Mutuáli, Estação Experimental do C.I.C.A., 24.iv.1961, *Balsinhas & Marrime* 441 (K; LISC; PRE). T: Cahora Bassa Distr., Chicoa, 26.ii.1972, *Macêdo* 4928 (K; LISC; PRE).

Pantropical. Rocky outcrops, woodland and seasonally damp grassland; 550–2085 m.

Closely related to *A. textilis*, but is an annual with conspicuous sterile, papery pedicelled spikelets much larger than the sessile, and shorter ligules. Features common to *A. fastigiatus* and *A. textilis*, such as single racemes and acute ligules, also occur in other species (*A. festuciformis* and *A. ligulatus* respectively).

19. **Andropogon textilis** Welw. ex Rendle in Hiern et al., Cat. Afr. Pl. Welw. **2**: 144 (1899). — Stapf in Prain, F.T.A. **9**: 255 (1919). —Bennett in Kirkia **11**: 175 (1980). Type from Angola.
  Andropogon roseus Napper in Kirkia **3**: 123 (1963). —Napper, Grasses Tanganyika: 105 (1965). —Simon in Kirkia **8**: 16, 49 (1971). Type from Tanzania.

Caespitose perennial; culms up to 165 cm high, erect, branched above, tinged with dark red. Leaf sheaths often pilose; ligule 6–12(27) mm long, acute; laminas 25–63 cm long, 5–6 mm wide, loosely pilose near the ligule, tapering to a fine point at the apex. Racemes solitary, 3.2–3.7 cm long, exserted from the spatheoles, silvery with white hairs; peduncles longer than the racemes; internodes and pedicels clavate, ciliate on both margins with hairs c. 4 mm long. Sessile spikelet 6–8 mm long; inferior glume with a deep median groove, pilose; superior glume convex on the back, 2-toothed at the apex and with an awn 7–15 mm long; superior lemma bifid to the middle; awn geniculate, 21–24 mm long; anthers 2.3–2.8 mm long, dark red. Pedicelled spikelet male, 5.8–8 mm long, shortly pilose, reddish, with an awn up to 3.3 mm long.

Zambia. B: Kaoma Distr., Kafue Watershed road, 2.iv.1964, *Verboom* 1361 (K). N: Mporokoso Distr., 16 km SW of Mporokoso, 13.iv.1961, *Phipps & Vesey-FitzGerald* 3141 (BM). **Zimbabwe**. N: Mazowe Distr., Spelonken Farm, 18.iv.1981, *H.H. Burrows* 1739 (K). C: Shurugwi Distr., above Ferny Creek, 1.iv.1967, *Biegel* 2030 (K). **Malawi**. N: Rumphi Distr., Nyika foothills, on approach road, 25.vi.1952, *Jackson* 858 (K). C: Lilongwe Distr., Kampini, Dzalanyama foothills, 18.iv.1952, *Jackson* 765 (K; PRE).
  Also in Tanzania and Angola. Growing in grassland and *Brachystegia* woodland; 1250–1450 m. See comments under *A. fastigiatus*.

22. CYMBOPOGON Spreng.

By Fatíma Sales

Cymbopogon Spreng., Pl. Min. Cogn. Pug. **2**: 14 (1815). —Bor in J. Bombay Nat. Hist. Soc. **51**: 890–916 (1953); **52**: 149–183 (1954). —Roberty in Boissiera **9**: 172–180 (1960). —Clayton in Kew Bull. **19**: 451–456 (1965). —Soenarko in Reinwardtia **9**: 225–375 (1977).

Racemes short, in pairs, axillary, enclosed by the spatheoles, arranged in a dense, decompound, globose to elongated and ± interrupted spatheate false panicle; peduncles 0.5–2.5 mm long, flattened, reflexed at maturity; internodes and pedicels filiform, long-ciliate on both margins, the lowermost sometimes swollen, barrel-shaped and fused together; lowermost pair of spikelets in lower racemes homogamous. Sessile spikelet dorsally compressed, narrowly lanceolate to narrowly ovate; callus obtuse; inferior glume with a median groove, occasionally flat on the back, narrowly winged on the keels or sometimes with only scabrid flanges near the apex, awnless; inferior floret reduced to a hyaline lemma; superior floret fertile, the lemma deeply bilobed or entire, awned. Pedicelled spikelet male, ± the same length as the sessile spikelet, rounded on the back, awnless.

A genus of about 40 species in the Old World tropics and subtropics; introduced in tropical America.
  The affinities of *Cymbopogon* with *Andropogon* and more distantly with *Hyparrhenia* are suggested by a number of characters in common (Clayton, op. cit.: 454). Only a few, but nevertheless important, characters separate them. The aromatic and flavoured leaves of *Cymbopogon* set it aside from the other two, scentless, genera. Our species also differ from *Andropogon* by the elaborate compound false panicle and the always extremely short and deflexed raceme-bases.
  A monographic work is still much needed. In Soenarko's revision many species were too narrowly circumscribed; in a number of cases the range of variation in most characters is continuous. The apparently interesting disjunctions in the genus, based on present taxonomy, should be regarded with caution.

  C. citratus (DC.) Stapf is cultivated throughout the tropics for culinary and medicinal purposes. Its inflorescence is awnless but in cultivation it rarely flowers; in the vegetative state it can be recognized by the long, scabrid, lemon-scented leaf laminas.

1. Awn yellowish throughout, without a distinct brown column; inflorescence a dense
   subovoid to globose head · 1. *densiflorus*
 – Awn with a distinct brown column and yellowish bristle or spikelets awnless; inflorescence
   oblong to narrow · 2
2. Lowermost internode and pedicel of the raceme swollen and fused together; leaves
   distributed throughout the plant; laminas smooth on the edges · · · · · · · · · · · · 2. *caesius*
 – Lowermost internode and pedicel of the raceme filiform and not fused together; leaves
   mostly confined to the base of the plant; laminas scabrid on the edges · · · · · · · · · · · · · 3
3. Sessile spikelet awnless; cultivated · *C. citratus* (see above)
 – Sessile spikelet awned; not cultivated · 4
4. Ligule (2.5)3.5–7.5 mm long; inferior glume of the sessile spikelet with a shallow narrow
   V-shaped median groove or flat, the keels winged · 3. *nardus*
 – Ligule 0.5–1.5 mm long; inferior glume of the sessile spikelet with a broad U-shaped
   median groove, the keels merely flanged above · 4. *pospischilii*

1. **Cymbopogon densiflorus** (Steud.) Stapf in Prain, F.T.A. **9**: 289 (1919). —Robyns, Fl. Agrost.
   Congo Belge **1**: 148 (1929). —Jackson & Wiehe, Annot. Check List Nyasal. Grass.: 33
   (1958). —Napper, Grasses Tanganyika: 105 (1965). —Simon in Kirkia **8**: 17, 50 (1971).
   —Soenarko in Reinwardtia **9**: 327 (1977). —Bennett in Kirkia **11**: 176 (1980). —Clayton
   & Renvoize in F.T.E.A., Gramineae: 761 (1982). —Zon, Gramin. Cameroun **2**: 448 (1992).
   Type from Gabon.
   Andropogon densiflorus Steud., Syn. Pl. Glumac. **1**: 386 (1854).

Robust, caespitose perennial; culms up to 220 cm high. Leaves smooth; ligule
2.5–3.5 mm long, erose; laminas 30–40 cm × 8–35 mm, tapering to a fine point at
the apex. Inflorescence a dense subovoid to globose head 5–19(30) cm long,
greenish. Racemes 10–14 mm long; internodes and pedicels pilose with silvery-
white hairs. Sessile spikelet 3–4 mm long, glabrous; inferior glume with an often
shallow V-shaped median groove, narrowly winged on the keels; superior lemma
inconspicuous, entire; awn delicate, 3.5–5 mm long, without a column; anthers 1–1.5
mm long, purple. Pedicelled spikelet male, 2–3 mm long, awnless.

Zambia. B: Kaoma (Mankoya), 2.iv.1964, *Verboom* 1359 (K). N: Mbala Distr., Nakatali,
27.vi.1950, *Bullock* 2952 (K). W: Kitwe Distr., near Kitwe, 29.v.1963, *Loveridge* 683 (K). C:
Mkushi Distr., near Mkushi Hotel, 4.viii.1959, *West* 4001 (BM). E: Lundazi Distr., Magodi,
17.viii.1938, *Greenway & Trapnell* 5617 (K; PRE). S: near Bowood, iii.1933, *Trapnell* 1668 (K).
Zimbabwe. W: Nkayi Distr., Gwampa Forest Reserve, iii.1956, *Goldsmith* 136/56 (K; PRE).
Malawi. N: Mzimba Distr., Kasitu Valley, 29.i.1938, *Fenner* 267 (K). **Mozambique**. Z: Gurué,
1942, *A.J.W. Hornby* 3369 (PRE).
Central Africa from Gabon and Dem. Rep. Congo to Angola; also in Tanzania; introduced in
Brazil. In woodland, swamps, dambos and fallow fields, in sandy and gravelly soils and in heavily
grazed areas; 680–1700 m.
A distinctive species on account of its dense globose inflorescence. It has a strong aroma of
turpentine. Some atypical specimens with more elongated inflorescences can be recognized by
their columnless awns. In native medicine the heads are dried and smoked in pipes as a remedy
for bronchial complaints.

2. **Cymbopogon caesius** (Hook. & Arn.) Stapf in Bull. Misc. Inform., Kew **1906**: 360 (1906).
   —Stapf in Prain, F.T.A. **9**: 287 (1919). —Robyns, Fl. Agrost. Congo Belge **1**: 147 (1929).
   —Soenarko in Reinwardtia **9**: 331 (1977). —Bennett in Kirkia **11**: 176 (1980). —Clayton
   & Renvoize in F.T.E.A., Gramineae: 761, fig. 179 (1982). —Müller, Grasses of Southwest
   Africa/Namibia: 106 (1984). Type from India.
   Andropogon caesius Hook. & Arn., Bot. Beechey's Voy.: 244 (1838).
   Andropogon excavatus Hochst. in Flora **29**: 116 (1846). Type from South Africa.
   Cymbopogon excavatus (Hochst.) Stapf in Prain, F.T.A. **9**: 285 (1919). —Sturgeon in
   Rhodesia Agric. J. **51**: 17 (1954). —Jackson & Wiehe, Annot. Check List Nyasal. Grass.: 33
   (1958). —Napper, Grasses Tanganyika: 105 (1965). —Launert in Merxmüller, Prodr. Fl.
   SW. Africa, fam. 160: 51 (1970). —Simon in Kirkia **8**: 17, 50 (1971). —Soenarko in
   Reinwardtia **9**: 325 (1977). —Gibbs Russell et al., Grasses South. Africa [Mem. Bot. Surv.
   S. Africa No. 58]: 94 (1990).

Caespitose perennial; culms up to 200(300) cm high. Leaves smooth, sometimes
purplish; ligule 0.5–2(3) mm long, minutely dentate; laminas 9–37 cm × 7–28 mm
wide, amplexicaul, cordate or rounded at the base, broader below and tapering to a

very fine point at the apex. Inflorescence elongated and ± interrupted, 7.5–70 cm long, reddish. Racemes 11–20 mm long; internodes and pedicels pilose with silvery-white hairs, the lowermost swollen and fused together. Sessile spikelet 3–5.5 mm long, glabrous; inferior glume with a V-shaped median groove, winged on the keels in the upper $\frac{1}{2}$–$\frac{1}{3}$, entire or dentate at the apex; superior lemma linear, bifid to about the middle; awn delicate, 7–19.5 mm long, with a distinct column; anthers 1.4–2.6(3) mm long, straw-coloured. Pedicelled spikelet male, 3–4 mm long, longer than the sessile spikelet, glabrous.

C. *caesius* is at the core of a species-group which, *inter alia*, includes and C. *giganteus* in Africa and C. *martinii* (Roxb.) Stapf in Asia. The species-group is characterized by having distinct swollen and fused lowermost internodes and pedicels. C. *caesius* sens. lat. (subsp. *caesius* and subsp. *giganteus*) shows great range of variation in habit and leaf size. Such characters are not fully correlated with each other and are certainly not correlated with any other character. Nevertheless, a reasonable number of specimens can be separated into those with delicate wiry culms and those with more robust culms by using lamina width. But there is a continuous range of variation in these and all other characters. The more robust plants (subsp. *giganteus*) are restricted to the northern, more tropical areas. Further studies are needed to ascertain whether the differences in habit between the two extremes of the range of variation are merely those of phenotypic plasticity or not. Lamina width can vary greatly in an individual and measurements should be taken of the widest leaves in a specimen.

In Asia, C. *caesius* and C. *martinii* seem to have a similar range of variation. C. *martinii*, which bears the oldest name in this complex, is the Asiatic counterpart of C. *caesius* subsp. *giganteus* and is probably not specifically different from it.

Largest leaf laminas up to 8 mm wide; culms wiry · subsp. *caesius*
Largest leaf laminas (9)10–22 mm wide; culms robust · · · · · · · · · · · · · · · · subsp. *giganteus*

Subsp. **caesius** TAB. 25.

Botswana. N: Ngamiland Distr., c. 25 km north of Dobe, 27.iv.1982, *Mavi* 1696 (K; PRE). SW: Kgalagadi Distr., Tshabong, above Govt. Camp, 2.iii.1977, *Mott* 1107 (K; PRE). SE: Kweneng Distr., Matloakgama Ranch, 17.iii.1978, *Hansen* 3379 (K; PRE). **Zambia**. S: Kalomo, 1.iii.1963, *Astle* 2219 (K). **Zimbabwe**. N: Zvimba Distr., 64 km NW of Harare, on Great Dyke, 11.ii.1974, *Davidse, Simon & Bennett* 6476 (K). W: Hwange Distr., Kazuma Range, 9.v.1972, *Simon* 2177 (LISC). C: Gweru (Gwelo), 25.ii.1967, *Biegel* 1921 (K). E: Chipinge Distr., Lower Save (Sabi), east bank, near Hippo Mine, 12.iii.1957, *Phipps* 595 (LISC). S: Mberengwa (Belingwe), 4.v.1973, *Simon, Pope & Biegel* 2451 (K; PRE). **Malawi**. N: Nyika, vii.1894, *Sacleux* 2089 (K). **Mozambique**. Z: Mopeia, 15 km from Campo towards Raposa, 28.xii.1967, *Torre & Correia* 16769 (LISC). MS: between Espungabera and Chibabava, 10.xi.1943, *Torre* 6134 (K; LISC). GI: Massinga Distr., Pomene, xii.1971, *Tinley* 2268 (K). M: Namaacha, 20.i.1958, *Barbosa & Lemos* 8242 (COI; K; LISC).

Sudan and Yemen southwards to southern Africa and Namibia, extending to southern India and Sri Lanka. In open forest and grassland, pastures, cultivated fields and roadsides in a variety of soils; 30–1400 m.

Plants have a slight turpentine scent when crushed, and are used for thatching.

Subsp. **giganteus** (Chiov.) Sales, comb. et stat. nov. TAB. **26**. Type: Sudan, Kordofan, *Kotschy* 250 (BM; K, isotypes).
    Andropogon giganteus Hochst. in Flora **27**: 242 (1844), non Tenore (1811). Type from Sudan.
    Cymbopogon giganteus Chiov., Int. Alc. Gram. Essenze Col. Eritrea: 12 (1909), based on the above. —Stapf in Prain, F.T.A. **9**: 288 (1919). —Robyns, Fl. Agrost. Congo Belge **1**: 148 (1929). —Sturgeon in Rhodesia Agric. J. **51**: 18 (1954). —Jackson & Wiehe, Annot. Check List Nyasal. Grass.: 34 (1958). —Napper, Grasses Tanganyika: 105 (1965). —Simon in Kirkia **8**: 17, 50 (1971). —Clayton in F.W.T.A., ed. 2, **3**: 482 (1972). —Soenarko in Reinwardtia **9**: 323 (1977). —Bennett in Kirkia **11**: 176 (1980). —Clayton & Renvoize in F.T.E.A., Gramineae: 763 (1982). —Zon, Gramin. Cameroun **2**: 448 (1992).

Botswana. N: Ngamiland Distr., 22 km south of Nokaneng on road to Tsao, 11.iii.1965, *Wild & Drummond* 6862 (K). SW: Ghanzi Distr., just west of Kobe Pan, 20.iii.1978, *Skarpe* 287 (K; PRE). **Zambia**. N: Mbala Distr., Itembwe Gap, 24.iv.1959, *McCallum Webster* A360 (K). C: Luangwa Valley, South Luangwa National Park, near Mfuwe, 7.ii.1966, *Astle* 4487 (K). S: Choma Distr., Mapanza West, 14.iii.1954, *E.A. Robinson* 608 (K). **Zimbabwe**. N: Mazowe Distr., Mvurwi (Umvukwe), 4.x.1953, *Pollit* s.n. (K). W: Hwange Distr., Kazuma Range, 11.v.1972, *Simon* 2203 (K). C: Marondera (Marandellas), 1.vi.1924, *Eyles* 4014 (K; PRE). E: Nyanga Distr., near

Tab. 25. CYMBOPOGON CAESIUS subsp. CAESIUS. 1, habit (× ½); 2, ligule (× 1); 3,
portion of raceme, showing: a. base of spatheole, b. raceme base; 4, spikelet pair (× 8).
Drawn by W.E. Trevithick. From F.T.E.A.

Tab. 26. CYMBOPOGON CAESIUS subsp. GIGANTEUS. 1, habit (× ½); 2, ligule (× 1½); 3, sessile spikelet (× 6); 4, callus of sessile spikelet (× 9); 5, superior lemma (× 15). Drawn by Victoria Friis. From Ghana Gr.

Cheshire, 15.i.1931, *Norlindh & Weimarck* 4346 (K; PRE). S: Chiredzi Distr., near Chiredzi, 3.iii.1971, *P. Taylor* 137 (K). **Malawi**. N: Nkhata Bay Distr., Kilwa, 89 km south of Nkhata Bay on Lakeshore road, 18.vi.1973, *Pawek* 6904 (K). C: Lilongwe Distr., Chitedze, 28.iii.1950, *Wiehe* 477 (K). S: Mwanza Distr., Neno Hills, 30.vi.1949, *Wiehe* 153 (K). **Mozambique**. N: Lichinga Distr., andados 37 km de Lichinga (Vila Cabral) para Maniamba, c. 1400 m, 29.ii.1964, *Torre & Paiva* 10890 (COI; LISC). Z: Mogoé road, iv.1972, *Bowbrick* J84 (K; LISC). T: Moatize Distr., 40 km from Zóbuè to Ulónguè (Vila Coutinho), 10.iii.1964, *Correia* 191 (LISC). MS: Gondola Distr., Chimoio Plateau, Vandúzi region, road to Tete, 23.iii.1948, *Garcia* 705 (LISC). M: Namaacha, 9.i.1947, *Barbosa* 58 (COI).

Centred in tropical Africa with extensions further south. Growing in grassland, savanna, open forest, dambos, fallow fields, waste ground and old maize fields, and along roadsides; 50–1550 m.

Field notes remark on its strong smell.

3. **Cymbopogon nardus** (L.) Rendle in Hiern et al., Cat. Afr. Pl. Welw. **2**: 155 (1899). —Stapf in Bull. Misc. Inform., Kew **1906**: 314–318 (1906). —Soenarko in Reinwardtia **9**: 349 (1977). —Bennett in Kirkia **11**: 176 (1980). —Clayton & Renvoize in F.T.E.A., Gramineae: 764 (1982). Type from Sri Lanka.
   Andropogon nardus L., Sp. Pl. **2**: 1046 (1753).
   Andropogon nardus var. *validus* Stapf in F.C. **7**: 352 (1898). Type from South Africa.
   Cymbopogon validus (Stapf) Stapf ex Burtt Davy in Ann. Transvaal Mus. **3**: 121 (1912). —Stent & Rattray in Proc. & Trans. Rhodesia Sci. Assoc. **32**: 12 (1933). —Sturgeon in Rhodesia Agric. J. **51**: 17 (1954). —Simon in Kirkia **8**: 17 (1971). —Soenarko in Reinwardtia **9**: 358 (1977). —Gibbs Russell et al., Grasses South. Africa [Mem. Bot. Surv. S. Africa No. 58]: 95 (1990).
   Cymbopogon validus var. *lysocladus* Stapf in Prain, F.T.A. **9**: 282 (1919). —Sturgeon in Rhodesia Agric. J. **51**: 17 (1954). Type: Zimbabwe, Victoria Falls, *Rogers* 5672 (K).

Densely caespitose perennial; culms up to 240 cm high. Leaves often reddish at maturity; ligule (2.5)3.5–10 mm long, membranous, erose; laminas 20–55 cm × 3.5–9 mm, broadest at about the middle, scabrid on the margins, tapering at the apex into a thread-like point. Inflorescence 8–47 cm long, elongated and interrupted, reddish. Racemes 9–18 mm long; internodes and pedicels pilose with silvery-white hairs. Sessile spikelet 3.8–6.5 mm long, glabrous; inferior glume with a shallow V-shaped median groove, occasionally flat, winged in the upper half, notched at the apex; superior lemma linear, bifid in the upper third; awn delicate, 6.5–11 mm long, with a well developed column; anthers 1.2–3 mm long, straw-coloured to brown-red. Pedicelled spikelet male, 2.7–6.2 mm long, awnless.

Zimbabwe. N: Gokwe South Distr., Sengwa River, Charama Plateau, 3.v.1965, *Simon* 415 (K; PRE). W: Matobo Distr., Besna Kobila Farm, iv.1955, *Miller* 2764 (K). E: Chipinge Distr., Gungunyana Forest Reserve, vi.1964, *Goldsmith* 22/63 (BM; K; LISC; PRE; SRGH). S: Mberengwa Distr., north side of Buhwa Mt., 1.v.1973, *Simon, Pope & Biegel* 2406 (K; PRE). **Mozambique**. N: andados 11 km de Nampula para Meconta, c. 380 m, 4.iv.1964, *Torre & Paiva* 11638 (LISC; PRE). MS: serra da Gorongosa, monte Unora, a caminho das quedas de agua, na propriedade do Sr. M. Ferrão, c. 800 m, 6.v.1964, *Torre & Paiva* 12247 (COI; LISC). GI: Xai-Xai (Vila de João Belo) near light-house by Limpopo River mouth, 1.iv.1959, *Barbosa & Lemos* 8454 (K; LISC; PRE). M: Namaacha Mts., 10.i.1948, *Torre* 7090 (LISC).

Eastern Africa from Sudan to southern Africa, extending eastwards to southern India, Sri Lanka and Burma. Wooded grassland and forest margins, often in rocky stony and sandy soils or in clay or shale soils; 0–1525 m.

Similar to *C. pospischilii* in habit, especially in the usually dense basal mat of leaves, and leaf shape, but is readily separated by the much longer ligule and the V-shaped median groove on the inferior glume of the sessile spikelet.

4. **Cymbopogon pospischilii** (K. Schum.) C.E. Hubb. in Kew Bull. **4**: 175 (1949). —Soenarko in Reinwardtia **9**: 315 (1977). —Bennett in Kirkia **11**: 176 (1980). —Clayton & Renvoize in F.T.E.A., Gramineae: 765 (1982). —Müller, Grasses of South West Africa/Namibia: 108 (1984). Type from Tanzania.
   Andropogon pospischilii K. Schum. in Bot. Jahrb. Syst. **24**: 328 (1897). —Stapf in Prain, F.T.A. **9**: 265 (1919).
   Andropogon plurinodis Stapf in Dyer, F.C. **7**: 353 (1898). Type from South Africa.
   Cymbopogon plurinodis (Stapf) Burtt Davy in Ann. Transvaal Mus. **3**: 121 (1912). —Sturgeon in Rhodesia Agric. J. **51**: 17 (1954). —Launert in Merxmüller, Prodr. Fl. SW. Afrika, fam. 160: 51 (1970). —Simon in Kirkia **8**: 17 (1971). —Gibbs Russell et al., Grasses South. Africa [Mem. Bot. Surv. S. Africa No. 58]: 95 (1990).

Caespitose perennial; culms up to 150 cm high. Leaves pilose with a few long hairs near the ligule only, often tinged with red; ligule 0.5–1.5 mm long; laminas 12–21 cm long, 4–5 mm wide, broadest at about the middle, switch-like, scabrid on the margins, very gradually tapering at the apex to a thread-like point. Inflorescence 17–38 cm long, elongated and ± interrupted, reddish. Racemes 10–28 mm long, tinged with red; internodes and pedicels pilose with silvery-white hairs. Sessile spikelet 4–7 mm long; inferior glume with a broad U-shaped median groove and scabrid flanges in the upper $^1/_4$, entire at the apex; superior lemma linear, bifid to about the middle; awn 10–16 mm long, with well developed column; anthers 1.6–3(4) mm long, straw-coloured. Pedicelled spikelet 4.5–7.5 mm long.

Botswana. N: Ngamiland Distr., Bushman Pits, 97 km east of Maun, 26.iii.1961, *Vesey-FitzGerald* 3334 (BM). SW: Ghanzi Distr., Central Kalahari Game Reserve, Deception Pan, 1.iv.1975, *Owens* 90 (K). SE: Kweneng Distr., Ngware, 7.i.1956, *de Beer* 18 (K). **Zimbabwe**. N: Gokwe Distr., Sasame River Test Herd, below Sasame River Gorge, 22.iv.1964, *Bingham* 1252 (K; PRE). W: Matobo Distr., Matopos Research Station, 14.ii.1954, *Rattray* 1572 (K; PRE). C: Gweru (Gwelo), 3.ii.1967, *Biegel* 1859 (K). S: Masvingo (Fort Victoria), 4.i.1948, *D.A. Robinson* 170 (K). **Mozambique**. GI: Massingir, 14.iv.1972, *Myre, Lousã & Rosa* 5774 (K; PRE). M: Namaacha Distr., Goba Station, 10.i.1980, *de Koning* 7931 (K; PRE).

Namibia; eastern Africa from Ethiopia to southern Africa, extending eastwards to northern Pakistan and Nepal. Growing in lush grassland and wooded grassland, in sandy, rocky or hard loamy soils, and along roadsides; 150–1400 m.

Similar to *C. nardus*, but more delicate.

23. SCHIZACHYRIUM Nees

By Fatíma Sales

Schizachyrium Nees, Agrost. Bras.: 331 (1829). —Roberty in Boissiera **9**: 172–180 (1960).

Racemes single, usually arranged in a spatheate false panicle, rarely solitary on the branches, usually not enclosed by the spatheoles; peduncles usually longer than the racemes, rarely shorter; internodes and pedicels ± clavate, glabrous or ciliate on both margins. Sessile spikelet dorsally or laterally compressed, narrowly ovoid to subterete; callus rounded; inferior glume chartaceous to subcoriaceous, with a ± deep and narrow median groove or ± flat or convex, with several intercarinal nerves, these distinct at least at the apex, awnless; superior glume usually keeled, with ± ciliate hyaline margins; inferior floret reduced to a hyaline lemma; superior floret bisexual, the lemma usually keeled, ± deeply bifid, rarely entire, awned. Pedicelled spikelet male or vestigial, usually smaller than the sessile spikelet, rarely bisexual or female, mucronate or with a terminal awn.

A genus of c. 50 species, distributed throughout the tropics.

The genus is closely related to, and merges with, *Andropogon*. The single raceme separates it from *Andropogon* sect. *Leptopogon*; the intercarinal nerves and the irregularly undulate rim at the internode apex separate it from those species of *Andropogon* with single racemes. Species delimitation in most of our species of *Schizachyrium* is reasonably clear-cut.

1. Racemes solitary on the culms · 2
 – Racemes arranged in a spatheate false panicle · 3
2. Racemes with a conspicuous ring of long white hairs at the base of the spikelet, otherwise glabrous; leaf sheaths glabrous throughout or thinly pilose with straight hairs · · · 8. *thollonii*
 – Racemes quite glabrous; leaf sheaths at the base of the plant covered with a dense indumentum of long curly white hairs · 5. *claudopus*
3. Spatheole narrowly lanceolate; mature raceme enclosed by the spatheole for most of its length · 4. *exile*
 – Spatheole linear or absent; mature raceme fully exserted · 4
4. Peduncles as long as or shorter than the racemes; leaf lamina obtuse at the apex; plants annual (very rarely perennial) · 5
 – Peduncles longer than the racemes; leaf lamina tapered to a fine point at the apex; plants perennial · 6

5. Delicate plants up to c. 40(80) cm high; leaf sheaths always shorter than the internodes; culms much branched from below; anthers small, c. 0.6 mm long; plants annual · 1. *brevifolium*
– Robust plants up to 120 cm high; leaf sheaths both shorter and longer than the internodes on the same plant; culms much branched above; anthers large, c. 1.7–2.5 mm long; plants annual, occasionally perennating · 2. *platyphyllum*
6. Racemes with short hairs only on the internodes and pedicels, occasionally glabrous; inferior glume of the sessile spikelet strongly convex or clearly concave, glabrous · 3. *sanguineum*
– Racemes silvery or yellowish with long hairs; inferior glume of the sessile spikelet only slightly convex, pilose on the back · 7
7. Slender plant 30–50 cm high; leaves mostly basal, the laminas up to 3.5 cm long; racemes 1.5–3.5 cm long; pedicelled spikelet awnless; fertile lemma bifid to less than $^{1}/_{3}$ its length, the lobes slender and often contiguous to the awn · 7. *lopollense*
– Robust plant (45)50–120 cm high; leaves both basal and cauline, the laminas 11–23 cm long; racemes 4–8 cm long; pedicelled spikelet awned; fertile lemma bifid to $^{1}/_{3}$–$^{1}/_{2}$ its length · 6. *jeffreysii*

1. **Schizachyrium brevifolium** (Sw.) Nees ex Büse in Miquel, Pl. Jungh.: 359 (1854). —Stapf in Prain, F.T.A. **9**: 187 (1917). —Robyns, Fl. Agrost. Congo Belge **1**: 114 (1929). —Sturgeon in Rhodesia Agric. J. **51**: 11 (1954). —Chippindall in Meredith, Grasses & Pastures S. Africa: 504 (1955). —Jackson & Wiehe, Annot. Check List Nyasal. Grass.: 57 (1958). — Napper, Grasses Tanganyika: 101 (1965). —Simon in Kirkia **8**: 18, 52 (1971). —Clayton in F.W.T.A., ed. 2, **3**: 478 (1972). —Bennett in Kirkia **11**: 184 (1980). —Clayton & Renvoize in F.T.E.A., Gramineae: 754 (1982). —Gibbs Russell et al., Grasses South. Africa [Mem. Bot. Surv. S. Africa No. 58]: 289 (1990). —Zon, Gramin. Cameroun **2**: 452. Type from Jamaica.

Andropogon brevifolius Sw., Prodr.: 26 (1788).
Andropogon flaccidus A. Rich., Tent. Fl. Abyss. **2**: 452 (1851). Type from Ethiopia.
Schizachyrium brevifolium var. *flaccidum* (A. Rich.) Stapf in Prain, F.T.A. **9**: 188 (1917). —Jackson & Wiehe, Annot. Check List Nyasal. Grass.: 57 (1958). —Clayton in F.W.T.A., ed. 2, **3**: 478 (1972). —Bennett in Kirkia **11**: 184 (1980).

Delicate, caespitose, erect or ± scrambling and decumbent annual, tinged with bright purplish-red; culms up to 40(80) cm high, much branched from below, often in fascicles; internodes delicate, linear, compressed. Leaf sheaths shorter than the internodes, glabrous; laminas 0.8–5 cm × up to 5.5 mm, keeled, obtuse at the apex. Racemes borne throughout the plant, 0.5–3 cm long, filiform, glabrous or sometimes with short hairs throughout, shortly exserted from and partially enclosed by the linear spatheole; peduncle shorter than the raceme; internodes and pedicels filiform and flattened below, clavate above. Sessile spikelet 2.3–3.5 mm long, $^{1}/_{5}$ as long again as the pedicel, linear-ovate, dorsally compressed, rounded at the base; inferior glume slightly convex; superior lemma bifid almost to the base; awn slender, 6–13 mm long, geniculate; anthers very small, 0.5–0.6 mm long. Pedicelled spikelet reduced to an awned vestigial glume 0.3–1.2 mm long; awn 2.5–3.5 mm long, straight.

Zambia. N: Mbala Distr., Simanwe Farm, 1520 m, 16.iv.1959, *McCallum Webster* A230 (K; PRE). W: Mwinilunga Distr., Chitunta River, 29 km north of Mwinilunga, 1400 m, 23.xii.1969, *Simon & Williamson* 1969 (K). C: Lusaka Distr., Mt. Makulu, Chilanga, 1250 m, 15.iv.1956, *E.A. Robinson* 1497 (K). E: Chipata (Fort Jameson), 25.vi.1962, *Verboom* 739 (K; PRE). S: Victoria Falls, 910 m, 4.iv.1956, *E.A. Robinson* 1427 (K). **Zimbabwe**. N: Guruve Distr., Nyamunyeche Estate, 18.iv.1979, *Nyariri* 800 (K; PRE). W: Matobo Distr., Besna Kobila Farm, iv.1955, *Miller* 2810 (K; PRE). C: Harare (Salisbury), 1460 m, 15.iv.1922, *Eyles* 3386 (K). E: Mutasa Distr., Honde Valley, 790 m, 17.iv.1958, *Phipps* 1067 (K; PRE). **Malawi**. N: Chitipa Distr., Misuku, 28.vi.1951, *Jackson* 557A (K). C: Salima, 7.iv.1952, *Jackson* 752 (K; PRE). S: Blantyre Distr., Matenje road, 2 km north of Limbe, 1190 m, 1.v.1970, *Brummitt* 10305 (K; LISC; MAL; PRE; SRGH). **Mozambique**. N: Ngauma Distr., Massangulo, 1100 m, iv.1933, *Gomes e Sousa* 1368 (K; PRE). T: Angónia, margens do rio Mauè, 12.v.1948, *Mendonça* 4198 (LISC). MS: near Manica (Macequece), 17.vii.1948, *Fisher & Schweickerdt* 238 (K; PRE).

Pantropical. Damp and waterlogged, grazed and ungrazed meadows, open sandy ground, river- and roadsides, and cultivated ground; 500–1530 m.

S. brevifolium is the most delicate species in the Flora Zambesiaca area. It is closely related to *S. platyphyllum*, from which it is usually separated by the small size of most of its parts, especially the anthers. *S. brevifolium* var. *flaccidum*, with ciliate peduncles and pedicels, and with a ± hairy inferior glume, does not merit formal recognition because the indumentum is extremely variable.

2. **Schizachyrium platyphyllum** (Franch.) Stapf in Prain, F.T.A. **9**: 188 (1917). —Robyns, Fl. Agrost. Congo Belge **1**: 111 (1929). —Jackson & Wiehe, Annot. Check List Nyasal. Grass.: 58 (1958). —Napper, Grasses Tanganyika: 101 (1965). —Simon in Kirkia **8**: 19, 52 (1971). —Clayton in F.W.T.A., ed. 2, **3**: 478 (1972). —Clayton & Renvoize in F.T.E.A., Gramineae: 755 (1982). —Zon, Gramin. Cameroun **2**: 452 (1992). Type from Dem. Rep. Congo.

Andropogon brevifolius var. *platyphyllus* Franch. in Bull. Soc. Hist. Nat. Autun **8**: 324 (1895).

Andropogon platyphyllus (Franch.) Pilg. in Engler & Prantl, Nat. Pflanzenfam., ed. 2, **14e**: 166 (1940).

Erect or ± scrambling annual, occasionally perennial; culms up to 120 cm high, much branched only above; internodes wide (2.5–3.5 mm), compressed. Leaf sheaths both shorter and longer than the internodes, glabrous; laminas 4.5–8.5 cm × 4.4–9 mm, keeled, obtuse at the apex. Racemes 2.5–4.5 cm long, often exserted from the linear spatheole, glabrous; peduncles both shorter than and as long as the racemes; internodes and pedicels flattened. Sessile spikelet 4.5–5.5 mm long, $\frac{1}{5}$ as long again as the pedicel, linear-ovate, dorsally compressed, rounded at the base; inferior glume slightly convex; superior lemma bifid almost to the base; awn 6–9.6 mm long, geniculate; anthers 1.7–2.5 mm long. Pedicelled spikelet reduced to an awned vestigial glume 0.7–1.3 mm long; awn 0.9–1.5 mm long.

Malawi. N: Nkhata Bay Distr., 8 km east of Mzuzu, 1220 m, 5.vi.1973, *Pawek* 6824 (K). S: Mt. Mulanje, Ruo River, 26.viii.1951, *Jackson* 591 (K). **Mozambique.** Z: serra de Gurué, 23.ix.1944, *Mendonça* 2221 (LISC).

Senegal to Sudan, Dem. Rep. Congo, Uganda, Kenya and Tanzania. Moist situations in shade near evergreen forest.

Although basically an annual species, perennial forms are not rare, a common occurrence in tropical grasses. See also comments under *S. brevifolium.*

3. **Schizachyrium sanguineum** (Retz.) Alston in Thwaites, Enum. Pl. Zeyl., Suppl.: 334 (1931). —Napper, Grasses Tanganyika: 101 (1965). —Simon in Kirkia **8**: 19, 52 (1971). —Clayton in F.W.T.A., ed. 2, **3**: 479 (1972). —Bennett in Kirkia **11**: 184 (1980). —Clayton & Renvoize in F.T.E.A., Gramineae: 756, fig. 178 (1982). —Gibbs Russell et al., Grasses South. Africa [Mem. Bot. Surv. S. Africa No. 58]: 290 (1990). —Zon, Gramin. Cameroun **2**: 458 (1992). TAB. **27**. Type from China.

Rottboellia sanguinea Retz., Observ. Bot. **3**: 25 (1783).

Schizachyrium semiberbe Nees, Agrost. Bras.: 336 (1829). —Stapf in Prain, F.T.A. **9**: 195 (1919). —Robyns, Fl. Agrost. Congo Belge **1**: 112 (1929). —Sturgeon in Rhodesia Agric. J. **51**: 12 (1954). —Chippindall in Meredith, Grasses & Pastures S. Africa: 502 (1955). —Launert in Merxmüller, Prodr. Fl. SW. Afrika, fam.160: 167 (1970). Type from Brazil.

Andropogon semiberbis (Nees) Kunth, Enum. Pl. **1**: 489 (1833).

Schizachyrium lindiense Pilg. in Notizbl. Bot. Gart. Berlin-Dahlem **14**: 100 (1938). Type from Tanzania.

Loosely caespitose perennial, eventually reddish throughout; culms up to 160 cm high, erect, branched; internodes terete. Leaf sheaths both shorter and longer than the internodes, occasionally villous towards the top; laminas 12.5–30 × 2.5–6 mm, keeled, tapering to a fine point at the apex. Racemes 4–18 cm long, exserted from the linear spatheoles, occasionally long-exserted; peduncles longer than the racemes; internodes and pedicels flattened, usually ciliate along both margins with hairs 2.5–4 mm long. Sessile spikelet 5.5–10 mm long, $\frac{1}{2}$ as long again as the pedicel, narrowly lanceolate, laterally compressed, pointed at the base; inferior glume strongly convex, glabrous; superior lemma bifid in the upper $\frac{3}{4}$; awn 10–22 mm long, geniculate; anthers 2.5–3.8 mm long. Pedicelled spikelet male or occasionally sterile, 4–7 mm long; inferior glume with an awn 1–4 mm long; anthers 1.3–2.3 mm long.

Botswana. N: Kwando–Movombi (Movombe) road, 9.iv.1982, *P.A. Smith* 3824 (K; PRE). SE: Serowe, Makoba–Lepalapala Flats, 26.iii.1957, *de Beer* 59 (K). **Zambia.** B: Kaoma Distr., 45 km west of Kaoma (Mankoya), 5.iv.1966, *E.A. Robinson* 6908 (K). N: Mbala Distr., Ndundu, 4.iv.1959, *McCallum Webster* A268 (K). W: Chizela Distr., 7 km east of Chizela (Chizera), 27.iii.1961, *Drummond & Rutherford-Smith* 7434 (K; LISC; PRE). C: Serenje Distr., Kundalila Falls, c. 22.5 km east of Kanona, 17.iii.1974, *Davidse & Handlos* 7247 (K; PRE). S: Choma, 15.iv.1963, *van Rensburg* 1950 (K; PRE; SRGH). **Zimbabwe.** N: Gokwe South Distr., near Sasame Gorge, 16 km from Gokwe, 21.iii.1963, *Bingham* 530B (K; PRE). W: Nkayi Distr., Gwampa Forest Reserve, 7.iii.1955, *Goldsmith* 86/55 (K; PRE). C: Harare (Salisbury), Mexico Rd. 83 km south of Harare, 15.ii.1974, *Davidse, Simon & Pope* 6675 (K; PRE). E: Chipinge Distr., 10 km south of Chimanimani (Melsetter),

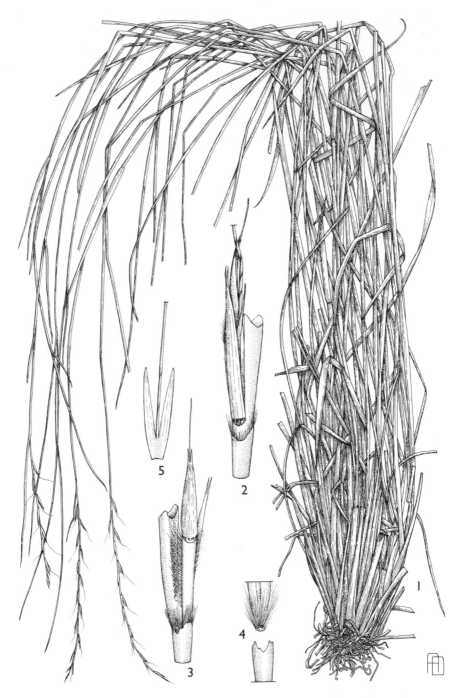

Tab. 27. SCHIZACHRYIUM SANGUINEUM. 1, habit (× ½); 2, detail of raceme, showing sessile spikelet (× 5); 3, detail of raceme, rear view, showing pedicelled spikelet (× 5); 4, raceme joint, disarticulated (× 5); 5, superior lemma (× 5), 1–5 from *Eichinger* 5774. Drawn by Ann Davies. From F.T.E.A.

Chipinge junction, c. 1300 m, 21.v.1972, *Simon* 2235 (K; PRE). S: Masvingo Distr., Makaholi Experimental Station (Farm), 20.ii.1948, *D.A. Robinson* 258 (K). **Mozambique**. N: Montepuez Distr., Montepuez, andados 48 km de Montepuez para Nairoto (Nantulo), 360 m, 8.iv.1964, *Torre & Paiva* 11780 (LISC). Z: Maganja da Costa Distr., Maganja da Costa (Vila da Maganja), 16.ii.1966, *Torre & Correia* 14709 (LISC). T: Tsangano Distr., Ntengo wa Mbalame (Metengobalame), 1350 m, 8.xii.1980, *Stefanesco & Nyongani* 1391 (K). MS: Gorongosa Distr., Parque Nacional da Gorongosa (Gorongosa National Park), Nhamussengere (Nhamasengeri) Falls, vii.1972, *Tinley* 2640 (K). M: between Boane and Namaacha, 20.xii.1944, *Torre* 6887 (K; LISC).

Pantropical. Sandy, gravelly and rocky soils in open woodland and grassland, and as a weed of cultivated fields; 1000–1700 m.

The red colouring of the whole plant seems to be a good spot character in the field but not in the herbarium. The red pigment is general in all *Schizachyrium* species but it becomes more evident in dry specimens. The specimens of *S. sanguineum* with hairier racemes can be mistaken for *S. jeffreysii*, but the very distinct shape of the sessile spikelet is an excellent differentiating character.

4. **Schizachyrium exile** (Hochst.) Pilg. in Bot. Jahrb. Syst. **54**: 284 (March 1917). —Stapf in Prain, F.T.A. **9**: 191 (July 1917). —Chippindall in Meredith, Grasses & Pastures S. Africa: 504 (1955). —Jackson & Wiehe, Annot. Check List Nyasal. Grass.: 57 (1958). —Launert in Merxmüller, Prodr. Fl. SW. Afrika, fam. 160: 167 (1970). —Simon in Kirkia **8**: 18, 52 (1971). —Clayton in F.W.T.A., ed. 2, **3**: 478 (1972). —Bennett in Kirkia **11**: 184 (1980). —Clayton & Renvoize in F.T.E.A., Gramineae: 756 (1982). —Müller, Grasses of South West Africa/Namibia: 222 (1984). —Gibbs Russell et al., Grasses South. Africa [Mem. Bot. Surv. S. Africa No. 58]: 289 (1990). —Zon, Gramin. Cameroun **2**: 459, t. 101/5–9 (1992). Type from Sudan.

Andropogon exilis Hochst. in Flora **27**: 241 (1844).

Schizachyrium inclusum Stent in Bothalia **1**: 172 (1923). —Sturgeon in Rhodesia Agric. J. **51**: 11 (1954). —Vesey-FitzGerald in Kirkia **3**: 103 (1963). —Simon in Kirkia **8**: 18, 52 (1971). Type: Zimbabwe, Enterprise, *Eyles* 1646 (K, isotype).

Loosely caespitose annual; culms up to 85(110) cm high, erect branched. Leaf sheaths much shorter than the internodes, glabrous; laminas 4–17 cm × 1–4 mm, tapering to a fine point at the apex. Racemes 2–7 cm long, ± stout, cylindrical, subsessile, usually enclosed by the broad spatheole for most of their length; peduncles shorter than the racemes; internodes and pedicels broad and canaliculate; spikelet base, internode, pedicel and inferior glume often with hairs up to 2.5 mm long. Sessile spikelet 4.7–7 mm long, $\frac{1}{4}$ as long again as the pedicel, narrowly lanceolate, laterally compressed, very pointed at the base; inferior glume strongly convex, the margins strongly inrolled; superior lemma bifid almost to the base; awn 9–20 mm long, geniculate; anthers 0.4–0.6 mm long. Pedicelled spikelet reduced to an awned vestigial glume 0.8–2.4 mm long; awn 3.5–6 mm long.

Botswana. N: Ngamiland Distr., Aha Hills, 22–28.iv.1980, *P.A. Smith* 3411A, B (K; PRE). **Zambia**. B: Kaoma Distr., 45 km west of Kaoma (Mankoya), 5.iv.1966, *E.A. Robinson* 6905 (K). N: Mbala Distr., Kasikalawi, west of Mpulungu, 820 m, 10.iv.1961, *Phipps & Vesey-FitzGerald* 3046 (K; PRE). W: Ndola, 1300 m, iii.1932, *Duff* 2005 (K). C: South Luangwa National Park, Mfuwe, 760 m, 21.iii.1969, *Astle* 5639 (K). E: Chipata Distr., Ngoni Area, v.1962, *Verboom* 587 (K). S: Livingstone Distr., Victoria Falls, 5–9.iii.1956, *Rattray* 1797 (K). **Zimbabwe**. N: Gokwe South Distr., Sengwa Research Station, 18.iii.1975, *P.R. Guy* 2347 (K; PRE). W: Matobo Distr., Besna Kobila Farm, 1460 m, *Miller* 2822 (K). C: Chegutu Distr., Msengezi, iv.1955, *Conradie* 4 (K). E: Chimanimani Distr., 58 km south of Mutare (Umtali), 760 m, 23.iv.1969, *Plowes* 3198 (K). **Malawi**. N: Karonga Distr., 19.iii.1954, *Jackson* 1260A (K). C: Lilongwe, 1.iv.1952, *Jackson* 748 (K). S: Mangochi Distr., Namwera Escarpment, c. 910 m, 8.v.1950, *Wiehe* 518 (K). **Mozambique**. N: Meconta Distr., andados 16 km de Corrane para Liupo, c. 150 m, 28.iii.1964, *Torre & Paiva* 11422 (LISC). Z: Lugela Distr., Namagoa, viii.1948, *Faulkner* 44 (K). T: Magoé, 19.ii.1970, *Torre & Correia* 18065 (LISC). MS: Gondola Distr., Lion's Creek, 300 m, 7.iv.1898, *Schlechter* 12196 (K).

Old World tropics. Barren rocky hillsides, sometimes in damp soil, in cultivated fields, and along roadsides; 150–1450 m.

A very distinctive species with its racemes usually enclosed in a broad spatheole.

5. **Schizachyrium claudopus** (Chiov.) Chiov. in Nuovo Giorn. Bot. Ital., n.s. **26**: 73 (1919). —Robyns, Fl. Agrost. Congo Belge **1**: 114 (1929). —Clayton & Renvoize in F.T.E.A., Gramineae: 759 (1982). Type from Dem. Rep. Congo.

Andropogon claudopus Chiov. in Ann. Bot. (Rome) **13**: 37 (1914).

Schizachyrium tomentosum Clayton in Kew Bull. **32**: 581 (1978). Type: Zambia, Dobeka Bridge, 4.xi.1937 *Milne-Redhead* 3090 (K, holotype).

Densely caespitose perennial; culms up to 100 cm high, unbranched, slender, erect; internodes terete. Leaves mostly basal; sheaths shorter than the internodes, covered with a dense indumentum of long curly white hairs at the base; laminas 2–10 cm × 1.8–3.3 mm, often inrolled and tapering to a fine point at the apex. Racemes 5.5–11.5 cm long, stout, cylindrical, solitary on the culms, long exserted from the linear spatheole, glabrous; peduncles longer than the racemes; internodes and pedicels canaliculate, very broad at the base (1.7 mm). Sessile spikelet 4.5–6.8 mm long, $^1/_3$ ($^1/_2$) as long again as the pedicel, conical, laterally compressed, rounded at the base; inferior glume convex, the margins strongly inrolled; superior lemma shortly bifid at the apex (and awn stout and straight) or bifid almost to the base (and awn long and geniculate); awn up to10 mm long; anthers 1.5–3.3 mm long. Pedicelled spikelet male, similar to the sessile spikelet but awnless.

Zambia. B: Kabompo Distr., 55 km west of Kabompo, 1100 m, 26.xii.1969, *Simon & Williamson* 2043 (K). N: Kawambwa Distr., Chishinga Ranch, near Luwingu, 1400 m, 13.ix.1961, *Astle* 906 (K; SRGH). W: Mwinilunga Distr., Chitunta River, 30 km north of Mwinilunga on Kalene Hill road, 25.x.1969, *Drummond & Williamson* 9439 (K).

Also in Tanzania and Dem. Rep. Congo. Sandy grassland, riverbanks and dambos; 1100–1400 m.

Resembling *S. thollonii* in its habit, the unbranched culms and the structure of the pedicelled spikelet. The occasional specimens of *S. claudopus* that lack the dense indumentum on the leaf sheaths can be readily separated from *S. thollonii* by the glabrous racemes. Curiously, all herbarium specimens of *S. claudopus*, and also of *S. lopollense* and *S. thollonii*, are of plants that have developed after burning of most of their vegetative parts.

6. **Schizachyrium jeffreysii** (Hack.) Stapf in Prain, F.T.A. **9**: 198 (1919). —Stent & Rattray in Proc. & Trans. Rhodesia Sci. Assoc. **32**: 10 (1933). —Sturgeon in Rhodesia Agric. J. **51**: 12 (1954). —Chippindall in Meredith, Grasses & Pastures S. Africa: 503 (1955). —Jackson & Wiehe, Annot. Check List Nyasal. Grass.: 57 (1958), in part, excl. *Jackson* 1296 (=*Andropogon textilis*). —Launert in Merxmüller, Prodr. Fl. SW. Afrika, fam. 160: 167 (1970). —Simon in Kirkia **8**: 18, 52 (1971). —Bennett in Kirkia **11**: 184 (1980). —Gibbs Russell et al., Grasses South. Africa [Mem. Bot. Surv. S. Africa No. 58]: 289 (1990). Type: Zimbabwe, Bulawayo, *Jeffreys* 78 (whereabouts uncertain).

Andropogon jeffreysii Hack. in Proc. Rhodesia Sci. Assoc. **7**(2): 70 (1908); in Repert. Spec. Nov. Regni Veg. **6**: 324 (1909).

Andropogon arthropogon Pilg. in Fries, Wiss. Ergebn. Schwed. Rhod.-Kongo-Exped. **1**: 193 (1916). Type: Zimbabwe, Victoria Falls, *Fries* 14 (UPS, holotype).

Schizachyrium ursulum Stapf in Prain, F.T.A. **9**: 197 (1919). —Chippindall in Meredith, Grasses & Pastures S. Africa: 503 (1955). —Simon in Kirkia **8**: 19, 52 (1971). Type from Angola.

Loosely caespitose, wiry, rhizomatous perennial, tinged with reddish-orange; culms up to 120 cm high, branched; internodes terete. Leaf sheaths shorter than the internodes, occasionally pubescent towards the top; laminas 11–23 cm × 2.5–4 mm, tapering to a fine point. Racemes 4–8 cm long, silvery or yellowish with usually dense long hairs 4–5.5 mm long on the internodes, pedicels and inferior glumes, exserted from the linear spatheole, occasionally long-exserted; peduncles longer than the racemes; internodes and pedicels flattened, narrow at the base (0.3 mm). Sessile spikelet 7–9 mm long, $^1/_3$–$^1/_2$ as long again as the pedicel, narrowly ovoid, somewhat laterally compressed, pointed at the base; inferior glume slightly convex; superior lemma linear, bifid in the upper $^1/_2$–$^2/_3$; awn dorsal, 13–22 mm long; anthers 2.5–4 mm long. Pedicelled spikelet male or sterile, 5–7.4 mm long, similar to the sessile spikelet and with an awn 1–3.5 mm long.

Botswana. N: Ngamiland Distr., Tsodilo Hills, Female Hill, 'Rhino Valley', 1020 m, 3.iv.1987, *Long & Rae* 611 (E; K). SE: Gaborone, 1.iii.1930, *van Son* in Herb. Tvl. Mus. 28618 (K; PRE). **Zambia**. B: Senanga Distr., Ngonye (Sioma) Falls, 26.vi.1964, *Verboom* 1120 (K). N: Mpika Distr., south Bangweulu and upper Lwitikila River, 5.x.1969, *Verboom* 2533 (K). C: Lusaka, 30.iii.1963, *van Rensburg* 1878 (K; PRE). S: Namwala, 1160 m, 22.ii.1964, *Astle* 2954 (K). **Zimbabwe**. N: Centenary Distr., Muzarabani Teacher Training School, Msengezi R., 9.iv.1965, *Bingham* 1450 (K; PRE). W: Nkayi Distr., Gwampa Forest Reserve, iii.1955, *Goldsmith* 150/55 (K; PRE). C: KweKwe Distr., Sable Park, 8 km NE of KweKwe, 23.iii.1977, *Chipunga* G15 (K; PRE). E: Mutare, 760 m, 23.iv.1969, *Plowes* 3199 (K; PRE). S: Chivi Res., 24.iii.1948, *D.A. Robinson* 323 (K). **Malawi**. C: Lilongwe Agric. Research Station, 4.iv.1951, *Jackson* 450 (K; PRE). **Mozambique**. Z: Maganja da Costa Distr., Maganja da Costa (Vila da Maganja), 21.xi.1967, *Torre & Correia* 16180

(LISC). T: Magoe Distr., Mphende (Magoé), 71 km from Mukumbura (Mucumbura), 10.iii.1970, *Torre & Correia* 18238 (LISC). M: between Matutuíne (Bela Vista) and Umbelúzi, 15.iv.1944, *Torre* 6456 (LISC; PRE).

Southern tropical Africa. In open woodland, shrubland and grassland, rocky outcrops and dambos; 10–1400 m.

See comments under *S. sanguineum*.

7. **Schizachyrium lopollense** (Rendle) Sales, comb. nov.
  Andropogon lopollensis Rendle in Hiern et al., Cat. Afr. Pl. Welw. **2**: 143 (1899). —Stapf in Prain, F.T.A. **9**: 187 (1917), in syn. of *S. thollonii*. Type from Angola.

Densely caespitose, rhizomatous perennial, ± reddish-purple; culms slender, up to 50 cm high, branched; internodes terete, filiform at the base but wider at the top. Leaves mostly basal, often falcate, green; sheaths shorter than the internodes, glabrous; laminas 1–3.5 cm × c. 2 mm, keeled, tapering to a fine point. Racemes 1.5–3.5 cm long, silvery with usually dense long hairs 3–4 mm long on the internodes, pedicels and inferior glume, shortly exserted from the linear spatheole; peduncles longer than the racemes; internodes and pedicels terete. Sessile spikelet 5.5–7.5 mm long, $\frac{1}{3}$–$\frac{1}{2}$ as long again as the pedicel, linear-ovate, dorsally compressed, rounded at the base; inferior glume slightly convex; superior lemma bilobed to less than $\frac{1}{3}$ its length, the lobes slender and often contiguous to the awn (thus the lemma appearing entire), not keeled, with 2 purple nerves; awn terminal, 4–9 mm long, geniculate; anthers 2.7–3 mm long. Pedicelled spikelet male, 4.5–5.5 mm long, similar to the sessile spikelet but awnless.

Zambia. B: Kalabo Distr., west of Sikongo, near the Angolan border, 15.xi.1959, *Drummond & Cookson* 6499 (SRGH). N: Chinsali Distr., Mbesuma Ranch, Chambeshi River, 24.x.1961, *Astle* 995 (K). W: Kitwe Distr., Buchi (Uchi) Dambo, 25.ix.1947, *Brenan & Greenway* 7956 (K). **Mozambique**. Z: Pebane Distr., between Pebane and Mucubela, 25.x.1942, *Torre* 4689 (LISC). MS: Dondo region, 17.x.1944, *Mendonça* 2467 (LISC).

Southern Angola. In dambos, sandy grassland, wooded grassland and among quartzite outcrops; 20–1350 m.

A plant of scattered occurrence with few herbarium collections, especially in the Flora Zambesiaca area where it has usually been named as *S. thollonii*. After Rendle had described the species in 1899, Stapf (1919) relegated it to synonymy under *S. thollonii*; it has languished there ever since. Although there are general similarities in plant size and basal leaves between these two species, there are striking differences in culm and spikelet structure, spikelet hairiness and leaf lamina length. See also comments under *S. claudopus*.

8. **Schizachyrium thollonii** (Franch.) Stapf in Prain, F.T.A. **9**: 200 (1919). —Robyns, Fl. Agrost. Congo Belge **1**: 114 (1929). TAB. **28**. Syntypes from Dem. Rep. Congo.
  Andropogon thollonii Franch. in Bull. Soc. Hist. Nat. Autun **8**: 324 (1895).
  Andropogon mukuluensis Vanderyst in Bull. Agric. Congo Belge **11**: 142 (1920). Type from Dem. Rep. Congo.
  Schizachyrium mukuluense (Vanderyst) Vanderyst in Robyns, Fl. Agrost. Congo Belge **1**: 115 (1929). —Simon in Kirkia **8**: 19, 52 (1971).
  Schizachyrium monostachyon P.A. Duvign. in Bull. Soc. Roy. Bot. Belgique **90**: 187 (1958). Type from Dem. Rep. Congo (Shaba).

Densely caespitose perennial, often reddish-purple; culms up to 130 cm high, slender, erect, unbranched; internodes terete. Leaves mostly basal, often falcate; sheaths shorter than the internodes, glabrous or thinly pilose with straight hairs; laminas 4.5–24 cm × 1.2–4 mm, flat at the base, involute above, tapering to a fine point. Racemes 5–9 cm long, solitary, glabrous except for a conspicuous ring of 2–4 mm long hairs at the base of the spikelet, long exserted from the linear spatheole, dull purple; peduncles longer than the racemes; internodes and pedicels terete, linear, widened at the apex. Sessile and pedicelled spikelets similar, 5–7.5 mm long, (the pedicelled spikelets $\frac{1}{3}$–$\frac{1}{2}$ as long again as the pedicel), sterile, female or bisexual, linear-ovate, dorsally compressed, flushed with dull purple; inferior glumes slightly convex, entire or occasionally 2-mucronate at the apex; superior lemma of sessile spikelet bifid only at the apex, mucronate or with an awn up to 11 mm long; lemma of pedicelled spikelet awnless or mucronate; anthers 2.5–4 mm long.

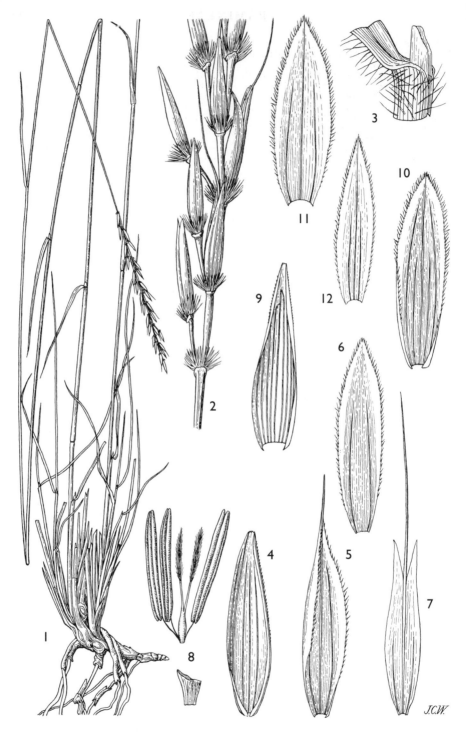

Tab. 28. SCHIZACHRYIUM THOLLONII. 1, habit ($\times$ $^2/_3$); 2, inflorescence ($\times$ 4); 3, ligule ($\times$ 6); 4–8 sessile spikclet. 4, inferior glume; 5, superior glume; 6, inferior lemma; 7, superior lemma; 8, flower; 9–12 pedicelled spikelet: 9, inferior glume; 10, superior glume; 11, inferior lemma; 12, superior lemma, 1–12 from *Vesey-FitzGerald* 2943. Drawn by J.C. Webb.

Zambia. B: road to Kaoma (Mankoya), 1.x.1957, *West* 3474 (SRGH). N: Mbala Distr., Zombe, 14.viii.1963, *Vesey-FitzGerald* 4174 (BM). W: Solwezi Distr., Solwezi to Chingola, 1.x.1947, *Greenway & Brenan* 8143 (K; LISC). C: Lusaka Distr., Chakwenga Headwaters, 100–129 km east of Lusaka, 27.x.1963, *E.A. Robinson* 5773 (K; PRE; SRGH). **Zimbabwe**. C: Marondera (Marandellas), 31.x.1975, *A.O. Crook* 2087 (K; PRE). E: Nyanga Distr., Mtarazi Falls, 30.x.1966, *Simon* 941 (K). **Malawi**. N: Mzimba Distr., Katoto, 5 km west of Mzuzu, 7.xi.1970, *Pawek* 3949 (K: MAL). C: Lilongwe Distr., Dzalanyama, Chiputu Dambo, 4.xii.1951, *Jackson* 694 (K).

Also in Congo-Brazzaville and Dem. Rep. Congo. Burnt perennial grassland and dambos on loam or sandy clay; 1050–1700 m.

The distribution of sexes in individual plants is rather interesting; some racemes are exclusively female, others bisexual throughout; some are bisexual below and female above or *vice versa*. There is some variation even in spikelets at individual nodes of the raceme: where sessile and pedicelled spikelets can be both bisexual or both female; the sessile spikelet bisexual and the pedicelled female or sterile. See also comments under *S. claudopus* and *S. lopollense*.

24. ARTHRAXON P. Beauv.

By Fatíma Sales

Arthraxon P. Beauv., Ess. Agrostogr.: 111, fig.11/6 (1812). —Roberty in Boissiera **9**: 293–296 (1960). —van Welzen in Blumea **27**: 255–300 (1981); in Rheedea **3**: 101 (1993).

Slender annuals; culms geniculate, often trailing and rooting at the decumbent nodes; internodes pilose. Leaf sheaths slightly inflated; ligule membranous; leaf laminas slightly cordate and semi-amplexicaul at the base, pilose with bulbous-based hairs at least below, usually becoming ciliate. Racemes in subdigitate groups, slender, ± exserted from the spatheoles; peduncles longer or shorter than the racemes, erect at maturity; internodes and pedicels delicate, filiform, long-ciliate on both margins. Sessile spikelet laterally compressed, with silvery-white appressed hairs at the base, the callus short, truncate; inferior glume chartaceous to coriaceous, rounded on the back, with or without lateral keels; inferior floret reduced to a hyaline lemma; superior floret fertile; fertile lemma bifid almost to the base and with a glabrous, geniculate sub-basal awn; anthers 2. Pedicelled spikelet absent or much reduced.

A genus of c. 10 species, occurring in the Old World tropics, especially in India.
Apparently a rather isolated genus of very narrowly circumscribed species.

Inferior glume of sessile spikelet narrowly ovate; superior glume convex on the back, acute or
    mucronate at the apex; pedicelled spikelet always absent · · · · · · · · · · · · · · · · 1. *micans*
Inferior glume of sessile spikelet linear; superior glume flat on the back, shortly awned at the
    apex; pedicelled spikelet present or absent · 2. *lancifolius*

1. **Arthraxon micans** (Nees) Hochst. in Flora **39**: 188 (1856). —Clayton & Renvoize in F.T.E.A., Gramineae: 742, fig. 173 (1982). —Zon, Gramin. Cameroun **2**: 461 (1992) in clavis et in t. 102 err. descript. sub. *A. hispidus* (Thunb.) Makino var. *hispidus.* Type from Bangladesh.
    Batratherum micans Nees in Edinburgh New Philos. J. **18**: 182 (1835).
    Alectoridia quartiniana A. Rich., Tent. Fl. Abyss. **2**: 448 (1850). Type from Ethiopia.
    Arthraxon quartinianus (A. Rich.) Nash, N. Amer. Fl. **17**: 99 (1912). —Stapf in Prain, F.T.A. **9**: 166 (1917). —Robyns, Fl. Agrost. Congo Belge **1**: 102 (1929). —Jackson & Wiehe, Annot. Check List Nyasal. Grass.: 30 (1958). —Bor, Grasses Burma Ceyl. Ind. Pak.: 102 (1960). —Napper, Grasses Tanganyika: 99 (1965). —Simon in Kirkia **8**: 49 (1971). —Clayton in F.W.T.A., ed. 2, **3**: 470 (1972).
    Arthraxon lancifolius sensu Vesey-FitzGerald in Kirkia **3**: 102 (1963), non (Trin.) Hochst.

Straggling, delicate annual; culms up to 80 cm long, much branched. Ligule 0.5 mm long; leaf lamina 2–6 cm × 4–12 mm, lanceolate to narrowly ovate, cordate and amplexicaul at the base, loosely pilose and often with bulbous-based hairs on the margins, acute at the apex. Racemes 2–14 in fascicles borne on long exserted peduncles, 1.5–5 cm long, ± terete, greenish-yellow or purplish; internodes filiform, with silvery-white appressed hairs 1–2 mm long at the summit; pedicels absent. Sessile spikelet 3.2–4 mm long; inferior glume rounded on the back, narrowly ovate, ± scabrid, awnless; superior glume strongly convex on the back with broad hyaline

Tab. 29. ARTHRAXON LANCIFOLIUS. 1, habit (× ⅔); 2, ligule (× 8); 3, ınflorescence (× 4); 4, sessile spikelet (× 16); 5, superior lemma (× 24). Drawn by Victoria Goaman. From Ghana Gr.

margins, acute to mucronate at the apex; awn of superior lemma 7–10 mm long; anthers 0.6–0.9 mm long, dark red. Pedicelled spikelet absent.

Zambia. N: Mbala Distr., bridge over Saisi River, 32 km from Mbala (Abercorn), 1520 m, 14.iv.1959, *McCallum Webster* A323 (K; LISC). W: Chingola Distr., banks of Kafue River, 11 km north of Chingola, 4.v.1960, *E.A. Robinson* 3686 (K; LISC; SRGH). E: Chama Distr., Nyika Plateau, Chowo, iv.1972, *G. Williamson* 2173A (K). **Malawi**. N: Rumphi Distr., Nyika Plateau, 18.v.1970, *Brummitt* 10893 (K; LISC; MAL; PRE; SRGH). C: Ntchisi Distr., Dowa, Mwera Hill, 8.v.1951, *Jackson* 492 (K; PRE). S: Zomba, 940 m, 17.vii.1949, *Wiehe* 166 (K). **Mozambique**. N: Lichinga (Vila Cabral), 16.v.1934, *Torre* 74 (COI; LISC).

Tropical Africa to India, Sri Lanka and Indonesia. In open *Brachystegia* woodland, along river banks and in shady rock crevices; 750–1520 m.

A. micans is very similar to *A. lancifolius*. Subtle differences in raceme, peduncle and inferior and superior glume shape can separate them. Only *A. lancifolius* sometimes has a pedicelled spikelet.

Van Welzen in his revision of the genus (in Blumea, loc. cit.) took a broad species concept and considered *A. micans* to be the same as the Asiatic *A. hispidus* (Thunb.) Makino, calling it var. *hispidus*.

2. **Arthraxon lancifolius** (Trin.) Hochst. in Flora **39**: 188 (1856). —Stapf in Prain, F.T.A. **9**: 165 (1917). —Jackson & Wiehe, Annot. Check List Nyasal. Grass.: 30 (1958). —Simon in Kirkia **8**: 16, 49 (1971). —Clayton in F.W.T.A., ed. 2, **3**: 470 (1972). —Bennett in Kirkia **11**: 175 (1980). —Clayton & Renvoize in F.T.E.A., Gramineae: 742 (1982). —Zon, Gramin. Cameroun **2**: 463 (1992). TAB. **29**. Type from Nepal.

Andropogon lancifolius Trin. in Mém. Acad. Imp. Sci. St.-Pétersbourg, Sér. 6, Sci. Math. **2**: 271 (1832).

Pleuroplitis lancifolius (Trin.) Regel in Bull. Acad. Imp. Sci. St.-Pétersbourg **10**: 375 (1866).

Similar to *A. micans* but differing by the following characters: culms 10–30 cm high; leaf laminas 1–3 cm × 3–7 mm; racemes 1–3, 0.5–2 cm long, not terete, greenish-yellow; internodes with hairs seldom over 0.5 mm long; pedicel present or absent; sessile spikelet 3–5 mm long; inferior glume flat on the back, linear, scaberulous; superior glume ± flat on the back, produced at the apex into a fine awn 0.5–1.5 mm long; awn of superior lemma 7–10 mm long; anthers 0.4–0.6 mm long; pedicelled spikelet lanceolate, 1.5–2 mm long or totally suppressed.

Zambia. C: Mazabuka Distr., Kafue Gorge, 975 m, 14.iv.1956, *E.A. Robinson* 1483 (K). S: Livingstone Distr., Victoria Falls, north bank, 28.viii.1947, *Greenway & Brenan* 8002 (K). **Zimbabwe**. N: Mount Darwin Distr., Msengesi Camp, 1220 m, 8.v.1955, *Whellan* 854 (K; LISC; PRE). E: Mutare Distr., Vumba Mts., 1130 m, 19.iv.1959, *Chase* 7096 (BM; K). **Malawi**. C: Ntchisi Distr., Ntchisi Forest Reserve, 27.iii.1970, *Brummitt* 9456 (K; LISC; MAL; PRE; SRGH). S: Zomba Distr., Zomba, by Malemia Hospital, 19.iv.1980, *Brummitt* 15512 (BR; C; K; MAL; P; SRGH; WAG). **Mozambique**. N: Malema Distr., Mutuáli, monte Cucuteia, junto as quedas do rio, c. 750 m, 16.iii.1964, *Torre & Paiva* 11223 (COI; K).

From Mali to Ethiopia southwards to Mozambique and eastwards through Arabia to China and Indonesia. Frequent in clearings in xerophytic forest; 560–1130 m.

Species doubtfully recorded

Van Welzen in his revision of the genus (Blumea, loc. cit.) recorded *A. lanceolatus* (Roxb.) Hochst. from Mozambique in a distribution map (p. 282), but no specimen was cited. No gathering of this species from the Flora Zambesiaca area was seen during preparation of this account.

25. DIHETEROPOGON Stapf

By Fatíma Sales

Diheteropogon Stapf in Hooker's Icon. Pl. **31**: t. 3093 (1922). —Clayton in Kew Bull. **20**: 73–76 (1966).

Annuals or caespitose perennials, often glaucous and tinged with red; culms not branched; ligules truncate; leaf laminas tapering to a fine point. Racemes in pairs, terminal or arranged in a scanty spatheate false panicle, rarely solitary, clearly exserted from the spatheoles; peduncles much longer than the racemes, erect at maturity; internodes and pedicels linear, long-ciliate on both margins with ±

ascending hairs. Sessile spikelet dorsally compressed, lanceolate; callus very pointed; inferior glume with a narrow median groove, coriaceous; inferior floret reduced to a hyaline lemma; superior floret bisexual, its lemma bilobed with a terminal awn; awn bigeniculate, shortly pilose. Pedicelled spikelet male or sterile, much longer than the sessile spikelet, dark red, coriaceous.

A genus of 5 species in tropical and southern Africa.

Diheteropogon in closely related to *Andropogon* and it has been suggested that it could be merged with it. Nevertheless, in my opinion it should be kept separate. The awns in *Diheteropogon* are always bigeniculate, pilose and extremely long; the lemmas are vestigial; the inferior glume of the pedicelled spikelets are always much longer than the sessile spikelet and ± coriaceous. *Andropogon schirensis* seems to be the closest species to *Diheteropogon* in the Flora Zambesiaca area because of the well developed inferior glume of the pedicelled spikelets which imbricates on one side of the raceme and because of its bigeniculate awns. *A. chinensis* also has bigeniculate awns.

1. Leaf laminas ± cordate at the base; awn of sessile spikelet 3–6 cm long · · · · · 2. *amplectens*
   - Leaf laminas not cordate at the base; awn of sessile spikelet 3.5–11 cm long · · · · · · · · · 2
2. Plants 60–160 cm high; pedicelled spikelet male, 14–20 mm long · · · · · · · · · · 1. *filifolius*
   - Plants 30–45 cm high; pedicelled spikelet sterile, 5.5–6.5 mm long · · · · · · · · 3. *microterus*

1. **Diheteropogon filifolius** (Nees) Clayton in Kew Bull. **20**: 75 (1966). —Clayton in F.W.T.A., ed. 2, **3**: 489 (1972). —Bennett in Kirkia **11**: 178 (1980). Type from South Africa.
   Heteropogon filifolius Nees, Fl. Afr. Austral. Ill.: 102 (1841).
   Andropogon filifolius (Nees) Steud., Syn. Pl. Glumac. **1**: 374 (1854).
   Andropogon grandiflorus Hack. in Flora **68**: 127 (1885). Type from Nigeria.
   Andropogon emarginatus De Wild. in Bull. Jard. Bot. État **6**: 33 (1919). Type from Dem. Rep. Congo.
   Diheteropogon grandiflorus (Hack.) Stapf in Hooker's Icon. Pl. **31**: t. 3093 (1922). —Simon in Kirkia **8**: 17, 50 (1971).
   Diheteropogon emarginatus (De Wild.) Robyns, Fl. Agrost. Congo Belge **1**: 152 (1929). —Napper, Grasses Tanganyika: 106 (1965).
   Diheteropogon maximus C.E. Hubb. in Bull. Misc. Inform., Kew **1936**: 296 (1936). Type from Angola.
   Heteropogon grandiflorus (Hack.) Roberty in Boissiera **9**: 137 (1960).

Caespitose perennial; culms 60–160 cm high, often glaucous. Leaf sheaths glabrous; ligule a minutely fringed rim 1–1.5 mm long; laminas 8–26 cm × 2.5–4 mm, inrolled, glabrous or with very short scattered hairs, tapering to a very fine point at the apex. Racemes in pairs, 7.5–9 cm long, robust, clearly exserted from the spatheoles; internodes and pedicels with silvery-white hairs. Sessile spikelet 6.5–9 mm long, with long silvery-white appressed hairs at the base; inferior glume with a deep median groove; lemma with an awn 7.5–11 cm long; anthers 4–6.5 mm long, straw-coloured. Pedicelled spikelet male, 14–20 mm long; inferior glume winged on the keels, purplish-green, coriaceous, acuminate at the apex and with a mucro 1–2 mm long.

Zambia. B: Kabompo Distr., 55 km west of Kabompo, 1100 m, 26.iii.1969, *Simon & Williamson* 2044 (BM; K; PRE). N: Mporokoso Distr., 42 km east of Mporokoso, 13.iv.1961, *Phipps & Vesey-FitzGerald* 3131 (BM). W: Mwinilunga Distr., 18 km east of Kalene Hill, 16.xii.1963, *E.A. Robinson* 6113 (K; PRE). C: Mkushi Distr., at junction of Munshiwemba and Mitupo, ii.1942, *Stöhr* 777 (K). S: 10 km SE of Choma, i.1969, *Heery* 36 (K). **Zimbabwe**. C: Chikomba Distr., Charter Estate, 21.iii.1969, *P.I. Thomas* 810 (K). E: Chimanimani Distr., Chimanimani Mts., south approach to Higher Valley, 25.ix.1966, *Simon* 833 (K; PRE). **Malawi**. C: Lilongwe Distr., Dzalanyama Forest Reserve, 27.iv.1958, *Jackson* 2215 (K).

Also in Nigeria, Dem. Rep. Congo, Tanzania, Angola and South Africa. Growing in dambos, swamps, sour grassland and along trodden paths; 1100–1700 m.

D. grandiflorus is only a form of *D. filifolius* with well developed leaves. *D. filifolius* is the earlier name when the two species are united (cf. Clayton's key in Kew Bull. **20**: 74 (1966)).

2. **Diheteropogon amplectens** (Nees) Clayton in Kew Bull. **20**: 75 (1966). —Simon in Kirkia **8**: 17, 50 (1971). —Clayton in F.W.T.A., ed. 2, **3**: 489 (1972). —Bennett in Kirkia **11**: 178 (1980). —Clayton & Renvoize in F.T.E.A., Gramineae: 784, fig. 182 (1982). —Gibbs Russell et al., Grasses South. Africa [Mem. Bot. Surv. S. Africa No. 58]: 115 (1990). Type from South Africa.

Andropogon amplectens Nees, Fl. Afr. Austral. Ill.: 104 (1841). —Stapf in Prain, F.T.A. **9**: 243 (1919). —Robyns, Fl. Agrost. Congo Belge **1**: 131 (1929). —Stent & Rattray in Proc. & Trans. Rhodesia Sci. Assoc. **32**: 11 (1933). —Chippindall in Meredith, Grasses & Pastures S. Africa: 498 (1955). —Jackson & Wiehe, Annot Check List Nyasal. Grass.: 28 (1958). —Napper, Grasses Tanganyika: 103 (1965).
Andropogon diversifolius Rendle in Hiern et al., Cat. Afr. Pl. Welw. **2**: 148 (1899). Type from Angola.
Cymbachne amplectens (Nees) Roberty in Boissiera **9**: 242 (1960).

Caespitose, rhizomatous perennial; culms 40–200(300) cm high, erect, glaucous, often flushed with purple.   Leaves glabrous, rarely densely pubescent; ligule a minutely fimbriate rim 0.5–1.5 mm long; laminas 8–30 cm × 2.5–26 mm wide at the base, flat, broadly or narrowly cordate at the base.   Racemes in pairs, 5–10 cm long, exserted from the spatheoles at maturity; internodes and pedicels with short, dense, silvery-white or rarely brown hairs, the internode broadening and hollow at the apex. Sessile spikelet 5.5–8.5 mm long, very pointed and with silvery-white appressed hairs at the base; inferior glume with a deep median groove; lemma with pilose awn 3–6 cm long; anthers 3–6 mm long, straw-coloured.   Pedicelled spikelet male, 8–13 mm long; inferior glume with an awn 2–11.5 mm long.

A very distinctive species on account of its leaf lamina shape and very long awns. The two varieties, based on leaf size and shape, are usually easy to separate. Intermediate specimens with all leaves broadly cordate are likely to be found.

Leaves all similar; lamina base 2.5–5.5 mm wide, narrowly cordate, ± parallel-sided, almost
    linear and gradually tapering to the apex · var. *amplectens*
Basal and upper leaves different, the basal leaves narrowly cordate, ± parallel-sided and
    gradually tapering to the apex, the upper leaves broadly cordate; lamina base 5.5–26 mm
    wide, abruptly tapering to a very fine point at the apex · · · · · · · · · · · · · var. *catangensis*

Var. **amplectens**   TAB. 30.

Zambia. N: Kasama Distr., Ntumba, near Kayambi, 8.viii.1958, *Vesey-FitzGerald* 1773 (BM). **Zimbabwe**. W: Matobo Distr., Matopos Research Station, 1370 m, 24.ii.1954, *Rattray* 1659 (K). **Malawi**. S: between Mandimba in Mozambique and Zomba, 6.iii.1964, *Correia* 172 (LISC). **Mozambique**. Z: between Namacurra (Inhamacurra) and Mocuba, 26.iii.1943, *Torre* 4995 (K; LISC). T: Angónia Distr., c. 20 km from Dómuè towards Chia, 19.iii.1980, *Macuácua & Mateus* 1214 (LISC). MS: Cheringoma Distr., 10 km south of Sengo, 16.xii.1971, *Müller & Pope* 2056 (K; LISC; PRE). GI: between Panda (Jacubécua) and Inharrime, *Mendonça* 3341 (K; LISC). M: entre Boane e Namaacha, 20.xii.1944, *Torre* 6888 (K; LISC).
Also in Dem. Rep. Congo and Kenya, and southwards to South Africa. In open forest, grassland and marshy areas in rocky, sandy and clay soils; 20–1800 m.
Var. *amplectens* is less common than var. *catangensis* in the Flora Zambesiaca area.

Var. **catangensis** (Chiov.) Clayton in Kew Bull. **20**: 75 (1966). —Zon, Gramin. Cameroun **2**: 468, t. 103/1–6 (1992). Type from Dem. Rep. Congo.
Andropogon amplectens var. *catangensis* Chiov. in Ann. Bot., Roma **13**: 38 (1914).
Diheteropogon amplectens sensu Simon in Kirkia **8**: 50 (1971).

Botswana. N: Chobe Distr., near the Leshomo–Ngwezumba road, 14.iv.1983, *P.A. Smith* 4318 (K; PRE). SE: South East Distr., c. 5 km north of Gaborone, 24°37'S, 25°56'E, 13.xi.1977, *Hansen* 3282 (K; PRE). **Zambia**. B: Mongu Distr., 48 km on Kaoma (Mankoya) road, 2.iv.1966, *Verboom* 1375 (BM; K). N: Mbala Distr., Simanwe Farm, 1520 m, 24.iii.1959, *McCallum Webster* A229 (K; LISC). W: Ndola–Kitwe road, 1190 m, 28.iv.1953, *Hinds* 130 (K). C: Mkushi Distr., Mulungushi, east rim of Mulungushi River Gorge, 17.iii.1973, *Kornaś* 3506 (K). E: Katete, 24.iii.1955, *Exell, Mendonça & Wild* 1157 (BM; LISC). S: Choma Distr., Dundwa Agriculture Station, 10 km SW of Mapanza, 2.vi.1953, *E.A. Robinson* 293 (K). **Zimbabwe**. N: Gokwe North Distr., Chirisa, Sengwa Research Area, 10.ii.1981, *Mahlangu* 426 (K; PRE). W: Hwange Distr., Victoria Falls National Park, 900 m, 26.iii.1975, *Gonde* 3 (K; PRE). C: Chegutu Distr., Poole Farm, 1.iii.1944, *H.I.E. Hornby* 2272 (K). E: Chimanimani (Melsetter), 1680 m, v.1934, *Brain* 10697 (K). S: Masvingo (Fort Victoria), 14.i.1948, *D.A. Robinson* 169 (K). **Malawi**. N: Mzimba Distr., Marymount, Mzuzu, 1370 m, 30.v.1971, *Pawek* 4877 (K). C: Dedza, 1430 m, 26.iii.1950, *Wiehe* 460 (K). S: Zomba Distr., road to Lake Chilwa (Shirwa), 610 m, 12.iii.1950, *Wiehe* 431 (K). **Mozambique**. N: Nampula Distr., andados 27 km na estr da antiga de Nampula para Muecate, c. 380 m, 2.iv.1964, *Torre & Paiva* 11579 (LISC). Z: Ile Distr., Ile (Errego), c. 3 km to monte Ile,

Tab. 30. DIHETEROPOGON AMPLECTENS var. AMPLECTENS. 1, habit (× ²/₃); 2, portion of raceme (× 2); 3, spikelet pair (× 4); 4, sessile spikelet, showing callus and rhachis internode (× 6); 5, tip of superior lemma (× 6), 1–5 from *Glover, Gwynne & Samuel* 1529. Drawn by Ann Davies. From F.T.E.A.

3.iii.1966, *Torre & Correia* 14986 (LISC). T: between Marueira and Bucha, 10.iii.1972, *Macêdo* 5048 (K; LISC). MS: Barué Distr., Catandica (Vila Gouveia), Chôa Mt., 30.iii.1966, *Torre & Correia* 15526 (LISC).

Also in Senegal and Dem. Rep. Congo to Tanzania and southwards to southern Africa and Angola. Growing in secondary forest, grassland and savanna in sandy and clay soils; 430–1680 m.

3. **Diheteropogon microterus** Clayton in Kew Bull. **21**: 485 (1968). —Simon in Kirkia **8**: 50 (1971). Type: Zambia, Mbala Distr., 20.iv.1963, *Vesey-FitzGerald* 4123 (K, holotype).

Caespitose annual; culms 30–45 cm high, tinged with red. Leaves glabrous; ligule a minutely fimbriate rim 0.5–1 mm long; laminas 7.5–13 cm × 1–2 mm, inrolled, tapering to a fine point at the apex. Racemes in pairs, occasionally solitary, 4–5 cm long; internodes and pedicels with short, dense, silvery-white hairs, clavate. Sessile spikelet 4.3–5.5 mm long, very pointed and with long appressed hairs at the base; inferior glume c. 4 mm long; lemma with awn 3.5–4.5 cm long; anthers 1–1.5 mm long. Pedicelled spikelet sterile, 5.5–6.5 mm long; inferior glume narrowly lanceolate-acuminate with narrow wings on the keels, purplish-green, with a mucro c. 1.5 mm long.

Zambia. N: Mbala Distr., Uningi Pans, lateritic outcrop between the pans, 20.iv.1963, *Vesey-FitzGerald* 4123 (K).

Known only from the type specimen.

This is the only species in the genus with sterile pedicelled spikelets.

26. HYPARRHENIA E. Fourn.

By T.A. Cope

Hyparrhenia E. Fourn., Mexic. Pl. **2**: 51 (1886). —Clayton, A revision of the genus *Hyparrhenia* in Kew Bull., Addit. Ser. 2 (1969).

Tall annuals or perennials. Ligule scarious; leaf laminas linear, never aromatic. Inflorescence composed of paired racemes, each pair supported on a peduncle and subtended by a sheathing spatheole, the latter crowded together in a large leafy false panicle; racemes short, slender, each borne upon a short stalk (raceme-base) which is often deflexed at maturity, and with up to 2 of the lowermost spikelet pairs (homogamous pairs) male or barren, awnless and tardily deciduous; internodes and pedicels linear. Sessile spikelet narrowly lanceolate to lanceolate-oblong, dorsally compressed or terete; callus obtuse to pungent, applied obliquely to the apex of the internode with its tip free; inferior glume coriaceous, broadly convex across the back and sides, without keels or these developed only in the uppermost third; superior glume awnless; inferior floret reduced to a hyaline lemma; superior lemma stipitiform, bidentate, passing between the teeth into a stout awn with pubescent or hirtellous column. Caryopsis oblong, subterete. Pedicelled spikelet male or barren, narrowly lanceolate, usually a little longer than the sessile spikelet, awnless or aristulate from the inferior glume.

A genus of c. 55 species; mainly African but with a few species extending into other tropical or warm temperate regions.

This genus is one of the most difficult to identify to species. As a result of hybridization, apomixy and polyploidy, *Hyparrhenia* consists of a mosaic of intergrading species. Sharp boundaries between these species are seldom present and intermediate specimens, difficult to place, are not uncommon. The form of the raceme-base is of great taxonomic significance and in general the different types are easy to recognize. There are, however, one or two anomalous species and those most likely to mislead are: *H. gazensis* (terete but pilose; terete raceme-bases are usually glabrous); *H. newtonii* (appendaged, but rather unequal and subterete); and *H. gossweileri* (flattened, but unequal).

It should be noted that 'glabrous' raceme-bases are usually pubescent in the fork, and that spikelet length includes the callus.

1. Callus of sessile spikelet broadly rounded, semicircular; spikelets glabrous; raceme-bases unequal, the superior 1.5–3 mm long (Sect. *Strongylopodia*) · · · · · · · · · · · · · · · · · 2
 – Callus of sessile spikelet acute to pungent, rarely obtuse but then the other characters not as above · 3
2. Plant robust, caespitose; leaf laminas up to 30 cm long; false panicle copious; awn puberulous · 1. *glabriuscula*
 – Plant slender, trailing; leaf laminas 4–8 cm long; false panicle scanty; awn glabrous · · · · ·
 · 2. *mobukensis*
3. Superior raceme-base terete or filiform, usually much longer than the inferior, glabrous or sometimes softly hirtellous (stiffly pilose in *gazensis*), not produced into a scarious appendage; spatheoles generally narrow (Sect. *Polydistachyophorum*) · · · · · · · · · · · · · · 4
 – Superior raceme-base flattened, usually not much longer than the inferior · · · · · · · · 15
4. Spikelets wholly or partly rufously pilose · 5
 – Spikelets white-pilose or glabrous · 8
5. Basal leaf sheaths densely pubescent to tomentose below · · · · · · · · · · · · · · · 3. *nyassae*
 – Basal leaf sheaths glabrous · 6
6. Sessile spikelets 5.5–7 mm long; pedicelled spikelets often with an awn-point up to 2 mm long · 6. *poecilotricha*
 – Sessile spikelets 3–5 mm long; pedicelled spikelets awnless, rarely mucronate · · · · · · · · 7
7. Peduncles exceeding the spatheoles at maturity, at least in the periphery of the false panicle; spatheoles linear, 3–5 cm long; superior raceme-base 2–5 mm long, glabrous; racemes 6–14-awned per pair; callus short and round to cuneate and narrowly truncate · · · · · · · 4. *rufa*
 – Peduncles half as long to as long as the spatheoles, mostly 1–2.5 cm long; spatheoles narrowly lanceolate, 2–3.5 cm long, embracing the racemes; superior raceme-base 1.5–2.5 mm long, usually hirtellous; racemes 6–9-awned per pair; callus cuneate, obtuse · · · · · · · · · 5. *dichroa*
8. Superior raceme with or without 1 homogamous pair (if with 2 homogamous pairs then raceme-pairs with at least 8 awns) · 9
 – Superior raceme with 2 homogamous pairs (if with only 1 then raceme-pairs with fewer than 8 awns) · 14
9. Plant annual · 12. *barteri*
 – Plants perennial · 10
10. Spikelets glabrous to hispidulous; racemes usually not deflexed (sometimes deflexed in *gazensis*) · 11
 – Spikelets pubescent to villous · 12
11. Culms slender; callus cuneate, 0.8–1.5 mm long; superior raceme-base (2)2.5–3.5 mm long, stiffly pilose · 7. *gazensis*
 – Culms robust; callus linear, 1–2 mm long; superior raceme-base 1.5–2.5 mm long, hirtellous · 8. *finitima*
12. Racemes not deflexed; sessile spikelet 4–6.5 mm long with awn 1–3.5 cm long · · · 9. *hirta*
 – Racemes, or some of them, deflexed (if spikelets glabrous to hispidulous see *gazensis*); sessile spikelet 4.5–5.5 mm long with awn 1.8–3.2 cm long, or sessile spikelet 6–7 mm long with awn 4–6 cm long · 13
13. Superior raceme-base 2–3.5 mm long; rhachis internodes 2–3 mm long; sessile spikelets 4.5–5.5 mm long; awn 1.8–3.6 cm long · 10. *quarrei*
 – Superior raceme-base 3.5–8 mm long; rhachis internodes 3.5–4.5 mm long; sessile spikelets 6–7 mm long; awn 4–6 cm long · 11. *griffithii*
14. Awn 3–5.5 cm long, hirtellous with hairs 0.7–1.2 mm long; callus 1.8–3 mm long; superior raceme-base (4)4.5–8(10) mm long; pedicelled spikelet with an awn 1–5 mm long; spikelets glabrous or white-villous; racemes 2–4-awned per pair · · · · · · · · · · · · · · 14. *filipendula*
 – Awn 2.5–4 cm long, pubescent with hairs 0.1–0.6 mm long; callus 1–1.8 mm long; superior raceme-base 3.5–6 mm long; pedicelled spikelet with or without an awn-point up to 2 mm long; spikelets white-villous; racemes 4–6(7)-awned per pair · · · · · · · · · · · · · 13. *anamesa*
15. Homogamous pairs of spikelets at the base of the inferior raceme only; raceme-bases barbate · 16
 – Homogamous pairs at the base of both racemes · 31
16. Raceme-bases unappendaged, though sometimes with a scarious rim (if with a short appendage then racemes more than 4-awned per pair); superior raceme-base seldom over 1.5 mm long (Sect. *Pogonopodia*) · 17
 – Raceme-bases extended into a scarious appendage 0.5–4 mm long (if the appendage short or obscure then raceme-pairs 2-awned); racemes 2–4-awned per pair; superior raceme-base up to 3(4) mm long (Sect. *Hyparrhenia*) · 27

17. Pedicelled spikelets glabrous to puberulous (sometimes sparsely pilose in *schimperi*) ·· 18
 – Pedicelled spikelets villous · 24
18. Plant annual; spatheoles 1.8–3.2 cm long; awns 3.2–4.5 cm long · · · · · · 15. *anthistirioides*
 – Plants perennial · 19
19. Callus square or broader than long, broadly obtuse; spatheoles 0.8–1.8(2.1) cm long; awns
    3–5(6) per raceme-pair, up to 1.6(2) cm long; peduncle 3–8 mm long; plant robust, rarely
    slender · 16. *cymbaria*
 – Callus oblong to cuneate, obtuse or acute (sometimes square but then spatheole longer
    than above) · 20
20. Awns 3–5 per raceme-pair; spatheoles 1.4–2.4 cm long; awns 1.8–3.2 cm long; peduncles
    3–9 mm long; callus cuneate; culms stout · 17. *variabilis*
 – Awns 6 or more per raceme-pair in at least part of the false panicle, rarely fewer but then
    plant slender and rambling · 21
21. Plant slender and rambling; spatheoles 2–3 cm long; awns 0.7–1.7 cm long; peduncles 9–30
    mm long, often sinuous, the racemes commonly exserted; callus square to oblong, broadly
    rounded · 18. *pilgeriana*
 – Plant robust · 22
22. Awns 10–25 per raceme-pair, seldom over 2 cm long; spatheoles 2.5–5 cm long; peduncles
    15–50 mm long; racemes dense, the internodes 1.5–2 mm long · · · · · · · · · · 21. *dregeana*
 – Awns 6–8 per raceme-pair, sometimes more but then the awns over 2.5 cm long · · · · · 23
23. Awns 0.8–1.8 cm long; spatheoles 1.8–2.6 cm long; peduncles 2–10 mm long; callus square
    to cuneate · 19. *formosa*
 – Awns 2–3.4 cm long; spatheoles 2.2–3.2 cm long; peduncles 10–15 mm long; callus cuneate
    · 20. *schimperi*
24. Basal leaf sheaths pubescent to tomentose, the plant densely caespitose; spatheoles 2.5–5
    cm long; awns up to 2 cm long, rarely more; pedicelled spikelets muticous or with a short
    awn-point · 25
 – Basal leaf sheaths glabrous; awns 4–7 per raceme-pair · 26
25. Awns 10–25 per raceme-pair · 21. *dregeana*
 – Awns 5–8 per raceme-pair · 22. *tamba*
26. Culms robust, supported by stilt-roots; awns 2.2–4 cm long; pedicelled spikelets usually with
    a bristle-like awn 2–6 mm long · 23. *rudis*
 – Culms slender, without stilt-roots, arising in clumps from a short rhizome; awns 1.5–2.5 cm
    long; pedicelled spikelets with a short awn-point 1–3 mm long · · · · · · · · · · · · 24. *collina*
27. Annuals; false panicle loose and leafy; awns 4–10.5 cm long · · · · · · · · · · · · · · · · · · · 28
 – Perennials; awns 1–5.5 cm long · 30
28. Peduncle barbate on one side near the apex, deflexed at maturity; raceme-base appendage
    obscure or absent; superior raceme-base 2–2.5 mm long, subterete, bearing a few scattered
    hairs, not deflexed · 25. *madaropoda*
 – Peduncle uniformly barbate near the apex, more or less straight; raceme-base appendage
    distinct; superior raceme-base 1–1.5 mm long, flat, densely barbate, deflexed at maturity
    · 29
29. Sessile spikelets 7–11 mm long; pedicelled spikelets 8–13 mm long · · · · · · · 26. *niariensis*
 – Sessile spikelets 5–7 mm long; pedicelled spikelets 6–8 mm long · · · · · · · · 27. *welwitschii*
30. Pedicel-tooth obtusely triangular, obscure; sessile spikelets 4–6 mm long, their awn 1–2.5
    cm long; pedicelled spikelets muticous or with a mucro up to 1 mm long; false panicle
    narrow, dense · 28. *bracteata*
 – Pedicel-tooth subulate, 0.2–1.5 mm long; sessile spikelets 6–10 mm long, their awn 2.2–5.5
    cm long; pedicelled spikelets with a bristle 1–5 mm long; false panicle loose, often scanty
    · 29. *newtonii*
31. Homogamous spikelets pectinate-ciliate on the margins, 1 pair on each raceme; peduncle
    densely barbate with yellow hairs above; raceme-bases subequal (Sect. *Arrhenopogonia*) · · ·
    · 30. *anemopaegma*
 – Homogamous spikelets scabrid on the margins, 2 pairs on each raceme, if only 1 pair then
    peduncle glabrous or shortly hirsute above and superior raceme-base much longer than the
    inferior (Sect. *Apogonia*) · 32
32. Column of awn with hairs 0.5–1.7 mm long, the awn 4.5–7.5 cm long; peduncles 1–3.5 cm
    long; spatheoles 3–7 cm long; pedicelled spikelets with a bristle 2–7 mm long · · · · · · · · ·
    · 31. *subplumosa*
 – Column of awn with hairs 0.2–0.4(0.5) mm long, the awn 2–5.5 cm long, rarely the awn
    quite absent · 33

33. Homogamous pairs 2 at the base of each raceme; superior raceme-base 1–2 mm long; racemes 1.5–2(2.5) cm long, (3)4–6(9)-awned per pair; peduncles 0.3–1.5 cm long; spatheoles 2–4.5 cm long · 32. *diplandra*
– Homogamous pairs 1 at the base of each raceme; superior raceme-base 3–4 mm long; racemes 2.5–3 cm long, 6–12-awned per pair; peduncles 1.5–3 cm long; spatheoles 4–7 cm long · 33. *gossweileri*

1. **Hyparrhenia glabriuscula** (Hochst. ex A. Rich.) Andersson ex Stapf in Prain, F.T.A. **9**: 372 (1919). —Clayton in Kew Bull., Addit. Ser. 2: 47 (1969); in F.W.T.A., ed. 2, **3**: 491 (1972). —Clayton & Renvoize in F.T.E.A., Gramineae: 792 (1982). TAB. **31**. Type from Ethiopia.
    Andropogon glabriusculus Hochst. ex A. Rich., Tent. Fl. Abyss. **2**: 468 (1851).
    Sorghum glabriusculum (Hochst. ex A. Rich.) Kuntze, Revis. Gen. Pl. **2**: 791 (1891).
    Hyparrhenia amoena Jacq.-Fél. in J. Agric. Trop. Bot. Appl. **1**: 46 (1954). Type from Senegal.
    Hyparrhenia gazensis sensu Jackson & Wiehe, Annot. Check List Nyasal. Grass.: 44 (1958), in part, non (Rendle) Stapf.

Caespitose perennial; culms up to 200 cm high, erect. Leaf sheaths glabrous; leaf laminas up to 30 cm × 5 mm. False panicle 15–30 cm long, narrow, congested; spatheoles linear-lanceolate, 2–3 cm long, green tinged with russet; peduncles ± half as long as the spatheole, glabrous or with a few hairs towards the summit; racemes 1.5–2.5 cm long, 5–7-awned per pair, seldom deflexed, laterally exserted; raceme-bases unequal, flattened, the superior 2.5–3 mm long, glabrous or sparsely ciliate on the margins, with a short scarious rim at the apex. Homogamous spikelets 4–6 mm long, a single pair at the base of the inferior raceme only, glabrous. Sessile spikelet c. 5 mm long; callus c. 0.5 mm long, square to oblong or almost semicircular, broadly obtuse; inferior glume lanceolate, glabrous; awns 1.5–2.5 cm long, the column minutely puberulous. Pedicelled spikelets c. 5 mm long, narrowly lanceolate, glabrous, acuminate; pedicel-tooth short, broadly triangular.

Malawi. C: Msekwe, near Salima, 26.iv.1951, *Jackson* 478 (K). **Mozambique**. T: Angónia Distr., margens do rio Mauè, 12.v.1948, *Mendonça* 4203 (K; LISC).
Also in Senegal to Cameroon, with scattered records from Ethiopia and Tanzania. Growing in seasonally swampy soils; c. 500 m.
The species bears a superficial resemblance to *Cymbopogon*, but the insertion of the callus is typical of *Hyparrhenia*, the inferior glume of the sessile spikelet is 2-keeled only at the apex, and the leaves have no aromatic oils.

2. **Hyparrhenia mobukensis** (Chiov.) Chiov. in Nuovo Giorn. Bot. Ital., n.s. **26**: 74 (1919). —Clayton in Kew Bull., Addit. Ser. 2: 48 (1969). —Simon in Kirkia **8**: 51 (1971). —Clayton & Renvoize in F.T.E.A., Gramineae: 793 (1982). TAB. **32**. Type from Uganda.
    Andropogon mobukensis Chiov. in Ann. Bot. (Rome) **6**: 147 (1907). —Stapf in Prain, F.T.A. **9**: 380 (1919).
    Andropogon scaettai Robyns in Bull. Jard. Bot. État **8**: 223 (1930). Type from Dem. Rep. Congo.
    Hyparrhenia absimilis Pilg. in Notizbl. Bot. Gart. Berlin-Dahlem **14**: 102 (1938). Type from Tanzania.
    Hypogynium absimile (Pilg.) Roberty in Boissiera **9**: 189 (1960).
    Cymbopogon tenuis Gilli in Ann. Naturhist. Mus. Wien **69**: 37 (1966). Type from Tanzania.

Trailing perennial; culms up to 1.6 m long, wiry, very slender. Leaf sheaths glabrous; leaf laminas 4–8 cm × 2–5 mm, linear-lanceolate, light green, flaccid. False panicle scanty, comprising up to 5 distant raceme-pairs (rarely the racemes single); spatheoles 4–5 cm long, linear-lanceolate, light green tinged with purple; peduncles mostly a little longer than the spatheoles, sometimes up to twice as long, rarely a little shorter, flexuous, scaberulous or pilose above; racemes 1.5–3 cm long, 7–15-awned per pair, not deflexed; raceme-bases unequal, the superior 1.5–3 mm long, terete or somewhat flattened, softly pilose or sometimes stiffly bristly, with a very short scarious rim at the apex. Homogamous spikelets 5–6 mm long, a single pair at the base of the inferior raceme only. Sessile spikelet 4–5 mm long; callus c. 0.5 mm long, square or semicircular, broadly obtuse; inferior glume lanceolate, glabrous or scaberulous, distinctly 9-nerved; awns 7–12 mm long, the column glabrous. Pedicelled spikelets c. 6 mm long, lanceolate, glabrous, acuminate; pedicel-tooth absent.

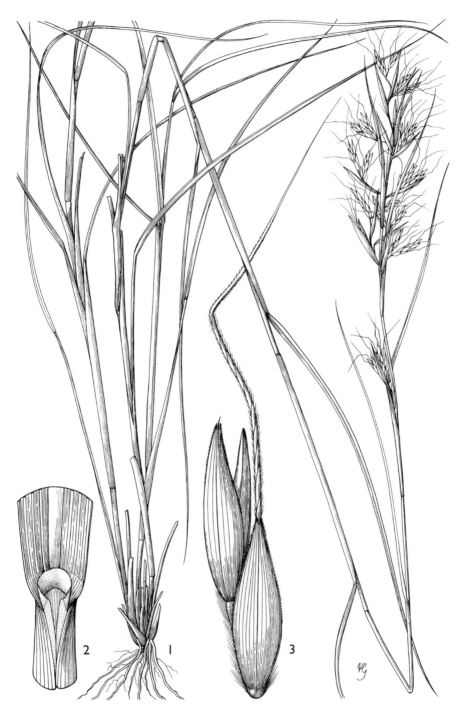

Tab. 31. HYPARRHENIA GLABRIUSCULA. 1, habit (× ¹/₂); 2, ligule (× 8); 3, spikelet pair (× 8). Drawn by Victoria Goaman. From Ghana Gr.

Tab. 32. HYPARRHENIA MOBUKENSIS. 1, inferior portion of plant (× ½), from *Kerfoot* 1953; 2, upper portion of culm (× ⅔); 3, raceme pair (× 3); 4–6 sessile spikelet: 4, inferior glume, outer view (× 6); 5, inferior glume, inner view (× 6); 6, superior lemma and awn (× 6), 2–6 from *E.A. Robinson* 3080. Drawn by Mary Grierson. From Kew Bull.

Zambia. N: Chama Distr., Nyika Plateau, Chire R., ix.1968, *G. Williamson* 1028 (K; SRGH).
Malawi. N: Chitipa Distr., 8 km east of Nganda, by tributary of Wovwe R., 1980 m, 2.viii.1972, *Brummitt, Munthali & Synge* WC128 (K).
Also in Uganda, Kenya, Tanzania and Dem. Rep. Congo. Growing in evergreen thickets, bogs and wet ground by rivers and waterfalls; 1900–2100 m.
The superior raceme-base is confusingly rather variable, ranging from terete to distinctly flattened. The species is best recognized by its trailing habit, broadly obtuse callus, often single racemes and the glabrous awn (it is the only species in the genus to have such an awn).

3. **Hyparrhenia nyassae** (Rendle) Stapf in Prain, F.T.A. **9**: 313 (1919). —Stent & Rattray in Proc. & Trans. Rhodesia Sci. Assoc. **32**: 13 (1933). —Sturgeon in Rhodesia Agric. J. **51**: 134 (1954). —Chippindall in Meredith, Grasses & Pastures S. Africa: 511 (1955). —Jackson & Wiehe, Annot. Check List Nyasal. Grass.: 45 (1958). —Clayton in Kew Bull., Addit. Ser. 2: 53 (1969). —Simon in Kirkia **8**: 18, 51 (1971). —Clayton in F.W.T.A., ed. 2, **3**: 491 (1972). —Clayton & Renvoize in F.T.E.A., Gramineae: 793 (1982). —Gibbs Russell et al., Grasses South. Africa [Mem. Bot. Surv. S. Africa No. 58]: 186 (1990). Type: Malawi, *Buchanan* 1423 (K, isotype).
   Andropogon nyassae Rendle in J. Bot. **31**: 358 (1893).
   Andropogon rufus var. *auricomus* Pilg. in Bot. Jahrb. Syst. **30**: 268 (1901). Type from Tanzania.
   Cymbopogon chrysargyreus Stapf in J. Bot. (Morot), sér. 2, **2**: 213 (1909). Type from Central African Republic.
   Andropogon chrysargyreus (Stapf) Stapf in A. Chevalier, Sudania **1**: 77 (1911).
   Andropogon lasiobasis Pilg. in Fries, Wiss. Ergebn. Schwed. Rhod.-Kongo-Exped. **1**: 197 (1916). Type: Zambia, Luwingu, x.1911, *Fries* 1089 (UPS, holotype).
   Cymbopogon nyassae (Rendle) Pilg. in Bot. Jahrb. Syst. **54**: 287 (1917).
   Hyparrhenia chrysargyrea (Stapf) Stapf in Prain, F.T.A. **9**: 312 (1919).
   Hyparrhenia vulpina Stapf in Prain, F.T.A. **9**: 310 (1919). —Stent & Rattray in Proc. & Trans. Rhodesia Sci. Assoc. **32**: 13 (1933). —Sturgeon in Rhodesia Agric. J. **51**: 134 (1954). Type from Angola.
   Cymbopogon vanderystii De Wild. in Bull. Jard. Bot. État **6**: 24 (1919). Type from Dem. Rep. Congo.
   Hyparrhenia vanderystii (De Wild.) Vanderyst in Bull. Agric. Congo Belge **11**: 144 (1920).
   Hyparrhenia rufa sensu Stent & Rattray in Proc. & Trans. Rhodesia Sci. Assoc. **32**: 13 (1933), in part, non (Nees) Stapf.
   Trachypogon spicatus sensu Jackson & Wiehe, Annot. Check List Nyasal. Grass.: 63 (1958), in part, non (L.f.) Kuntze.

Caespitose perennial; culms up to 200 cm high and 1–4 mm in diameter at the base. Leaf sheaths at the base of the plant mostly woolly-tomentose with white hairs, occasionally merely pubescent; leaf laminas up to 45 cm × 2–5 mm, rigid, often coarsely pilose near the base. False panicle 15–45 cm long, lax; spatheoles 3–6 cm long, linear, russet-coloured, eventually inrolled; peduncles usually longer than the spatheoles, pilose above with spreading white hairs; racemes 2–3 cm long, 8–13-awned per pair, with white-pilose internodes and pedicels, terminally exserted on the flexuous peduncle, tardily and imperfectly deflexed; raceme-bases unequal, the superior 2–3 mm long, terete, glabrous or sometimes thinly pilose, the articulation with the peduncle apex marked by a ring of white hairs or these sometimes brown or absent. Homogamous spikelets 5–6 mm long, a single pair at the base of the inferior raceme only. Sessile spikelets 5–6 mm long; callus 0.8–1.2 mm long, linear to narrowly cuneate, acute to very narrowly truncate at the apex; inferior glume narrowly lanceolate, yellowish-green to violet, densely pubescent with fulvous hairs; awn 2–4 cm long, the column fulvously pubescent with hairs 0.2–0.5 mm long. Pedicelled spikelets 4.5–7 mm long, linear-lanceolate, acute and muticous at the apex; callus absent; pedicel-tooth 0.2–0.5 mm long, subulate.

Botswana. N: 40 km NE of Maun, 12.vi.1930, *van Son* in Herb. Tvl. Mus. 28619 (K). **Zambia**. B: Kaoma Distr., 56 km along Kafue Watershed, 2.iv.1964, *Verboom* 1368 (K). N: Mbala Distr., Luombe River, 103 km from Mbala (Abercorn) on Kasama road, 1370 m, 15.iv.1958, *Vesey-FitzGerald* 1648 (K). W: Mwinilunga Distr., Matonchi, 1370 m, 21.xii.1969, *Simon & Williamson* 1936 (K). C: Chisamba–Kabwe (Broken Hill) road, 1220 m, 5.vi.1932, *Trapnell* 2082 (K). E: near Chadiza, 850 m, 27.xi.1958, *Robson* 754 (K). S: Choma Distr., Choma National Forest (Siamambo Forest Reserve), 4.ii.1960, *White* 6806 (K). **Zimbabwe**. N: Guruve Distr., Nyamunyeche Estate, near Great Dyke East, 4.i.1979, *Nyariri* 616 (K; PRE). W: Matobo Distr., Besna Kobila Farm, 1460 m, xii.1954, *Miller* 2578 (K). C: Chegutu Distr., Poole Farm, 1190 m, 7.iv.1954, *Wild* 4547 (K; PRE). E: Nyanga Distr., St. Swithin's C.L. (Tribal Trust Land), Mica

(Mika) Hill, 20.iv.1972, *Wild & Pope* 7943 (K; PRE). S: Masvingo Distr., Makaholi Experimental Station, 15.i.1948, *Newton & Juliasi* 16 (K; PRE). **Malawi**. N: Chitipa Distr., Nyika Plateau, Rumphi Bridge, 2300 m, 16.iv.1975, *Pawek* 9257 (K). C: Dedza Distr., Chongoni Forestry College, base of Ciwawo (Chiwao) Hill, 1650 m, 4.ii.1959, *Robson* 1446 (K). S: neighbourhood of Zomba, c. 900 m, 1936, *Cormack* 472 (K). **Mozambique**. N: Marrupa Distr., andados 36 km de Maúa para Marrupa, c. 600 m, 19.ii.1964, *Torre & Paiva* 10677 (K; LISC). T: Angónia Distr., between Vila Coutinho and the frontier, 11.v.1948, *Mendonça* 4147 (K; LISC). MS: Gorongosa Distr., Parque Nacional da Gorongosa (Parque Nacional de Caça), ao km 1 do Acampamento de Chitengo para a entrada do Parque, c. 40 m, 11.xi.1963, *Torre & Paiva* 9170 (K; LISC).

Tropical Africa from Cameroon and Ethiopia southwards to South Africa; also in Thailand and Vietnam. Growing in open and wooded grassland, deciduous bushland, in moist places, and at the edges of swamps and dambos; 40–2300 m.

Usually, the species can be recognized quite readily by the felt of white hairs on the basal leaf sheaths. It merges with *H. rufa* in most other features, and if the basal parts are missing the two can be very difficult to distinguish. *H. nyassae*, however, has a generally longer and more pointed callus to the sessile spikelet.

At one time narrow-leaved tussocky specimens (*H. nyassae*) were separated from taller specimens with leafy culms (*H. vulpina*), but the two variants intergrade so completely that a distinction is no longer tenable.

4. **Hyparrhenia rufa** (Nees) Stapf in Prain, F.T.A. **9**: 304 (1919). —Stent & Rattray in Proc. & Trans. Rhodesia Sci. Assoc. **32**: 13 (1933), in part. —Sturgeon in Rhodesia Agric. J. **51**: 134 (1954). —Chippindall in Meredith, Grasses & Pastures S. Africa: 511 (1955). —Jackson & Wiehe, Annot. Check List Nyasal. Grass.: 45 (1958). —Clayton in Kew Bull., Addit. Ser. 2: 60 (1969). —Simon in Kirkia **8**: 18, 51 (1971). —Clayton in F.W.T.A., ed. 2, **3**: 492, t. 455 (1972). —Clayton & Renvoize in F.T.E.A., Gramineae: 794 (1982). —Gibbs Russell et al., Grasses South. Africa [Mem. Bot. Surv. S. Africa No. 58]: 187 (1990). TAB. **33**. Type from Brazil.

> *Trachypogon rufus* Nees, Agrost. Bras.: 345 (1829).
> *Andropogon xanthoblepharis* Trin. in Mém. Acad. Imp. Sci. St.-Pétersbourg, Sér. 6, Sci. Math. **2**: 281 (1832); Sp. Gram. **3**: t. 330 (1836). Type from Congo-Brazzaville.
> *Andropogon rufus* (Nees) Kunth, Enum. Pl. **1**: 492 (1833).
> *Andropogon altissimus* Hochst. ex A. Braun in Flora **24**: 277 (1841), non Raspail (1825), *nec* Colla (1836). Type from Ethiopia.
> *Andropogon fulvicomus* Hochst. in Schimper, Iter. Abyss. Sched. **2**: 1118 (1842). Type from Ethiopia.
> *Andropogon fulvicomus* var. *approximatus* Hochst., op. cit.: 928 (1842). Type from Ethiopia.
> *Hyparrhenia fulvicoma* (Hochst.) Andersson in Schweinfurth, Beitr. Fl. Aethiop.: 310 (1867), nom. inval., publ. gen. ante.
> *Andropogon rufus* var. *fulvicomus* (Hochst.) Hack. in Bol. Soc. Brot. **5**: 213 (1867).
> *Andropogon bouangensis* Franch. in Bull. Soc. Hist. Nat. Autun **8**: 333 (1895). Type from Congo-Brazzaville.
> *Cymbopogon rufus* (Nees) Rendle in Hiern et al., Cat. Afr. Pl. Welw. **2**: 155 (1899).
> *Cymbopogon rufus* var. *fulvicomus* (Hochst.) Rendle, loc. cit. (1899).
> *Cymbopogon rufus* var. *major* Rendle, loc. cit. (1899). Type from Angola.
> *Andropogon rufus* var. *glabrescens* Chiov. in Ann. Bot. (Rome) **8**: 288 (1907). Type from Eritrea.
> *Hyparrhenia rufa* var. *major* (Rendle) Stapf in Prain, F.T.A. **9**: 306 (1919).
> *Hyparrhenia altissima* Stapf in Prain, F.T.A. **9**: 307 (1919), based on *Andropogon altissimus* Hochst. ex A. Braun.
> *Hyparrhenia rufa* var. *fulvicoma* (Hochst.) Chiov. in Nuovo Giorn. Bot. Ital., n.s. **26**: 74 (1919).
> *Hyparrhenia hirta* (L.) Stapf var. *brachypoda* Chiov. in Atti Reale Accad. Italia, Mem. Cl. Sci. **11**: 63 (1940). Type from Ethiopia.
> *Hyparrhenia parvispiculata* Bamps in Bull. Jard. Bot. État **25**: 391 (1955). Type from Dem. Rep. Congo.

Caespitose perennial or sometimes annual; culms up to 250 cm high, erect. Leaf sheaths glabrous; leaf laminas 30–60 cm × 2–8 mm, rigid. False panicle 5–80 cm long, lax or contracted and fasciculate; spatheoles 3–5 cm long, linear-lanceolate, at length reddish and rolled about the peduncle; peduncles usually longer than the spatheoles by up to 2 cm or more, rarely shorter, glabrous or pilose above with spreading white hairs; racemes (1.5)2–2.5 cm long, (7)9–14-awned per pair, fulvously to rufously pilose or with the callus and lower part of the internode and pedicel white-hirsute, usually terminally exserted, not or very rarely deflexed; raceme-bases unequal, sometimes ± connate, the superior 2–3.5(4) mm long, terete, glabrous

Tab. 33. **HYPARRHENIA RUFA.** 1, habit; 2, ligule; 3, spikelet pair: a. internode of raceme, b. pedicel of pedicelled spikelet, c. sessile spikelet from back, d. pedicelled spikelet; 4, spikelet pair: a. pedicel of pedicelled spikelet, b. sessile spikelet from front, c. pedicelled spikelet. Drawn by W.E. Trevithick. From F.W.T.A.

(very rarely with a few long hairs), the articulation with the peduncle tip glabrous. Homogamous spikelets similar to the pedicelled, a single pair at the base of the inferior raceme or both racemes. Sessile spikelets (3)3.5–4.5(5) mm long; callus 0.2–0.8 mm long, short and round to cuneate and narrowly truncate at the apex; inferior glume lanceolate, yellowish-brown to reddish-brown often tinged with violet, rarely green or glaucous, glabrous to pubescent but typically with a scanty covering of stiff rufous hairs, usually glossy; awn 2–3 cm long, the column rufously pubescent. Pedicelled spikelets 3–5 mm long, narrowly lanceolate, acute or rarely mucronate at the apex; callus absent; pedicel-tooth up to 0.3 mm long, triangular.

Botswana. N: Ngamiland Distr., Linyanti River bank (18°23'S, 23°46'E), 8.iv.1982, *P.A. Smith* 3819 (K; PRE). SE: near Kgale (Khale), along main road to Lobatse from Gaborone, 31.iii.1989, *Camerik* 1000 (K; PRE). **Zambia**. B: Kaoma Distr., Luena Flats, 14.vi.1964, *Verboom* 1394 (K). N: Kasama Distr., Chibutubutu, Lukulu River, 16.iv.1958, *Vesey-FitzGerald* 1657 (K). W: Ndola, 1220 m, v.1961, *Wilberforce* 108 (K). C: Lusaka Distr., near Lilayi, road from Lusaka to Kafue, 1230 m, 1.iv.1964, *Symoens* 10973 (K). S: Kafue Flats near Mazabuka, 1000 m, iv.1932, *Trapnell* 1091 (K). **Zimbabwe**. N: Gokwe South Distr., Sengwa River off southern border of Charama Platcau, 1220 m, 3.v.1966, *Simon* 413 (K; PRE). W: Hwange Distr., Victoria Falls, 18.x.1958, *West* 3735 (K). C: Shurugwi Distr., access road to Selukwe Mine, 1300 m, 30.iv.1973, *Simon & Biegel* 2402 (K; PRE). E: Chimanimani Distr., Melsetter Pasture Res. sub-station, 1520 m, 19.ii.1950, *F.R. Williams* 77 (K). S: Mberengwa Distr., SW of Mt. Buhwa on road to Masvingo Mission, 1000 m, 1.v.1973, *Simon, Pope & Biegel* 2413A (K). **Malawi**. N: Mzimba Distr., road to Lunyangwa from Marymount, Mzuzu, 1360 m, 10.vi.1971, *Pawek* 4887 (K). C: Nkhotakota Distr., 5 km north of Nkhotakota (Nkhota Kota), 490 m, 17.vi.1970, *Brummitt* 11522 (K; MAL). S: Thyolo Distr., Bvumbwe Research Station (Tung Station), 7.xii.1949, *Wiehe* 371 (K). **Mozambique**. N: Marrupa Distr., 11 km along road from Marrupa to Nungo, c. 750 m, 5.viii.1981, *Jansen, de Koning & de Wilde* 77 (K). Z: Arredores de Namarrói (Nhamarroi), 25.ix.1941, *Torre* 3499 (K; LISC). T: Angónia Distr., Posto Zootécnico, 12.v.1948, *Mendonça* 4188 (K; LISC). MS: Parque Nacional da Gorongosa (Parque Nacional da Caça), andados 17 km do Acampamento de Chitengo para batelao do rio Urema, c. 40 m, 2.v.1964, *Torre & Paiva* 12216 (K; LISC; PRE).

Tropical and South Africa; introduced in tropical America and Australia. Growing in deciduous bushland and wooded grassland, usually preferring damper soils but also found on roadsides and in other disturbed areas; 0–1520 m.

A very widespread and extremely common species in tropical Africa. Not surprisingly, therefore, it is also extremely variable. It is usually perennial but some specimens are quite clearly annual; it ranges from 30 cm to 2.5 m in height, the smaller plants usually being associated with trampled or grazed areas. The degree of exsertion of the racemes varies with their position in the false panicle; those towards the periphery are usually more exserted than those nearer the centre. The raceme-bases are usually separate, but may be connate for part of their length, and the superior, while usually glabrous, may be pubescent or pilose. Callus and spikelet lengths also cover remarkably wide ranges. The species is indistinctly, but probably justifiably, distinguished from *H. dichroa*. It is, for the greater part of its range, readily enough recognized by its longer racemes with more awns per pair, its longer and narrower spatheoles that do not embrace the raceme, and the usually glabrous superior raceme-base. There is, however, an area of overlap where specimens are difficult to assign to *H. rufa*, *H. dichroa* or even *H. poecilotricha*.

5. **Hyparrhenia dichroa** (Steud.) Stapf in Prain, F.T.A. **9**: 302 (1919). —Stent & Rattray in Proc. & Trans. Rhodesia Sci. Assoc. **32**: 12 (1933). —Sturgeon in Rhodesia Agric. J. **51**: 134 (1954). —Jackson & Wiehe, Annot. Check List Nyasal. Grass.: 43 (1958). —Clayton in Kew Bull., Addit. Ser. 2: 68 (1969). —Simon in Kirkia **8**: 17, 51 (1971). —Clayton & Renvoize in F.T.E.A., Gramineae: 796 (1982). —Gibbs Russell et al., Grasses South. Africa [Mem. Bot. Surv. S. Africa No. 58]: 184 (1990). Type of *Andropogon bicolor* from South Africa.
    Andropogon bicolor Nees, Fl. Afr. Austral. Ill. **1**: 113 (1841), non Roxb. (1820).
    Andropogon dichroos Steud., Syn. Pl. Glumac. **1**: 389 (1854). —Hackel in A. & C. de Candolle, Monogr. Phan. **6**: 622 (1889). —Stapf in Dyer, F.C. **7**: 360 (1898). Based on *Andropogon bicolor* Nees.
    Sorghum bicolor (Nees) Kuntze, Revis. Gen. Pl. **2**: 790 (1891), non (L.) Moench (1794).
    Cymbopogon dichroos (Steud.) Pilg. in Bot. Jahrb. Syst. **54**: 286 (1917).
    Cymbopogon luembensis De Wild. in Bull. Jard. Bot. État **6**: 14 (1919). Type from Dem. Rep. Congo.
    Hyparrhenia luembensis (De Wild.) Robyns , Fl. Agrost. Congo Belge **1**: 183 (1929).
    Hyparrhenia fastigiata Robyns, tom. cit.: 164 (1929); in Bull. Jard. Bot. État **8**: 231 (1930). Type from Dem. Rep. Congo.
    Hyparrhenia gazensis sensu Jackson & Wiehe, Annot. Check List Nyasal. Grass.: 44 (1958), in part, non (Rendle) Stapf.

Hyparrhenia vulpina sensu Jackson & Wiehe, Annot. Check List Nyasal. Grass.: 45 (1958), non Stapf.

Caespitose perennial; culms up to 300 cm high, stout, erect. Leaf sheaths glabrous; leaf laminas up to 60 cm × 8 mm, rigid. False panicle 20–60 cm long, copiously branched; spatheoles 2–3.5 cm long, narrowly lanceolate, at length reddish; peduncles half as long to almost as long as the spatheoles, glabrous or pilose above with spreading white hairs; racemes 1–1.5(2) cm long, 6–9-awned per pair, glabrescent to fulvously pilose, embraced by the spatheole, terminally exserted, not or very rarely deflexed; raceme-bases unequal, the superior 1.5–2.5 mm long, terete, usually softly hirtellous, sometimes glabrous, the articulation with the peduncle tip glabrous. Homogamous spikelets similar to the pedicelled, a single pair at the base of the inferior raceme or both racemes. Sessile spikelets 4–5 mm long; callus 0.4–0.8 mm long, cuneate, obtuse; inferior glume lanceolate, yellowish-brown or violet-tinged, the hairs usually fulvous but often pallid and scanty, the back usually glossy; awn (1)2–3 cm long, the column rufously pubescent. Pedicelled spikelets 3–5 mm long, narrowly lanceolate, acute or mucronate (or with a bristle not more than 1 mm long) at the apex; callus absent; pedicel-tooth short, broadly triangular.

Botswana. N: Ngamiland Distr., 11 km south of Gumore on road to Nokaneng, 18.iii.1976, *Ellis* 2679 (K; PRE). **Zambia**. B: Kabompo Distr., 113 km south of Mwinilunga on Kabompo road, 1250 m, 25.xii.1969, *Simon & Williamson* 2023 (K; PRE). N: Mansa Distr., Mote area, road to Kapalala from Kasanka, Lumanya Dambo, 1.iii.1996, *Renvoize* 5745 (K). W: Ndola, 1250 m, v.1945, *Trapnell* 1900 (K). C: Kabwe Distr., Mutendere Game Ranch, Kamaila, 1173 m, 1.iii.1996, *Bingham* 10975 (K). S: Monze Distr., 10 km south of Pemba, 1220 m, 19.vii.1963, *Astle* 2609 (K). **Zimbabwe**. N: Zvimba Distr., Darwendale, 21.iii.1963, *Wild* 6091 (K). C: Harare (Salisbury), near Royal Salisbury Golf Course, 31.iii.1954, *Sturgeon* in *GHS* 57589 (K; PRE). E: Chipinge Distr., Rupise (Rupisi) Hot Springs, 26.vii.1960, *Wier* 7308 (K). S: Masvingo Distr., Mushandike River, 30.iii.1984, *Mahlangu* 1098 (K). **Malawi**. N: Karonga, 480 m, 15.vii.1970, *Brummitt* 12147 (K). C: Kasungu Distr., Lisasadzi Dambo, 23.xi.1951, *Jackson* 682 (K). S: Zomba, 940 m, 14.v.1949, *Wiehe* 105 (K). **Mozambique**. N: Mecúfi Distr., andados 15 km de Namapa para Pemba (Porto Amélia), c. 300 m, 6.iv.1964, *Torre & Paiva* 11653 (LISC). Z: Morrumbala Distr., M'bobo Hospital, 3.viii.1942, *Torre* 4475 (K; LISC). MS: Barué Distr., Catandica (Vila Gouveia), 27.x.1943, *Torre* 6093 (K; LISC).

Eastern tropical Africa southwards to South Africa, but more common towards the south. Growing in deciduous bushland and wooded grassland, especially on the moister soils; also in weedy places along roadsides and on old cultivated land; 50–1250 m.

The species is tolerably well distinguished from *H. rufa* but the relationship is extremely close and separation at species level is sometimes hard to justify. It differs from *H. rufa* in little more than the slightly shorter and broader spatheoles, the enclosed racemes and the hirtellous superior raceme-base.

6. **Hyparrhenia poecilotricha** (Hack.) Stapf in Prain, F.T.A. **9**: 309 (1919). —Clayton in Kew Bull., Addit. Ser. 2: 68 (1969). —Simon in Kirkia **8**: 18, 51 (1971). —Clayton in F.W.T.A., ed. 2, **3**: 492 (1972). —Clayton & Renvoize in F.T.E.A., Gramineae: 796 (1982). —Gibbs Russell et al., Grasses South. Africa [Mem. Bot. Surv. S. Africa No. 58]: 186 (1990). Type from Angola.

Andropogon poecilotrichus Hack. in Bol. Soc. Brot. **3**: 138 (1885); in A. & C. de Candolle, Monogr. Phan. **6**: 638 (1889).
Andropogon buchananii Stapf in Dyer, F.C. **7**: 362 (1898). Type from South Africa.
Andropogon pleiarthron Stapf in Dyer, F.C. **7**: 364 (1898). Type from South Africa.
Cymbopogon pleiarthron (Stapf) Stapf ex Burtt Davy in Ann. Transvaal Mus. **3**: 121 (1912).
Hyparrhenia buchananii (Stapf) Stapf ex Stent in Bothalia **1**: 249 (1923). —Chippindall in Meredith, Grasses & Pastures S. Africa: 511 (1955).
Hyparrhenia familiaris (Steud.) Stapf var. *pilosa* Robyns, Fl. Agrost. Congo Belge **1**: 176 (1929); in Bull. Jard. Bot. État **8**: 236 (1930). Type from Dem. Rep. Congo.

Perennial; culms up to 130 cm high. Leaf sheaths glabrous or very rarely loosely pilose; leaf laminas up to 30 cm × 3 mm, rigid. False panicle up to 30 cm long, lax and open; spatheoles 4–8 cm long, linear; peduncles about as long as the spatheoles, glabrous or with whitish or yellowish hairs towards the apex; racemes 1.5–2 cm long, 4–7-awned per pair, fulvously pilose, terminally exserted, not or tardily deflexed; raceme-bases unequal, the superior 3.5–7 mm long, terete, glabrous or rarely loosely and thinly pilose, the articulation with the peduncle glabrous. Homogamous spikelets similar to the pedicelled spikelets, 1 or very rarely

2 pairs at the base of the inferior raceme and 2 (or sometimes only 1) at the base of the superior, the number usually variable within the same panicle. Sessile spikelets 5.5–7 mm long; callus 1–2 mm long, acute to pungent; inferior glume narrowly lanceolate, pubescent with fulvous or yellow hairs; awn 2.5–4 cm long, the column rufously pubescent. Pedicelled spikelets 4–7 mm long, narrowly lanceolate, usually with an awn-point up to 2 mm long at the apex; callus absent; pedicel-tooth up to 0.2 mm long, triangular.

Zambia. B: Kabompo Distr., Kusokweji Dambo, 27 km north of Manyinga, 1130 m, 26.xii.1969, *Simon & Williamson* 2034 (K; PRE). N: Mporokoso Distr., 16 km WSW of Mporokoso, 13.iv.1961, *Phipps & Vesey-FitzGerald* 3142a (K). C: 8 km east of Lusaka, 1220 m, 24.ii.1954, *Best* 58 (K). S: Mazabuka, 1220 m, 31.i.1964, *Astle* 2925 (K). **Zimbabwe**. W: Matobo Distr., Matopos Research Station, 1370 m, 16.ii.1954, *Rattray* 1608 (K). C: Marondera Distr., Digglefold Vlei, 1520 m, 7.vi.1948, *Corby* 123 (K; SRGH). E: Nyanga Distr., Juliasdale, 1800 m, 16.iii.1975, *Crook* 2076 (K). **Malawi**. N: Mzimba Distr., Marymount, Mzuzu, 1370 m, 30.v.1971, *Pawek* 4881 (K). **Mozambique**. T: Angónia Distr., between Posto Zootécnico and the frontier, 12.v.1948, *Mendonça* 4187A (K). MS: entre Nhamatanda (Vila Machado) e Chimoio (Vila Pery), 22.ix.1943, *Torre* 5925 (K).

Tropical and South Africa, but mostly on the eastern side. Growing in deciduous bushland and in dambos; 1120–1800 m.

A variable species that may be of hybrid origin. It bridges the gap between *H. nyassae* and *H. rufa* on one hand, and *H. filipendula* on the other. The characters that separate these taxa are not wholly reliable, but those that work best for distinguishing *H. poecilotricha* from *H. rufa* are the long superior raceme-base, the longer sessile spikelet and the more pointed callus of the former. *H. filipendula* has glabrous or white-villous spikelets and racemes only 2–4-awned per pair.

7. **Hyparrhenia gazensis** (Rendle) Stapf in Prain, F.T.A. **9**: 301 (1919). —Stent & Rattray in Proc. & Trans. Rhodesia Sci. Assoc. **32**: 12 (1933). —Sturgeon in Rhodesia Agric. J. **51**: 134 (1954). —Chippindall in Meredith, Grasses & Pastures S. Africa: 513 (1955). —Clayton in Kew Bull., Addit. Ser. 2: 71 (1969). —Simon in Kirkia **8**: 18 (1971). —Clayton & Renvoize in F.T.E.A., Gramineae: 797 (1982). —Gibbs Russell et al., Grasses South. Africa [Mem. Bot. Surv. S. Africa No. 58]: 185 (1990). Type: Zimbabwe, Chirinda, *Swynnerton* 1637 (BM, holotype; K, isotype).

   Cymbopogon gazensis Rendle in J. Linn. Soc., Bot. **40**: 226 (1911).
   Andropogon gazensis (Rendle) Eyles in Trans. Roy. Soc. South Africa **5**: 295 (1916).
   Hyparrhenia snowdenii C.E. Hubb. in Bull. Misc. Inform., Kew **1928**: 38 (1928). Type from Uganda.

Loosely caespitose perennial; culms up to 180 cm high and up to 2 mm in diameter, weakly erect or geniculately ascending, sometimes sprawling. Leaf sheaths glabrous, or pubescent above; leaf laminas 8–20 cm × 2–5 mm, glabrous or thinly pilose, long-acuminate at the apex. False panicle 10–35 cm long, narrow, lax; spatheoles 3–4 cm long, linear-lanceolate, glabrous, reddish-brown; peduncles $^1/_2$–$^3/_4$ as long as the spatheoles, pilose above with white hairs; racemes 1–1.5 cm long, 4–5-awned per pair, laterally exserted, not or sometimes tardily deflexed; raceme-bases unequal, divergent, the superior (2)2.5–3.5 mm long, terete, stiffly pilose with white bulbous-based hairs, the articulation with the peduncle glabrous. Homogamous spikelets c. 5 mm long, narrowly lanceolate, a single pair at the base of the inferior or both racemes. Sessile spikelets 4.5–5 mm long; callus 0.8–1.5 mm long, cuneate, acute at the apex; inferior glume linear-oblong to lanceolate-oblong, sparsely hispidulous; awn 2–3 cm long, the column pubescent with fulvous hairs. Pedicelled spikelets 5–6 mm long, linear-lanceolate, glabrous, terminating in a short mucro 1–2 mm long; callus absent; pedicel-tooth short, broadly triangular.

Zambia. N: Samfya Distr., Lake Bangweulu, southern part, 1062 m, 13.ii.1996, *Renvoize* 5591 (K). **Zimbabwe**. E: Chimanimani Distr., Belmont West Farm, c. 16 km south of Chimanimani (Melsetter), 16.i.1954, *Crook* 501 (K; PRE). S: Masvingo Distr., Great Zimbabwe, 19.x.1930, *Fries, Norlindh & Weimarck* 2074 (K).

Also in Uganda, Kenya, Tanzania, Dem. Rep. Congo and South Africa (Transvaal). Growing in flooded grassland, *Brachystegia* woodland, and as a ruderal on poor soils; 1000–1100 m.

The species closely resembles *H. dichroa* in overall appearance, but is distinguished by its weak culms, fewer awns per raceme-pair, almost glabrous spikelets and longer superior raceme-base clothed in stiff hairs. It differs little from *H. finitima* except in its weaker habit, slightly shorter and broader callus and slightly longer superior raceme-base.

GRAMINEAE

8. **Hyparrhenia finitima** (Hochst.) Andersson ex Stapf in Prain, F.T.A. **9**: 299 (1919). —Jackson & Wiehe, Annot. Check. List Nyasal. Grass.: 44 (1958). —Clayton in Kew Bull., Addit. Ser. 2: 72 (1969). —Simon in Kirkia **8**: 18, 51 (1971). —Clayton in F.W.T.A., ed. 2, **3**: 492 (1972). —Clayton & Renvoize in F.T.E.A., Gramineae: 797 (1982). —Gibbs Russell et al., Grasses South. Africa [Mem. Bot. Surv. S. Africa No. 58]: 185 (1990). Type from Ethiopia.
Andropogon finitimus Hochst. in Schimper, Iter. Abyss. Sched. **2**: 1797 (1844); in A. Richard, Tent. Fl. Abyss. **2**: 465 (1851). —Hackel in A. & C. de Candolle, Monogr. Phan. **6**: 637 (1889).
Cymbopogon finitimus (Hochst.) Thomson in Speke, J. Discov. Source Nile: 652 (1863).
Sorghum finitimum (Hochst.) Kuntze, Revis. Gen. Pl. **2**: 791 (1891).
Hyparrhenia rhodesica Stent & Rattray in Proc. & Trans. Rhodesia Sci. Assoc. **32**: 14 (1933). —Sturgeon in Rhodesia Agric. J. **51**: 136 (1954). Type: Zimbabwe, *Perrott* in *Herb. Eyles* 3069 (K, isotype).
Hyparrhenia hirta var. *garambensis* Troupin, Fl. Garamba **1**: 47 (1956). Type from Dem. Rep. Congo.
Hyparrhenia gazensis sensu Jackson & Wiehe, Annot. Check List Nyasal. Grass.: 44 (1958), in part, non (Rendle) Stapf.

Caespitose perennial; culms up to 200 cm high and up to 6 mm in diameter, robust, erect. Leaf sheaths at the base of the plant pubescent to hirsute along the margins and sometimes also on the back; leaf laminas up to 60 cm × 8 mm. False panicle up to 60 cm long, copiously branched and usually rather dense; spatheoles 2.5–4 cm long, narrowly lanceolate, enclosing the racemes; peduncles $^1/_3$–$^2/_3$ the length of the spatheoles, pilose above with white hairs; racemes 1–1.5 cm long, 2–6-awned per pair, yellowish, laterally exserted, not deflexed; raceme-bases unequal, the superior 1.5–2.5 mm long, terete, softly hirtellous to pilose with white tubercle-based hairs, glabrous at the foot, the articulation with the peduncle glabrous. Homogamous spikelets 6–8 mm long, narrowly lanceolate, a single pair at the base of the inferior or both racemes. Sessile spikelet 5.5–6 mm long; callus 1–2 mm long, linear, acute to very narrowly truncate at the apex; inferior glume yellowish, narrowly lanceolate, glabrous to hispidulous with short white hairs; awn 2.5–4 cm long, the column fulvously puberulous. Pedicelled spikelets 5–6 mm long, narrowly lanceolate, glabrous or rarely hispidulous, terminating in an awnlet 2–5 mm long; callus absent; pedicel-tooth up to 0.2 mm long, subulate.

Zambia. N: Chinsali Distr., 25.v.1964, *Astle* 3054 (K). W: Mwinilunga Distr., east of Matonchi Farm, 11.xi.1937, *Milne-Redhead* 3198 (K; PRE). C: Chisamba–Kabwe (Broken Hill) road, 1220 m, 14.vi.1932, *Trapnell* 2034 (K). E: Chipata (Fort Jameson), near St. Paul's Church, 30.viii.1929, *Burtt Davy* 1023 (K). S: Choma Distr., Mochipapa, 1280 m, 25.iv.1963, *Astle* 2993 (K). **Zimbabwe**. N: Guruve Distr., Rukovakuona (Rukowakuona) Mt., 18.x.1962, *Wild* 5905 (K; PRE). C: Harare (Salisbury), near Royal Salisbury Golf Course, 31.iii.1954, *Sturgeon* in *GHS* 57586 (K; PRE). E: Mutare Distr., c. 32 km south of Odzi, 1370 m, 28.iii.1966, *Simon* 774 (K; SRGH). S: Mberengwa Distr., SW of summit of Mt. Buhwa, on road to Masvingo Mission, 1000 m, 1.v.1973, *Simon, Pope & Biegel* 2413 (K). **Malawi**. N: Mbawa, near Mzimba, 7.iv.1955, *Jackson* 1610 (K). C: Lilongwe Agric. Research Station, 20.iv.1951, *Jackson* 460 (K; PRE). S: Chiradzulu, 915 m, 17.iii.1950, *Wiehe* 443 (K). **Mozambique**. MS: between Catandica (Vila Gouveia) and Macossa, 1.vii.1941, *Torre* 2962B (LISC). M: Namaacha Hills, 3.ix.1940, *M.G. Hornby* 1052 (K).

Sierra Leone and Ethiopia southwards to South Africa, but mainly in the southern part of the range. Growing in deciduous bushland and wooded grassland, or on disturbed soils in farmland and along roadsides; 600–1370 m.

Closely allied to *H. gazensis* from which it can be distinguished by its robust culms, slender callus, more positively awned pedicelled spikelets and shorter raceme-bases.

9. **Hyparrhenia hirta** (L.) Stapf in Prain, F.T.A. **9**: 315 (1919). —Chippindall in Meredith, Grasses & Pastures S. Africa: 510 (1955). —Clayton in Kew Bull., Addit. Ser. 2: 75 (1969). —Simon in Kirkia **8**: 18, 51 (1971). —Clayton in F.W.T.A., ed. 2, **3**: 492 (1972). —Clayton & Renvoize in F.T.E.A., Gramineae: 798 (1982). —Gibbs Russell et al., Grasses South. Africa [Mem. Bot. Surv. S. Africa No. 58]: 185 (1990). TAB. **34**. Type from Italy.
Andropogon hirtus L., Sp. Pl. **2**: 1046 (1753).
Andropogon podotrichus Hochst. in Schimper, Iter. Abyss. Sched. **2**: 1056 (1842). Type from Ethiopia.
Cymbopogon hirtus (L.) Thomson in Speke, J. Discov. Source Nile: 652 (1863).
Hyparrhenia podotricha (Hochst.) Andersson in Schweinfurth, Beitr. Fl. Aethiop.: 310 (1867), *nom. inval.*, *publ. gen. ante.*
Andropogon hirtus var. *podotrichus* (Hochst.) Hack. in A. & C. de Candolle, Monogr. Phan. **6**: 620 (1889).

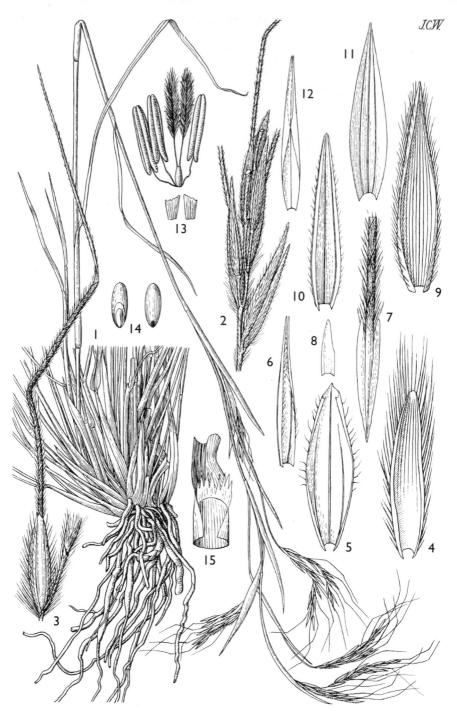

Tab. 34. HYPARRHENIA HIRTA. 1, habit (× ²/₃); 2, portion of inflorescence, showing the lower homogamous spikelets (× 5 ¹/₂); 3, spikelet pair (× 5 ¹/₂); 4–8 sessile spikelet: 4, inferior glume (× 8); 5, superior glume (× 8); 6, inferior lemma (× 8); 7, superior lemma with base of awn (× 8); 8, its palea (× 8); 9–15 pedicelled spikelet: 9, inferior glume (× 8); 10, superior glume (× 8); 11, lemma (× 8); 12, palea (× 8); 13, flower (× 8); 14, grain (× 4); 15, ligule (× 4), 1–15 from *Gillett* 8232. Drawn by J.C. Webb. From Fl. Iraq.

Andropogon transvaalensis Stapf in Dyer, F.C. **7**: 363 (1898). Type from South Africa.
Cymbopogon transvaalensis (Stapf) Burtt Davy in Ann. Transvaal Mus. **3**: 122 (1912).
Cymbopogon modicus De Wild. in Bull. Jard. Bot. État **6**: 16 (1919). Type from Dem. Rep. Congo.
Hyparrhenia modica (De Wild.) Robyns, Fl. Agrost. Congo Belge **1**: 172 (1929).
Hyparrhenia hirta var. *podotricha* (Hochst.) Pic. Serm., Ric. Bot. Lago Tana **1**: 174 (1951).

Caespitose perennial arising from short rhizomes; culms up to 60 cm high (up to 1 m in exceptionally robust specimens), wiry, standing above a dense leafy tussock 10–20 cm high. Leaf sheaths compressed and keeled, glabrous or rarely obscurely puberulous; leaf laminas 2–15(30) cm × 1–2(4) mm, narrowly linear to conduplicate and filiform, flexuous, glaucous, harshly scaberulous. False panicle up to 30 cm long, typically scanty with only 2–10 raceme-pairs but sometimes a little fuller with more raceme-pairs; spatheoles 3–8 cm long, linear-lanceolate, at length reddish; peduncles about as long as the spatheoles, glabrous or with white bulbous-based hairs above; racemes 2–4 cm long, 8–13(16)-awned per pair, white-villous, terminally exserted, never deflexed; raceme-bases unequal, the superior 2.5–5 mm long, filiform, glabrous or more usually pubescent to hirsute, with or without a white beard at the foot. Homogamous spikelets similar to the pedicelled, a single pair at the base of the inferior or both racemes. Sessile spikelets 4–6.5 mm long; callus 0.5–1.5 mm long, subacute to acute; inferior glume linear-elliptic, yellowish-green to violet, white-villous but occasionally the hairs rather sparse; awn 10–35 mm long, puberulous with white hairs 0.1–0.3 mm long. Pedicelled spikelets 3–7 mm long, narrowly lanceolate, white-villous, acute and muticous at the apex; callus absent; pedicel-tooth 0.2–1 mm long, subulate.

Botswana. N: east slope of Ntwetwe Pan in Makgadikgadi Pans (Makgadikgadi Depression), 15.2 km north of main Maun–Francistown road at 20°01'S, 25°48'E, 16.iv.1978, *P.A. Smith* 2392 (K; PRE). SE: 9 km east of Kanye at 24°57'S, 25°22'E, 1200 m, 15.iv.1978, *Hansen* 3411 (K; PRE). **Zambia**. W: Mwinilunga Distr., Chitunta R., 29 km north of Mwinilunga, 1400 m, 23.xii.1969, *Simon & Williamson* 1968 (K). C: Serenje Distr., Lusiwashi Dambo near Kanona, 1430 m, 6.iv.1961, *Phipps & Vesey-FitzGerald* 2965 (K; SRGH). **Zimbabwe**. E: Nyanga Distr., Bonda Mission, 10.xii.1967, *Biegel* 2365 (K; SRGH).
Mainly in the Mediterranean region and NE tropical Africa, and extending eastwards through Arabia and SW Asia to Pakistan. Absent from much of the rest of tropical Africa, except for isolated records from Niger and Angola as well as the Flora Zambesiaca area. Reappearing in South Africa and probably introduced in Australia and Central America. Growing mostly in upland dambos; 1200–1700 m.
A widespread and variable species which tends to intergrade with several others. It is best recognized by its scanty panicle of white-villous racemes which never deflex, by the many-awned racemes with 0–1 homogamous pairs at the base of the superior, and by the harsh narrow leaves forming a basal tussock.

10. **Hyparrhenia quarrei** Robyns, Fl. Agrost. Congo Belge **1**: 171 (1929); in Bull. Jard. Bot. État **8**: 234 (1930). —Clayton in Kew Bull., Addit. Ser. 2: 82 (1969). —Simon in Kirkia **8**: 18, 51 (1971). —Clayton in F.W.T.A., ed. 2, **3**: 492 (1972). —Clayton & Renvoize in F.T.E.A., Gramineae: 799 (1982). —Gibbs Russell et al., Grasses South. Africa [Mem. Bot. Surv. S. Africa No. 58]: 186 (1990). Type from Dem. Rep. Congo.

Caespitose perennial; culms up to 200 cm high. Leaf sheaths at the base of the plant usually white-pubescent, sometimes glabrous, rarely tomentose; leaf laminas up to 40 cm × 5 mm, rigid, glaucous, harsh. False panicle c. 30 cm long, narrow, moderately dense; spatheoles 3–5 cm long, linear, russet coloured; peduncles flexuous, a little longer than the spatheoles, with spreading white tubercle-based hairs above; racemes 1.5–2 cm long, 6–10-awned per pair, white-hirsute, terminally exserted, at least some of them deflexed; internodes 2–3 mm long; raceme-bases unequal, the superior 2–3.5 mm long, hirsute, sometimes with a few stiff bristles or sometimes glabrous. Homogamous spikelets similar to the pedicelled spikelets, a single pair at the base of the inferior or both racemes. Sessile spikelets 4.5–5.5 mm long; callus 0.7–1.2 mm long, linear to narrowly cuneate; inferior glume narrowly lanceolate, pubescent to villous with white hairs; awn 1.8–3.6 cm long, the column pubescent with hairs 0.2–0.5 mm long. Pedicelled spikelets 5–7 mm long, linear-lanceolate, pubescent to villous, acute and muticous at the apex; callus absent; pedicel-tooth 0.1–0.5 mm long, subulate.

Zambia. N: Kawambwa Distr., Chishinga Ranch, 1430 m, 14.v.1964, *Astle* 3022 (K). W: Ndola, 1250 m, v.1945, *Trapnell* 1901 (K). S: Choma Distr., Mochipapa, 25.iv.1964, *Astle* 2991 (K). **Zimbabwe**. C: Chegutu Distr., Mondoro C.L., vi.1949, *Davies* 32696 (K). **Malawi**. N: Mzimba Distr., Marymount, Mzuzu, 1370 m, 2.vi.1973, *Pawek* 6787 (K). C: Lilongwe Distr., Dzalanyama Forest Reserve, 27.iv.1958, *Jackson* 2220 (K). **Mozambique**. T: Angónia Distr., Posto Zootécnico, 12.v.1948, *Mendonça* 4185 (K; LISC).

Throughout tropical and South Africa, but mainly in the southern half of the continent and rather scattered elsewhere; probably introduced in Australia. Growing in deciduous bushland, wooded grassland and dambos, at high altitude; 1230–1600 m.

The species sits within a cluster of others and may represent introgression products between them. It is most similar to *H. hirta*, differing only by the deflexion of some (but by no means all) racemes, the more robust habit and, less reliably, the pubescent basal sheaths.

11. **Hyparrhenia griffithii** Bor in Indian Forest Rec., Bot. **1**: 92 (1938). —Clayton in Kew Bull., Addit. Ser. 2: 84 (1969). —Simon in Kirkia **8**: 51 (1971). —Clayton & Renvoize in F.T.E.A., Gramineae: 799 (1982). Type from India.

Caespitose perennial; culms up to 200 cm high. Leaf sheaths glabrous or sparsely pilose; leaf laminas up to 40 cm × 8 mm. False panicle up to 35 cm long, narrow, lax; spatheoles 4–7 cm long, linear, convolute; peduncle from $^2/_3$ as long as to slightly longer than, the spatheole, pilose with white hairs above; racemes 1.5–3.5 cm long, 5–10-awned per pair, white-hirsute, strongly deflexed; internodes 3.5–4.5 mm long; raceme-bases unequal, the superior 3.5–8 mm long, terete, glabrous. Homogamous spikelets similar to the pedicelled spikelets, a single pair at the base of each raceme. Sessile spikelets 6–7 mm long; callus 1.5–2 mm long, pungent; inferior glume lanceolate, brownish to dark violet, white-hirsute; awn 4–6 cm long, the column pubescent with rufous hairs 0.4–0.6 mm long. Pedicelled spikelets 6–8 mm long, linear-lanceolate, villous, with or without a terminal awnlet up to 4 mm long; callus absent; pedicel-tooth c. 0.5 mm long, subulate.

Zambia. N: Mporokoso Distr., 42 km east of Mporokoso, 1400 m, 13.iv.1961, *Phipps & Vesey-FitzGerald* 3128 (K; PRE; SRGH).

Abundant but localized in India in the Khasi Hills; scattered records in Africa from Sudan, Kenya, Tanzania and Madagascar. Growing in a dambo; c. 1400 m.

The species is distinguished from *H. quarrei* by its longer awns, longer superior raceme-base and longer sessile spikelets. It was originally thought to be an Indian endemic, but the Indian material is identical to the African extremes of *H. quarrei*. It is probably the better option to retain a narrow view of the species and extend the range of the Indian representative to Africa than to amalgamate them into a single species. Because of its close link with *H. quarrei* it can be assumed that *H. griffithii* arose in Africa and migrated to India where it is restricted to, but quite common in, the Khasi Hills.

12. **Hyparrhenia barteri** (Hack.) Stapf in Prain, F.T.A. **9**: 321 (1919). —Vesey-FitzGerald in Kirkia **3**: 106 (1963). —Clayton in Kew Bull., Addit. Ser. 2: 90 (1969). —Simon in Kirkia **8**: 50 (1971). —Clayton in F.W.T.A., ed. 2, **3**: 492 (1972). —Clayton & Renvoize in F.T.E.A., Gramineae: 800 (1982). Type from Nigeria.
    Andropogon barteri Hack. in Flora **68**: 124 (1885).

Annual; culms up to 200 cm high. Leaf sheaths glabrous; leaf laminas up to 30 cm × 4 mm, scaberulous. False panicle 30–40 cm long, narrow, dense, of up to 8 fastigiate tiers one above the other; spatheoles 3–4 cm long, linear, at length reddish-brown, embracing the base of the racemes; peduncle up to half as long as the spatheole, with or without spreading white hairs above; racemes 8–10 mm long, 2-awned per pair, greenish or pallid, laterally exserted, not deflexed; raceme-bases unequal, the superior 4–7 mm long, filiform, glabrous. Homogamous spikelets 3–5 mm long, narrowly lanceolate, a single pair at the base of each raceme. Sessile spikelets 5.5–7 mm long; callus 1.5–2 mm long, rarely longer, pungent; inferior glume narrowly lanceolate, 9-nerved with the inner 3–5 nerves ± raised and scaberulous, glabrous; awn 4–4.5 cm long, the column coarsely hirsute with rufous hairs 3–5 mm long. Pedicelled spikelets 3.5–5 mm long, linear-lanceolate, glabrous, terminating in an awn-point 1–2 mm long; callus absent; pedicel-tooth c. 0.1 mm long, subulate.

Zambia. N: North Luangwa National Park, Lubanga River, 11°46'S, 32°10'E, 700 m, 27.ii.1994, *P.P. Smith* 969 (K). E: Chipata Distr., Luangwa Valley, Msandile R, Nsefu, 5.iv.1968, *R. Phiri* 132 (K). **Malawi**. N: Karonga Distr., Fort Hill, 19.iii.1954, *Jackson* 1257 (K).

Around the periphery of the Dem. Rep. Congo Basin from Togo and Tanzania to Zambia. Growing in old farmland and along roadsides on poor soils, and in riverine grassland; 700–1280 m.

An annual species recognized at once by the remarkably long hairs on the awn, and by the fastigiate false panicle with non-overlapping tiers comprising tight bundles of spathes, spatheoles and racemes bundled around the axis.

13. **Hyparrhenia anamesa** Clayton in Kew Bull., Addit. ser. 2: 85 (1969). —Simon in Kirkia **8**: 17, 50 (1971). —Clayton & Renvoize in F.T.E.A., Gramineae: 800, fig. 184 (1982). TAB. **35**. Type from Kenya.

   Hyparrhenia filipendula (Hochst.) Stapf var. *pilosa* sensu Sturgeon in Rhodesia Agric. J. **51**: 135 (1954), in part, non (Hochst.) Stapf.

   Hyparrhenia filipendula sensu Stent & Rattray in Proc. & Trans. Rhodesia Sci. Assoc. **32**: 13 (1933), in part (*Eyles* 2174), non (Hochst.) Stapf.

Densely caespitose perennial; culms up to 120 cm high. Leaf sheaths glabrous or sometimes thinly hirsute; leaf laminas up to 40 cm × 4 mm, but often much shorter, harsh, glaucous. False panicle 15–45 cm long, lax; spatheoles 4–6 cm long, linear; peduncles slender and flexuous, usually longer than the spatheoles, with or without spreading white hairs above; racemes 1.5–2.5 cm long, 4–6(7)-awned per pair, white-villous, terminally exserted, not deflexed; raceme-bases very unequal, the superior 3.5–6 mm long, filiform, glabrous or sometimes softly hirsute. Homogamous spikelets 5–6 mm long, linear-lanceolate, glabrous or sometimes softly hirsute, 1 or 2 pairs at the base of the inferior raceme, 2 pairs (rarely only 1) at the base of the superior raceme. Sessile spikelet 5–6.5 mm long; callus 1–1.8 mm long, acute; inferior glume linear-oblong, white-villous; awn 2.5–4 cm long, the column pubescent with hairs 0.1–0.6 mm long. Pedicelled spikelets 4–6 mm long, linear-lanceolate, white-villous, muticous or occasionally with an awn-point 1–2 mm long; callus absent; pedicel-tooth c. 0.6 mm long, triangular.

Botswana. SE: Gaborone University Campus, 990 m, 29.xi.1973, *Mott* 31 (K). **Zambia**. B: Kaoma Distr., 56 km from Kafue Watershed, 2.iv.1969, *Verboom* 1366 (K). N: Mbala Distr., road to Itembwe Gorge, near Mbala (Abercorn), 1710 m, 24.iv.1959, *McCallum Webster* A357 (K). W: Mufulira, Copperbelt Exp. Farm, 21.iv.1966, *Lawton* 1395 (K). C: North Luangwa National Park, Muchinga Escarpment, 11°25'S, 32°01'E, 1200 m, 29.iv.1994, *P.P. Smith* 993 (K). S: Kafue National Park, 80 km west of Mumbwa on Kaoma (Mankoya) road, 23.iv.1964, *B.L. Mitchell* 25/39 (K; PRE; SRGH). **Zimbabwe**. N: Zvimba Distr., Darwendale, 14.iii.1931, *Brain* 2357 (K). W: Matobo Distr., Matopos Research Station, 1370 m, 12.ii.1954, *Rattray* 1590 (K; PRE). C: Chegutu Distr., Poole Farm, 1190 m, 7.iv.1954, *Wild* 4548 (K). E: Chimanimani Distr., Melsetter Pasture Res. sub-station, 1520 m, 19.ii.1950, *Williams* 76 (K). **Malawi**. N: Mzimba Distr., Marymount, toward Army base, 1370 m, 10.vi.1971, *Pawek* 4891 (K). **Mozambique**. N: Ngauma Distr., Massangulo, 13°55'S, 35°35'E, 1100 m, iv.1955, *Gomes e Sousa* 1363 (K). MS: Beira, ii.1912, *Rogers* 5945 (K). M: Marracuene, 26.ii.1940, *A.J.W. Hornby* 3122 (LISC).

In eastern Africa from Sudan southwards to South Africa (Cape). Growing on dry slopes and in grassy plains; 0–1710 m.

The species sits somewhat between *H. hirta* and *H. filipendula*. The latter has only 2 awns per raceme-pair while the former has at least eight; *H. filipendula* has two homogamous spikelet-pairs at the base of the superior raceme while *H. hirta* has only one. Furthermore, *H. filipendula* has longer hairs on the column of the awn (0.7–1.2 mm). *H. hirta* and *H. anamesa* share the characteristic of short hairs on the column of the awn (0.1–0.6 mm) and in many respects are closer together than either is to *H. filipendula*. They are distinguished from each other by number of awns per raceme-pair and number of homogamous spikelet-pairs on the superior raceme. Occasionally, specimens of *H. hirta* have two homogamous pairs on the superior raceme but these can be assigned to the species on account of their awn-number (8 or more per raceme-pair). Likewise, there are rare specimens of *H. anamesa* that have only one homogamous pair on the superior raceme, but these can be assigned to the appropriate species again by awn-number (always fewer than 8). The distinction works well enough north of the equator, but in southern Africa it is somewhat less than convincing with the species merging much more than they do elsewhere. Nevertheless, a distinction seems to be worth maintaining since the specimens from a Mediterranean-type climate (whether in the Mediterranean region or in South Africa) do seem to be different from those of the tropics.

Tab. 35. HYPARRHENIA ANAMESA. 1, habit (× ²/₃); 2, inflorescence (× ²/₃); 3, diagram of raceme pair, showing homogamous spikelets (shaded), sessile spikelets (black) and pedicelled spikelets (white) (× 3); 4, base of raceme pair, showing homogamous spikelet pairs and one heterogamous pair (× 6); 5, sessile spikelet, showing inferior glume and callus (× 6), 1–5 from *Bogdan* 2197. Drawn by Mary Grierson. From F.T.E.A.

14. **Hyparrhenia filipendula** (Hochst.) Stapf in Prain, F.T.A. **9**: 322 (1919). —Stent & Rattray in Proc. & Trans. Rhodesia Sci. Assoc. **32**: 13 (1933), pro parte excl. *Eyles* 2174. —Sturgeon in Rhodesia Agric. J. **51**: 135 (1954). —Chippindall in Meredith, Grasses & Pastures S. Africa: 510 (1955). —Jackson & Wiehe, Annot. Check List Nyasal. Grass.: 44 (1958) pro parte excl. *Jackson* 857 (=*Homozeugos eylesii*). —Vesey-FitzGerald in Kirkia **3**: 106 (1963). —Wild in Kirkia **5**: 54 (1965). —Clayton in Kew Bull., Addit. Ser. 2: 95 (1969). —Simon in Kirkia **8**: 18, 51 (1971). —Clayton in F.W.T.A., ed. 2, **3**: 494 (1972). —Clayton & Renvoize in F.T.E.A., Gramineae: 803 (1982). —Gibbs Russell et al., Grasses South. Africa [Mem. Bot. Surv. S. Africa No. 58]: 185 (1990). Type from South Africa (KwaZulu-Natal).

Andropogon filipendulus Hochst. in Flora **29**: 115 (1846).

Andropogon filipendulus var. *pilosus* Hochst., loc. cit. (1846). Type from South Africa.

Cymbopogon filipendulus (Hochst.) Rendle in Hiern et al., Cat. Afr. Pl. Welw. **2**: 157 (1899).

Cymbopogon filipendulus var. *angolensis* Rendle, loc. cit. (1899). Types from Angola.

Andropogon barteri var. *menyharthii* Hack. in Denkschr. Kaiserl. Akad. Wiss. Wien, Mat.-Naturwiss. Kl., **78**: 398 (1905). Type: Mozambique, *Menyharth* 894 (Z, holotype).

Hyparrhenia filipendula var. *pilosa* (Hochst.) Stapf in Prain, F.T.A. **9**: 324 (1919). —Sturgeon in Rhodesia Agric. J. **51**: 135 (1954), in part. —Chippindall in Meredith, Grasses & Pastures S. Africa: 510 (1955). —Jackson & Wiehe, Annot. Check List Nyasal. Grass.: 44 (1958). —Simon in Kirkia **8**: 18, 51 (1971). —Clayton in F.W.T.A., ed. 2, **3**: 494 (1972). —Gibbs Russell et al., Grasses South. Africa [Mem. Bot. Surv. S. Africa No. 58]: 185 (1990).

Hyparrhenia piovanii Chiov. in Atti Reale Accad. Italia, Mem. Cl. Sci. **11**: 63 (1940). Type from Ethiopia.

Hyparrhenia familiaris sensu Vesey-FitzGerald in Kirkia **3**: 106 (1963), non (Steud.) Stapf.

Caespitose perennial arising from short scaly rhizomes; culms up to 200 cm high, branched from the inferior nodes. Leaf sheaths glabrous or rarely sparsely pilose with stiff white hairs; leaf laminas up to 30 cm × 4 mm. False panicle 30–80 cm long, virgate, its branches slender and graceful; spatheoles 4.5–5.5 cm long, linear to almost filiform; peduncles about as long as the spatheoles, very fine and flexuous, with or without spreading white hairs above; racemes 10–12 mm long, 2–4-awned per pair, delicate, yellowish-green often tinged with violet, terminally exserted, not deflexed; raceme-bases very unequal, the superior (4)4.5–8(10) mm long, slender, glabrous. Homogamous spikelets 5–7 mm long, linear-lanceolate, glabrous, a single pair at the base of the inferior raceme and 2 pairs at the base of the superior. Sessile spikelets 5.5–7 mm long; callus 1.8–3 mm long, pungent; inferior glume linear-oblong, flat on the back or with the inner nerves ± raised towards the apex and with an indistinct median hollow towards the base, glabrous to villous with white hairs; awn 3–5.5 cm long, the column hirtellous with fulvous hairs 0.7–1.2 mm long. Pedicelled spikelets 5–6 mm long, linear-lanceolate, terminating in an awnlet 1–5 mm long; callus absent; pedicel-tooth very short, triangular.

Caprivi Strip. Katima Mulilo, 915 m, 28.xii.1958, *Killick & Leistner* 3175 (K; PRE). **Botswana**. N: Chobe Distr., near the Leshomo–Ngwezumba road, 17°58'S, 25°12'E, 14.iv.1983, *P.A. Smith* 4316 (K; PRE). SE: South East Distr., Mogobane, 1160 m, 3.iv.1957, *de Beer* Mog.5 (K). **Zambia**. N: Kasama Distr., Chibutubutu, Lukulu River, 16.iv.1958, *Vesey-FitzGerald* 1658 (K). W: Mwinilunga Distr., Matonchi River, by dam, 6.xi.1937, *Milne-Redhead* 3123 (K; PRE). C: Lusaka Distr., Quien Sabe, 1100 m, 7.ix.1929, *Sandwith* 49 (K). E: Chipata (Fort Jameson), iii.1962, *Verboom* 542 (K; PRE). S: Mazabuka Distr., Burdett's Farm, Monze to Magoye, km 9.7, 19.ii.1960, *White* 7229 (K). **Zimbabwe**. N: Gokwe South Distr., Charama Plateau, 1220 m, 2.v.1965, *Simon* 386 (K). W: Bulawayo, 20°09'S, 25°25'E, 1360 m, 9.iv.1912, *Rogers* 5869 (K). C: Harare (Salisbury), National Botanic Garden, 1500 m, 7.ii.1965, *Simon* 126 (K; SRGH). E: Nyanga Distr., Susurumba, 23.v.1968, *Wild* 7720 (K). S: Umzingwane Distr., Essexvale Ranch, 18.v.1969, *Wild Life Society, Bulawayo* 166 (K). **Malawi**. N: Karonga Distr., Vinthukutu Forest, 3 km north of Chilumba, 475 m, 26.iv.1975, *Pawek* 9585 (K). C: Lilongwe Agric. Research Station, 25.iv.1951, *Jackson* 470 (K). S: Blantyre Distr., Ndirande Mt., 1300 m, 10.vi.1970, *Brummitt* 11379 (K; LISC; MAL; PRE; SRGH). **Mozambique**. N: Malema, andados 17 km do Mutuáli para Lioma, 650 m, 10.ii.1964, *Torre & Paiva* 10497 (K; LISC). Z: Nas encostas dos montes Milange, 9.ix.1941, *Torre* 3380 (K; LISC). T: entre Furancungo e a serra de Pandarajala (Pandalanjala), 15.v.1948, *Mendonça* 4245 (K; LISC). MS: between Catandica (Vila Gouveia) and Macossa, 1.vii.1941, *Torre* 2962 (K). GI: Inharrime Distr., Nhacoongo, 5.iii.1963, *Macêdo & Balsinhas* 1099 (K; PRE). M: arredores de Moamba, 20.ii.1948, *Torre* 7381 (K; LISC).

Tropical and South Africa, but mainly on the eastern side of the continent; also in Madagascar and extending eastwards through Sri Lanka to Australia. A common species of open situations in a wide range of soil and vegetation types, particularly those subject to disturbance; 100–1500 m.

In its typical form *H. filipendula* is most readily recognized by its delicate and graceful appearance derived from the copious slender ascending inflorescence branches, the flexuous peduncles and the numerous small racemes with a long superior raceme-base and commonly only 2 awns per raceme-pair. It is most similar to *H. anamesa* but this has a pubescent rather than a hirtellous awn-column.

The form with pilose spikelets and up to 4 awns per raceme-pair tends to merge with *H. hirta*, but again the form of the hairs on the awn-column will serve to distinguish them, as will the higher number of awns in *H. hirta.*

15. **Hyparrhenia anthistirioides** (Hochst. ex A. Rich.) Andersson ex Stapf in Prain, F.T.A. **9**: 331 (1919). —Clayton in Kew Bull., Addit. Ser. 2: 106 (1969). —Simon in Kirkia **8**: 50 (1971). —Clayton & Renvoize in F.T.E.A., Gramineae: 804 (1982). Type from Ethiopia.

Andropogon anthistirioides Hochst. ex A. Rich., Tent. Fl. Abyss. **2**: 463 (1851).

Anthistiria pseudocymbaria Steud., Syn. Pl. Glumac. **1**: 399 (1854). Type from Ethiopia.

Anthistiria quinqueplex Hochst. ex Steud., Syn. Pl. Glumac. **1**: 400 (1854). Type from Ethiopia.

Hyparrhenia quinqueplex (Hochst. ex Steud.) Andersson in Schweinfurth, Beitr. Fl. Aethiop.: 300 (1867), *comb. inval. publ. gen. ante.*

Sorghum anthistirioides (Hochst. ex A. Rich.) Kuntze, Revis. Gen. Pl. **2**: 791 (1891).

Hyparrhenia pseudocymbaria (Steud.) Stapf in Prain, F.T.A. **9**: 329 (1919). —Jackson & Wiehe, Annot. Check List Nyasal. Grass.: 46 (1958) (as 'sp. aff. *pseudocymbaria*').

Annual; culms up to 200 cm high, often much less on poorer soils, erect or geniculately ascending, sometimes with stilt roots from the lowest nodes. Leaf sheaths glabrous; leaf laminas up to 30 cm × 10 mm, glabrous or very sparsely pilose, somewhat flaccid. False panicle 20–30 cm long, ample or sometimes reduced to just a few raceme-pairs, the coloured spathes and spatheoles contrasting with the dark awns; spatheoles 1.8–3.2 cm long, lanceolate, thinly scarious, often brightly streaked in shades of green, yellow and orange-brown; peduncles 5–10 mm long, $^{1}/_{4}$–$^{1}/_{2}$ the length of the spatheole, barbate above; racemes 1.1–1.3 cm long, 3–4(5)-awned per pair, exserted laterally, deflexed; raceme-bases subequal, the superior c. 1 mm long, flattened, stiffly barbate, with or without a scarious lobe up to 0.2 mm long at the summit. Homogamous spikelets 8–11 mm long, a single pair at the base of the inferior raceme only, glabrous except for the ciliolate margins. Sessile spikelets 3.5–6 mm long; callus 0.5–1 mm long, cuneate, subacute; inferior glume lanceolate-oblong, glabrous or sparsely pubescent; awn 3.2–4.5 cm long, the column usually dark brown and hirtellous with hairs 0.8–1 mm long. Pedicelled spikelets 5–6 mm long, glabrous except for the ciliolate margins, terminating in a fine straight awnlet 3–6 mm long; callus sometimes perceptible and up to 0.3 mm long (occasionally more); pedicel-tooth obscure.

Zambia. N: Mbala Distr., Kawimbe, 1710 m, 26.iii.1959, *McCallum Webster* A236 (K; SRGH). **Malawi**. C: Salima road, 56 km from Lilongwe, 1.iv.1950, *Wiehe* 485 (K).

Also in Sudan, Ethiopia, Somalia and Tanzania. A pioneer species of open or disturbed places; 1020–1710 m.

Although the best diagnostic character for this species is the annual habit, it should be noted that in perennial species with stilt roots the connection between the rhizome and the robust prop-rooted culm is often rather tenuous and easily overlooked. Another distinctive feature of the species is the brightly striped spatheoles contrasting with the dark awns.

16. **Hyparrhenia cymbaria** (L.) Stapf in Prain, F.T.A. **9**: 332 (1919). —Sturgeon in Rhodesia Agric. J. **51**: 136 (1954). —Chippindall in Meredith, Grasses & Pastures S. Africa: 514 (1955). —Jackson & Wiehe, Annot. Check List Nyasal. Grass.: 43 (1958). —Clayton in Kew Bull., Addit. Ser. 2: 110 (1969). —Simon in Kirkia **8**: 17, 50 (1971). —Clayton in F.W.T.A., ed. 2 **3**: 494 (1972). —Clayton & Renvoize in F.T.E.A., Gramineae: 804 (1982). —Gibbs Russell et al., Grasses South. Africa [Mem. Bot. Surv. S. Africa No. 58]: 184 (1990). TAB. **36**. Type said to be from India, but probably from Comoro Is.

Andropogon cymbarius L., Mant. Pl. **2**: 303 (1771). —Hackel in A. & C. de Candolle, Monogr. Phan. **6**: 629 (1889). —Stapf in Dyer, F.C. **7**: 360 (1898).

Cymbopogon elegans Spreng., Pl. Min. Cogn. Pug. **2**: 14 (1815), *nom. superfl.*, based on *Andropogon cymbarius* L.

Anthistiria cymbaria (L.) Roxb., Fl. Ind. **1**: 255 (1820).

Andropogon lepidus Nees, Fl. Afr. Austral. Ill. **1**: 113 (1841). —Hackel in A. & C. de Candolle, Monogr. Phan. **6**: 624 (1889). Type from South Africa.

Andropogon intonsus Nees, op. cit.: 114 (1841). Type from south Africa.

Cymbopogon cymbarius (L.) Thomson in Speke, J. Discov. Source Nile: 652 (1863).

Tab. 36. **HYPARRHENIA CYMBARIA.** 1, habit ($\times \frac{1}{2}$); 2, ligule ($\times 1$); 3, spikelet pair: a. joint of raceme, b. sessile spikelet, c. pedicelled spikelet; 4, spikelet pair: a. joint of raceme, b. sessile spikelet, c. pedicelled spikelet, d. pedicel of pedicelled spikelet. Drawn by W.E. Trevithick. From E. Afr. Pasture Gr.

Anthistiria latifolia Andersson in Peters, Naturw. Reise Mossambique, Bot. part 2: 562 (1864). Type from Comoro Is.

Andropogon lepidus var. *intonsus* (Nees) Hack. in A. & C. de Candolle, Monogr. Phan 6: 625 (1889).

Sorghum cymbarium (L.) Kuntze, Revis. Gen. Pl. **2**: 791 (1891).

Sorghum lepidum (Nees) Kuntze, tom. cit.: 792 (1891).

Andropogon cymbarius var. *lepidus* (Nees) Stapf in Dyer, F.C. **7**: 361 (1898).

Cymbopogon lepidus (Nees) Chiov. in Monogr. Rapp. Colon. **24**: 66 (1912).

Hyparrhenia formosa sensu Sturgeon op. cit.: 136 (1954), in part, non Stapf.

Hyparrhenia variabilis sensu Stent & Rattray in Proc. & Trans. Rhodesia Sci. Assoc. **32**: 14 (1933), in part, non (Stapf).

Hyparrhenia gazensis sensu Jackson & Wiehe, Annot. Check List Nyasal. Grass.: 44 (1958), in part, non (Rendle) Stapf.

Robust caespitose perennial arising from a slender rhizome clad in small cataphylls; culms up to 400 cm high, initially slender and rambling, but subsequently erect and stout (up to 8 mm in diameter) and supported by stilt roots. Leaf sheaths usually glabrous, though often ciliate on the margins and sometimes pubescent at the base; leaf laminas up to 45 cm × 6–20 mm, rigid to subflaccid, dull green, glabrous or hirsute at the base. False panicle typically 20–40 cm long, large, dense, decompound; spatheole 0.8–1.8(2.1) cm long, boat-shaped (narrowly ovate in profile), glabrous, turning bright russet-red at maturity; peduncles 3–8 mm long, $^1/_3$–$^1/_2$ the length of the spatheole, barbate above with white or yellowish hairs; racemes 0.7–1.3 cm long, 3–5(6)-awned per pair, partially laterally exserted, deflexed; raceme-bases subequal, very short, the superior up to 0.5 mm long, flattened, stiffly barbate, with or without a scarious frill up to 0.2 mm long at the apex. Homogamous spikelets 4–6(7) mm long, a single pair at the base of the inferior raceme only, glabrous to puberulous on the back, ciliate on the margins. Sessile spikelets 3.8–4.5 mm long; callus 0.2–0.3 mm long, square or broader than long, broadly rounded at the apex; inferior glume oblong-lanceolate, glabrescent to sparsely and shortly pubescent, often becoming purplish; awn 0.5–1.6(2) cm long, rarely almost suppressed, the column pubescent or puberulous with pallid hairs. Pedicelled spikelets 4–5 mm long, glabrous to puberulous on the back, ciliate on the margins, acuminate or sometimes with an awn-point up to 1.5 mm long at the apex; callus scarcely developed; pedicel-tooth obscure.

Zambia. N: North Luangwa National Park, 11°40'S, 31°57'E, 1200 m, 28.iv.1994, *P.P. Smith* 987 (K). W: Copperbelt Distr., Mufulira, 28.v.1934, *Eyles* 8412 (K; SRGH). C: Kabwe (Broken Hill), c. 1220 m, viii.1934, *Trapnell* 1526 (K). **Zimbabwe**. W: Bulawayo area, *Jeffreys* in GHS 41283 (K). C: Hwedza Distr., SE slopes of Hwedza (Wedza) Mt., 64 km south of Marondera (Marandellas), 1680 m, 21.v.1968, *Simon, Rushworth & Mavi* 1828 (K). E: Chimanimani Distr., Melsetter Pasture Research Station, 22.v.1953, *West* 3312 (K). S: Bikita Reserve, 25.v.1948, *D.A. Robinson* 343 (K; PRE). **Malawi**. N: Chitipa Distr., Misuku Hills, Mugesse (Mughesse), 1590 m, 6.vii.1973, *Pawek* 7052 (K). C: Ntchisi Distr., Ntchisi Forest Reserve, 1370 m, 27.iii.1970, *Brummitt* 9454 (K; LISC; MAL; PRE; SRGH). S: Blantyre Distr., Ndirande Mt., 1220 m, 3.v.1970, *Brummitt* 10344 (K; MAL; SRGH). **Mozambique**. N: Ribáuè, viii.1931, *Gomes e Sousa* 837 (LISC). Z: Serra de Gurué, 29.ix.1944, *Mendonça* 2214 (K; LISC). T: Moatize Distr., Zóbuè, 17.vi.1941, *Torre* 2867 (K; LISC). MS: west face of serra de Gorongosa, c. 1370 m, 10.vii.1969, *Leach & Cannell* 14307 (K).

Eastern Africa from Eritrea southwards to South Africa (KwaZulu-Natal), reaching the west coast in the Cameroon area and in northern Angola; also in Madagascar and the Comoro Islands. Growing in wooded grassland, on open hillsides, on the edges of evergreen forest and along stream banks; 480–1680 m.

When mature the plant is usually encountered as tall and robust. However, the initial shoot from the rhizome is slender and rambling, eventually turning upwards, thickening and becoming supported by large stilt roots; the original connection between the culm and the rhizome can become very obscure. Under favourable conditions the rambling culms can flower, thus simulating the habit of *H. pilgeriana*.

The most conspicuous feature of the species is its beautiful panicle of short, broad, brightly coloured spatheoles that almost completely enclose the shortly awned racemes. Unfortunately, these characters cannot be used with absolute assurance to distinguish the species from *H. variabilis*. There is a tendency for the panicle described to be associated with the short broadly rounded callus, but the latter is itself liable to a degree of subjective interpretation. A combination of callus and awn-length seems to work best to discriminate the taxa.

17. **Hyparrhenia variabilis** Stapf in Prain, F.T.A. **9**: 334 (Jan. 1919). —Stent & Rattray in Proc. & Trans. Rhodesia Sci. Assoc. **32**: 14 (1933), in part. —Sturgeon in Rhodesia Agric. J. **51**: 136 (1954). —Jackson & Wiehe, Annot. Check List Nyasal. Grass.: 45 (1958), in part. —Clayton in Kew Bull., Addit. Ser. 2: 113 (1969). —Simon in Kirkia **8**: 18, 51 (1971). —Clayton & Renvoize in F.T.E.A., Gramineae: 805, fig. 185 (1982). —Gibbs Russell et al., Grasses South. Africa [Mem. Bot. Surv. S. Africa No. 58]: 188 (1990). TAB. **37**. Type: Zambia, *Macauley* 62 (K, lectotype).

    Hyparrhenia spectabilis Stapf in Prain, F.T.A. **9**: 338 (1919). Type from Angola.
    Cymbopogon acutispathaceus De Wild. in Bull. Jard. Bot. État **6**: 6 (Jan. 1919). Type from Dem. Rep. Congo.
    Hyparrhenia acutispathacea (De Wild.) Robyns, Fl. Agrost. Congo Belge **1**: 181 (1929).
    Hyparrhenia iringensis Pilg. in Notizbl. Bot. Gart. Berlin-Dahlem **14**: 101 (1938). Type from Tanzania.

Robust perennial arising from a short rhizome clad in hard cataphylls; culms up to 300 cm high, supported by stilt roots, initially decumbent. Leaf sheaths glabrous or sometimes pubescent at the base; leaf laminas up to 45 cm × 15 mm, firm, glabrous or rarely hirsute at the base. False panicle 20–40 cm long, large, lax, decompound; spatheoles 1.4–2.4 cm long, boat-shaped (lanceolate in profile), glabrous, russet-red tinged with yellow and green at maturity; peduncles 3–9 mm long, up to c. ¹/₃ the length of the spatheole, barbate above with white hairs; racemes 0.8–1.3 cm long, 3–5-awned per pair, projecting laterally from the spatheole, deflexed; raceme-bases subequal, short, the superior c. 1 mm long, flattened, stiffly barbate, with a scarious rim up to 0.4 mm long at the apex. Homogamous spikelets 7–9 mm long, a single pair at the base of the inferior raceme only, glabrous to puberulous on the back, ciliate on the margins. Sessile spikelets 4–5 mm long; callus 0.5–1 mm long, cuneate, narrowly obtuse to subacute at the apex; inferior glume oblong-lanceolate, glabrescent to sparsely and shortly pubescent; awn 1.8–3.2 cm long, the column pubescent with pallid or tawny hairs. Pedicelled spikelets 5–8 mm long, glabrous to puberulous on the back, ciliate on the margins, terminating in an awn-point up to 6 mm long; callus scarcely developed; pedicel-tooth obscure.

Zambia. B: Kaoma Distr., Luampa River Bridge, 17.iv.1964, *Verboom* 1385 (K). N: Kasama Distr., Chibutubutu, Lukulu River, 16.iv.1958, *Vesey-FitzGerald* 1654 (K). W: Mwinilunga Distr., Lisombo River tributary, 14.vi.1963, *Loveridge* 992 (K). C: west of Lusaka, 1200 m, v.1932, *Trapnell* 2017 (K). S: Choma Distr., Mapanza, 24.v.1953, *E.A. Robinson* 274 (K). **Zimbabwe**. N: Gokwe South Distr., Sengwa River off south border of Charama Plateau, 1220 m, 24.iv.1965, *Simon* 219A (K; SRGH). W: Matobo Distr., near Absent Farm, Konkora Stream, 18.iv.1954, *Plowes* 1713 (K; PRE). C: Goromonzi Distr., Ruwa, Tanglewood Farm, v.1960, *Miller* 7295 (K; PRE). E: Chipinge Distr., Cheredza Valley, near Mt. Selinda, 4.vii.1934, *Michelmore* 295 (K). S: Masvingo Distr., Zimbabwe National Park, 2.vi.1973, *Vernon* 40 (K). **Malawi**. N: Mzimba Distr., Mzuzu, Marymount, 1370 m, 13.vi.1973, *Pawek* 6851 (K). C: Lilongwe Agric. Research Station, 16.v.1966, *Salubeni* 448 (K). S: Mangochi Distr., Namwera highlands east of Mangochi (Fort Johnston), 910 m, 20.iv.1937, *Lawrence* 385 (K). **Mozambique**. N: Lichinga (Vila Cabral), Meponda, 30.viii.1958, *Rui Monteiro* 18 (LISC). Z: Morrumbala Distr., encostas da serra da Morrumbala, 1.v.1943, *Torre* 5259 (K; LISC). T: Angónia Distr., arredores do Posto Zootécnico de Angónia, 12.v.1948, *Mendonça* 4188A (K; LISC). MS: Gondola Distr., Chimoio (Vila Pery), hills, 4.vi.1941, *Torre* 2793 (K; LISC).
    The eastern half of Africa from Ethiopia to South Africa (Transvaal); also in Madagascar, the Comoro Is. and Java. Growing in tall grassland and open woodland on a variety of soils; also common on roadsides and in areas of abandoned cultivation; 700–1560 m.
    Very similar to, and intergrades with, *H. cymbaria*. See under that species for the best discriminatory characters.

18. **Hyparrhenia pilgeriana** C.E. Hubb. in Bull. Misc. Inform., Kew **1928**: 39 (1928), based on *Cymbopogon stolzii* Pilg. —Clayton in Kew Bull., Addit. Ser. 2: 53 (1969), as "*pilgerana*". —Simon in Kirkia **8**: 18, 51 (1971). —Clayton & Renvoize in F.T.E.A., Gramineae: 807 (1982). —Gibbs Russell et al., Grasses South. Africa [Mem. Bot. Surv. S. Africa No. 58]: 186 (1990). TAB. **38**. Type from Tanzania.

    Cymbopogon stolzii Pilg. in Bot. Jahrb. Syst. **54**: 286 (1917), non *Hyparrhenia stolzii* Stapf (1918). Type as above.
    Hyparrhenia claessensii Robyns, Fl. Agrost. Congo Belge **1**: 180 (1929). Type from Dem. Rep. Congo.
    Hyparrhenia formosa auct. non Stapf. —Stent & Rattray in Proc. & Trans. Rhodesia Sci. Assoc. **32**: 15 (1933). —Sturgeon in Rhodesia Agric. J. **51**: 136 (1954), in part. —Jackson & Wiehe, Annot. Check List Nyasal. Grass.: 44 (1958).

Tab. 37. HYPARRHENIA VARIABILIS. 1, inflorescence (× ½); 2, spatheole and raceme pair
(× 4); 3, raceme pair (× 5); 4, lower raceme, showing one homogamous pair, one
heterogamous pair and a terminal triad (× 5); 5, tip of superior lemma (× 20), 1–5 from
Tanner 4520. Drawn by Ann Davies. From F.T.E.A.

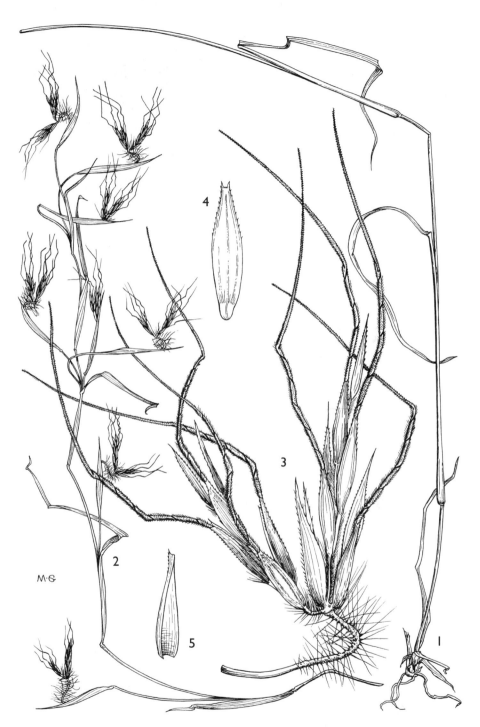

Tab. 38. HYPARRHENIA PILGERIANA. 1, lower portion of plant (× ⅔), from *Gillett* 14479; 2, inflorescence (× ⅔); 3, raceme pair (× 5); 4 & 5, sessile spikelet: 4, inferior glume, including callus (× 6); 5, superior glume (× 6), 2–5 from *Mooney* 8618. Drawn by Mary Grierson. From Kew Bull.

Laxly caespitose perennial arising from a short rhizome clad in scaly white cataphylls; culms up to 60 cm high and 1–2 mm in diameter, slender, weakly ascending or untidily rambling. Leaf sheaths glabrous; leaf laminas 5–10(15) cm × 2–4 mm, usually rather thin, glabrous. False panicle 5–30 cm long, open and scanty, often only with 10–20 raceme-pairs but sometimes denser with more numerous pairs; spatheoles 2–3 mm long, narrowly lanceolate, reddish-brown, glabrous; peduncles 0.9–3 cm long, from half as long as to slightly longer than the spatheole, pilose with yellow or sometimes white hairs above; racemes 1–1.5 cm long, (4)6–7-awned per pair, exserted laterally with or without a sinuous hook in the peduncle, not or tardily deflexed; raceme-bases subequal and up to 0.5 mm long, or rather unequal with the superior up to 1.5 mm long, subterete, typically with a scarious rim up to 0.2 mm long at the summit, or rarely with a short appendage up to 0.5 mm long. Homogamous spikelets 4–5.5 mm long, a single pair at the base of the inferior raceme only, glabrous except for the ciliolate margins. Sessile spikelets 3.5–5 mm long; callus 0.4–0.7 mm long, oblong or sometimes square, broadly rounded at the apex; inferior glume lanceolate, glabrous or white-puberulous; awn 0.7–1.7 cm long, the column white-puberulous. Pedicelled spikelets 4–4.5 mm long, purple, glabrous, awnless or with a short awn-point up to 1.5 mm long; callus scarcely developed; pedicel-tooth obscure.

Zambia. N: Samfya Distr., Lake Bangweulu, Mboyalubambe Is., 10 km south of Lake Chali, 1062 m, 7.iii.1996, *Renvoize* 5766 (K). W: Copperbelt Distr., Mufulira, 23.v.1934, *Eyles* 8393 (K; SRGH). **Zimbabwe**. C: Harare (Salisbury), 1460 m, 21.iii.1927, *Eyles* 4768 (K; PRE; SRGH). E: Camp site near Chimanimani (Melsetter), 1500 m, 22.v.1972, *Simon* 2233 (K). **Malawi**. N: Nyika, 16.vi.1951, *Jackson* 529 (K). C: Mchinji (Fort Manning), Bua River, 31.i.1952, *Jackson* 721 (K).

Eastern Africa from Ethiopia southwards to Malawi and then sporadically to South Africa (KwaZulu-Natal). Growing in the uplands in seasonal swamps, on forest margins and on grassy slopes, sometimes descending into the lowlands in deciduous bushland, wooded grassland and old fallow fields; mostly 1050–1500 m.

The species is best distinguished from *H. cymbaria* and *H. formosa* by its slender straggling habit and relatively long peduncle. The longer spatheoles will separate it from depauperate (and similarly straggling) specimens of *H. cymbaria*. It bears a superficial resemblance to *H. gazensis*, but this has longer raceme-bases and awns.

19. **Hyparrhenia formosa** Stapf in Prain, F.T.A. **9**: 340 (1919). —Clayton in Kew Bull., Addit. Ser. 2: 117 (1969). —Clayton & Renvoize in F.T.E.A., Gramineae: 807 (1982). Type from Ethiopia.

Hyparrhenia variabilis sensu Jackson & Wiehe, Annot. Check List Nyasal. Grass.: 45 (1958), in part, non Stapf.

Robust, coarsely caespitose perennial; culms up to 200 cm high, sometimes supported by stilt roots. Leaf sheaths glabrous; leaf laminas up to 50 cm × 12 mm, scaberulous. False panicle 30–40 cm long, dense; spatheoles 1.8–2.6 cm long, lanceolate, glabrous or thinly pilose, russet or tinged with purple and yellow; peduncles 2–10 mm long, $\frac{1}{8}$–$\frac{1}{2}$ as long as the spatheole, barbate above; racemes 1–1.5 cm long, 6–8-awned per pair, exserted laterally, deflexed; raceme-bases subequal, the superior up to 0.5 mm long, flattened, stiffly barbate. Homogamous spikelets 4.5–5.5 mm long, a single pair at the base of the inferior raceme only, glabrous except for the ciliate margins. Sessile spikelets 4–4.5 mm long; callus 0.5–1 mm long, oblong to cuneate, rarely square, broadly rounded at the apex; inferior glume oblong-lanceolate, glabrous to shortly pilose; awn 0.8–1.8 cm long, the column minutely pubescent. Pedicelled spikelets 4.5–6 mm long, glabrous to puberulous, acuminate at the apex; callus scarcely developed; pedicel-tooth obscure.

Malawi. N: Mzimba Distr., Mzuzu, Lunyangwa, 1370 m, 27.vi.1973, *Pawek* 6957 (K). S: Zomba Plateau, 1520 m, 27.iii.1937, *Lawrence* 318 (K).

Scattered throughout eastern Africa and tropical Arabia (Yemen), especially in Uganda, but nowhere common. Growing in upland marshes, along rivers and streams, on roadsides and in savanna grassland; 1370–1520 m.

The species is scarcely distinct from *H. cymbaria*, separation being justified by the longer spatheole and callus and the more numerous awns; there is also a difference in chromosome number (*H. formosa* 2n – 30, *H. cymbaria* 2n = 20) that gives some further support. Specimens cited by Sturgeon in Rhodesia Agric. J. **51**: 136–137 (1954) are a mixture of *H. pilgeriana* (*Eyles* 4768) and *H. cymbaria* (*D.A. Robinson* 343 and 351).

20. **Hyparrhenia schimperi** (Hochst. ex A. Rich.) Andersson ex Stapf in Prain, F.T.A. **9**: 341
     (1919). —Sturgeon in Rhodesia Agric. J. **51**: 137 (1954). —Clayton in Kew Bull., Addit. Ser.
     2: 118 (1969). —Simon in Kirkia **8**: 18, 51 (1971). —Clayton & Renvoize in F.T.E.A.,
     Gramineae: 808 (1982). —Gibbs Russell et al., Grasses South. Africa [Mem. Bot. Surv. S.
     Africa No. 58]: 187 (1990). Types from Ethiopia.
        Andropogon schimperi Hochst. ex A. Rich., Tent. Fl. Abyss. **2**: 466 (1851).
        Cymbopogon schimperi (Hochst. ex A. Rich.) Rendle in Hiern et al., Cat. Afr. Pl. Welw. **2**:
     155 (1899).
        Hyparrhenia viridescens Robyns, Fl. Agrost. Congo Belge **1**: 182 (1929). Type from Dem.
     Rep. Congo.
        Hyparrhenia variabilis sensu Stent & Rattray in Proc. & Trans. Rhodesia Sci. Assoc. **32**: 14
     (1933), in part, non Stapf.
        Hyparrhenia gazensis sensu Jackson & Wiehe, Annot. Check List Nyasal. Grass.: 44 (1958),
     in part, non (Rendle) Stapf.

Robust, coarsely caespitose perennial arising from a short rhizome; culms up to
400 cm high and up to 8 mm in diameter, often supported by stilt roots. Leaf
sheaths glabrous or rarely pubescent at the base; leaf laminas up to 60 cm × 20 mm,
glabrous or with a few hairs beneath. False panicle 30–60 cm long, large,
decompound; spatheoles 2.2–3.2 cm long, lanceolate, glabrous, turning russet-
brown; peduncles 10–15 mm long, up to half as long as the spatheole, pilose above;
racemes 1.2–1.6 cm long, up to a maximum of 6–8-awned per pair in any given
panicle, exserted laterally, at length deflexed; raceme-bases subequal, the superior
0.5–1.5 mm long, flattened to subterete, stiffly barbate, with a scarious lobe up to 0.6
mm long at the apex. Homogamous spikelets 5–8 mm long, a single pair at the base
of the inferior raceme only, glabrous to sparsely pilose on the back, ciliate on the
margins. Sessile spikelets 4–5 mm long; callus 0.5–0.8 mm long, cuneate, narrowly
obtuse to acute at the apex; inferior glume oblong-lanceolate, glabrescent to
sparsely pubescent; awn 2–3.3 cm long, the column pubescent with white or tawny
hairs. Pedicelled spikelets 5–7 mm long, glabrous to shortly and sparsely pilose,
with an awn-point up to 6 mm long, occasionally muticous; callus scarcely
developed; pedicel-tooth obscure.

Zambia. B: 72 km west of Nangweshi, 1040 m, 6.viii.1952, *Codd* 7420 (K; PRE). N: Mpika
Distr., North Luangwa National Park, Muchinga Escarpment, 11°40'S, 32°04'E, 950 m, 6.v.1994,
P.P. Smith 996 (K). W: Ndola Forest Reserve, 1950, *Jackson* 1 (K). C: Lusaka Distr., Quien Sabe
Farm, 1100 m, 9.ix.1929, *Sandwith* s.n. (K). E: Chipata (Fort Jameson) (13°40'S, 32°40'E), 1160
m, 27.xii.1957, *D.R.M. Stewart* 95 (K; SRGH). S: Gwembe, 520 m, iv.1934, *Trapnell* 1490 (K).
Zimbabwe. N: Shamva Distr., Chipoli, Mazowe (Mazoe) River, 14.v.1972, *Simon* 2220 (K; PRE).
C: Harare Distr., Harare University Campus, 14.iii.1985, *Bennett* in *GHS* 283138 (K; PRE). E:
Chipinge Distr., Mt. Selinda, 4.vii.1934, *Michelmore* 297 (K). S: Masvingo Distr., Great Zimbabwe,
11.vii.1955, *Plowes* 1871 (K). **Malawi**. N: Rumphi Distr., Nyika Plateau, Dembo R. Bridge, 2300
m, 17.iv.1975, *Pawek* 9302 (K). C: Lilongwe Agric. Research Station, 30.iii.1953, *Jackson* 1193
(K). S: neighbourhood of Zomba, 910 m, 1936, *Cormack* 393 (K). **Mozambique**. Z:
Morrumbala Distr., M'bobo area, 27.iv.1943, *Torre* 5210 (K; LISC; PRE). MS: Manica, v.1970,
Horrocks in *GHS* 205771 (K; LISC).
     On the eastern side of Africa from Ethiopia southwards to South Africa (KwaZulu-Natal); also
in Madagascar. Growing in open, wooded and riverine grasslands, deciduous bushland,
dambos and old areas of cultivation; 500–2300 m.
     The species is rather indistinctly separated from *H. variabilis* of which it is likely to be a
tetraploid derivative (2n = 40); the two share the same geographical distribution. Specimens
with both glabrous and sparsely pilose pedicelled spikelets are included, but the two sorts
intergrade so gradually that it is scarcely possible, or worthwhile, to separate them. The scarious
rim at the apex of the superior raceme-base is variable and on occasion approaches the
condition found in the truly appendaged species; these, however, can be distinguished by their
few-awned raceme-pairs.

21. **Hyparrhenia dregeana** (Nees) Stapf ex Stent in Bothalia **1**: 249 (1923). —Clayton in Kew
     Bull., Addit. Ser. 2: 124 (1969). —Simon in Kirkia **8**: 18 (1971). —Clayton & Renvoize in
     F.T.E.A., Gramineae: 809 (1982). —Gibbs Russell et al., Grasses South. Africa [Mem. Bot.
     Surv. S. Africa No. 58]: 184 (1990). Type from South Africa.
        Andropogon dregeanus Nees, Fl. Afr. Austral. Ill. **1**: 112 (1841). —Stapf in Dyer, F.C. **7**: 359
     (1898).
        Andropogon pilosissimus Hack. in A. & C. de Candolle, Monogr. Phan. **6**: 690 (1889). Type
     from South Africa.
        Andropogon acutus Stapf in Dyer, F.C. **7**: 357 (1898). Type from South Africa.

Hyparrhenia elongata Stapf in Prain, F.T.A. **9**: 343 (1919). Type from Ethiopia.
Hyparrhenia phyllopoda Stapf in Prain, F.T.A. **9**: 346 (1919). —Jackson & Wiehe, Annot. Check List Nyasal. Grass.: 45 (1958), as 'sp. aff. *phyllopoda.*'
Hyparrhenia acuta (Stapf) Stapf ex Stent in Bothalia **1**: 249 (1923). —Chippindall in Meredith, Grasses & Pastures S. Africa: 514 (1955).
Cymbopogon micratherus Pilg. in Notizbl. Bot. Gart. Berlin-Dahlem **10**: 596 (1929). Type from Kenya.
Hyparrhenia micrathera (Pilg.) Pilg. ex Peter in Repert. Spec. Nov. Regni Veg. **40**(1): 379 (1936).
Hyparrhenia subaristata Peter in Repert. Spec. Nov. Regni Veg. **40**(1): 375; Anh.: 120, t. 77 (1936). Type from Tanzania.
Hyparrhenia brachychaete Peter, tom. cit.: 376; Anh.: 121, t. 78 (1936). Type from Tanzania.
Hyparrhenia pilosissima (Hack.) J.G. Anderson in Bothalia **9**: 130 (1966).
Hyparrhenia cymbaria sensu Stent & Rattray in Proc. & Trans. Rhodesia Sci. Assoc. **32**: 14 (1933), in part (*Fitt* 156), non (L.) Stapf.

Robust, densely caespitose perennial; culms up to 200 cm high (or more) and up to 4 mm in diameter at the base, simple or scantily branched. Basal leaf sheaths silky pubescent below, otherwise all sheaths glabrous; leaf laminas up to 60 cm × 3–8 mm, flat or inrolled and narrower, glabrous, usually stiff and glaucous with scaberulous nerves and harsh to the touch, sometimes soft and green. False panicle 20–50 cm long, narrow, fairly dense; spatheoles 2.5–5 cm long, narrowly lanceolate, glabrous or thinly hirsute (rarely the whole panicle copiously villous), becoming reddish-brown; peduncles 1.5–5 cm long, from half as long as to rather longer than the spatheole, pilose above with yellowish hairs; racemes 2–3 cm long, 10–25-awned per pair, usually exserted terminally, tardily deflexed; raceme-bases subequal, the superior 1–1.5 mm long, flattened, stiffly barbate, the apex with a scarious rim and usually produced into a definite oblong appendage up to 0.5 mm long. Homogamous spikelets 5–7 mm long, a single pair at the base of the inferior raceme only, villous or rarely glabrous. Sessile spikelets 4–5 mm long; callus c. 1 mm long, cuneate, subacute to narrowly obtuse at the apex; inferior glume oblong-lanceolate, yellowish-green to light brown or purple, densely villous to hispidulous, rarely glabrous; awn (1)8–20(28) mm long, the column pubescent with tawny hairs. Pedicelled spikelets 5–6 mm long, similar to the sessile spikelets in indumentum and colouring, muticous or with a short awn-point up to 1.5 mm long at the apex; callus scarcely developed; pedicel-tooth obscure.

Zimbabwe. C: Harare Distr., Gletwyn Farm, Enterprise Road, 1545 m, 8.vi.1972, *Beamish* in *GHS* 231613 (K; PRE; SRGH). E: Mutasa Distr., Stapleford Forest Reserve, 12.vi.1934, *Gilliland* 275 (K; PRE; SRGH). **Malawi**. N: Rumphi Distr., Nyika Plateau, Nganda base-camp, 2255 m, 27.iii.1997, *Patel & Ludlow* 5003 (K). C: Dowa Distr., Bua River on Lilongwe–Kasungu road, 1000 m, 9.viii.1949, *Wiehe* 180 (K; PRE). S: Zomba Mt., 21.viii.1950, *Jackson* 140 (K).
The eastern side of Africa from Ethiopia southwards to South Africa (Cape), but most abundant in the northern and southern extremities of this range. Growing in upland grassland, often in marshy ground and alongside streams; 1000–2300 m.
The species may be recognized from the densely packed spikelets in the racemes, the short awns (almost absent in some spikelets) and the silky pubescent basal leaf sheaths, but there is a great deal of variation in other characters, particularly with regard to spikelet indumentum and leaf texture. Less constant, but still helpful characters, are the rather long peduncles, the muticous pedicelled spikelets, and the scarious rim at the apex of the raceme-base that is sometimes extended into a short appendage.
Species 15–20 in this account have glabrous pedicelled spikelets and have no exact counterpart among those species that have villous spikelets (species 22–24); the character is therefore accorded some considerable taxonomic significance. *H. dregeana* is an exception since it displays an identical range of variation in both glabrous and villous forms.

22. **Hyparrhenia tamba** (Hochst. ex Steud.) Andersson ex Stapf in Prain, F.T.A. **9**: 336 (1919). —Clayton in Kew Bull., Addit. Ser. 2: 126 (1969). —Clayton & Renvoize in F.T.E.A., Gramineae: 810 (1982). —Gibbs Russell et al., Grasses South. Africa [Mem. Bot. Surv. S. Africa No. 58]: 187 (1990). Type from Ethiopia.
Andropogon tamba Hochst. ex Steud., Syn. Pl. Glumac. **1**: 385 (1854).
Andropogon lepidus Nees var. *tamba* (Hochst. ex Steud.) Hack. in A. & C. de Candolle, Monogr. Phan. **6**: 625 (1889).

Cymbopogon tamba (Hochst. ex Steud.) Rendle in J. Linn. Soc., Bot. **40**: 227 (1911).
Hyparrhenia glauca Stent in Bothalia **1**: 251 (1923). —Sturgeon in Rhodesia Agric. J. **51**: 134 (1954). —Chippindall in Meredith, Grasses & Pastures of South Africa: 514 (1955). Type from South Africa.

Stout, caespitose perennial; culms up to 200 cm high and up to 4 mm in diameter at the base, without conspicuous stilt-roots. Basal leaf sheaths silky pubescent at the base, otherwise the sheaths glabrous; leaf laminas up to 80 cm × 3–7 mm, glabrous, stiff and harsh, scabrid on the margins and scaberulous on the nerves, strongly glaucous, rarely soft and green. False panicle 30–60 cm long, linear-oblong, lax; spatheoles 2.6–4 cm long, narrowly lanceolate, glabrous or sparsely pilose, turning glaucous brown; peduncles 2–3 cm long, $^1\!/_2$–$^3\!/_4$ as long as the spatheole, pilose above with white hairs; racemes c. 1.5 cm long, 5–8-awned per pair, exserted laterally on the curved peduncles, tardily deflexed; raceme-bases subequal, the superior up to 1 mm long, flattened, stiffly barbate, the apex produced into a lobed scarious rim 0.2–0.5 mm long. Homogamous spikelets 8–9 mm long, a single pair at the base of the inferior raceme only, villous with white hairs. Sessile spikelets 4–5.5 mm long; callus c. 0.5 mm long, cuneate, subacute at the apex; inferior glume oblong-lanceolate, villous with white hairs, becoming dark purplish-grey at maturity; awn 16–25 mm long, the column shortly pubescent with tawny hairs. Pedicelled spikelets 7–8 mm long, villous with white hairs, muticous or with a short bristle up to 2 mm long at the apex; callus scarcely developed; pedicel-tooth obscure.

Zimbabwe. C: Harare Distr., Harare Nat. Bot. Garden, 19.xi.1982, *Browning & Bennett* in *GHS* 281214 (K).
Also in Sudan, Ethiopia, Kenya, Dem. Rep. Congo and South Africa.
This species lacks the dense many-spiculate racemes and relatively long peduncles of *H. dregeana*; it resembles *H. collina* but has a more robust habit and pubescent basal leaf sheaths.

23. **Hyparrhenia rudis** Stapf in Prain, F.T.A. **9**: 344 (1919). —Jackson & Wiehe, Annot. Check List Nyasal. Grass.: 45 (1958). —Clayton in Kew Bull., Addit. Ser. 2: 128 (1969). —Simon in Kirkia **8**: 18, 51 (1971). —Clayton in F.W.T.A., ed. 2, **3**: 494 (1972). —Clayton & Renvoize in F.T.E.A., Gramineae: 811 (1982). —Gibbs Russell et al., Grasses South. Africa [Mem. Bot. Surv. S. Africa No. 58]: 186 (1990). Type from Angola.
Hyparrhenia acutispathacea (De Wild.) Robyns var. *pilosa* Bamps in Bull. Jard. Bot. État **25**: 392 (1955). Type from Dem. Rep. Congo.

Robust, coarsely caespitose perennial arising from a short slender rhizome clad in white cataphylls; culms up to 300 cm high and up to 8 mm in diameter at the base, erect and usually supported by stilt roots. Leaf sheaths glabrous; leaf laminas 30–60 cm × 3–18 mm, pale green to glaucous, stiff, harsh, scabrid on the margins and scabrid or scaberulous on the nerves. False panicle 30–50 cm long, narrowly oblong, lax or somewhat contracted, decompound; spatheoles 2.5–4 cm long, narrowly lanceolate, glabrous or thinly pilose, turning reddish-brown at maturity; peduncles 1–2 cm long, $^1\!/_3$–$^1\!/_2$ as long as the spatheole but sometimes almost as long, pilose towards the apex with off-white hairs; racemes c. 1.5 cm long, 4–7-awned per pair, laterally exserted, deflexed; raceme-bases subequal, the superior c. 1.5 mm long, flattened, stiffly barbate, the apex produced into a scarious lobe 0.2–0.5 mm long. Homogamous spikelets 5–9 mm long, a single pair at the base of the inferior raceme only, villous. Sessile spikelets 5–6 mm long; callus c. 0.6 mm long, narrowly oblong, narrowly obtuse at the apex; inferior glume oblong-lanceolate, pale or reddish-brown, silky villous with white (rarely fulvous) hairs; awn 2.2–4 cm long, the column pubescent with pallid hairs. Pedicelled spikelets 6–7 mm long, villous with white (rarely fulvous) hairs, terminating in a bristle-like awn 2–6 mm long or sometimes merely acuminate.

Zambia. B: Kaoma Distr., Luena River, 2.iv.1964, *Verboom* 1370 (K). N: 19 km south of Samfya, between escarpment and Lake Bangweulu, on edge of mushitu, 13.vi.1957, *Seagrief* 3110 (K). C: Mumbwa–Nangoma road, 20.iii.1963, *van Rensburg* 1722 (K). E: Chipata (Fort Jameson), iv.1962, *Verboom* 702 (K). S: Mazabuka, Kafue Flats, 1000 m, iv.1932, *Trapnell* 1093 (K). **Zimbabwe**. N: Gokwe South Distr., Sengwa/Korwe confluence, 19.iii.1984, *Mahlangu* 980 (K). W: Hwange Distr., Kazuma Range, bank of Katsatetsi River, c. 1000 m, 10.v.1972, *Simon* 2195 (K; PRE). C: Harare (Salisbury), Gwebi source, 25.iv.1975, *Wild* 7995 (K; PRE). **Malawi**. C: Mchinji (Fort Manning), Nyoka, 21.iv.1952, *Jackson* 776 (K). **Mozambique**. T: Angónia

Distr., Posto Zootécnico, 12.v.1948, *Mendonça* 4191 (K; LISC). MS: Sussundenga Distr., Revué Valley, 11.iv.1905, *Vasse* 209 (K).

Mostly in southern tropical Africa and South Africa, but extending northwards through Dem. Rep. Congo and Tanzania to Nigeria and Sudan; also in Madagascar. Growing in upland grassland and open woodland, favouring moist soils especially along streams and in dambos; 900–1200 m.

Very similar to *H. collina* but of more robust habit and with longer awns.

24. **Hyparrhenia collina** (Pilg.) Stapf in Prain, F.T.A. **9**: 337 (1919). —Stent & Rattray in Proc. & Trans. Rhodesia Sci. Assoc. **32**: 14 (1933). —Sturgeon in Rhodesia Agric. J. **51**: 136 (1954). —Jackson & Wiehe, Annot. Check List Nyasal. Grass.: 43 (1958). —Clayton in Kew Bull., Addit. Ser. 2: 130 (1969). —Simon in Kirkia **8**: 17, 50 (1971). —Clayton in F.W.T.A., ed. 2, **3**: 494 (1972). —Clayton & Renvoize in F.T.E.A., Gramineae: 811 (1982). —Gibbs Russell et al, Grasses South. Africa [Mem. Bot. Surv. S. Africa No. 58]: 184 (1990). Types from Tanzania and Rwanda.

Andropogon collinus Pilg. in Mildbraed (ed.), Wiss. Ergebn. Deutsch. Zentr.-Afrika Exped., Bot. **2**: 43 (1910), non Lojac. (1909).

Cymbopogon collinus Pilg. in Bot. Jahrb. Syst. **54**: 287 (1917), based on *Andropogon collinus* Pilg. Types as above.

Cymbopogon scabrimarginatus De Wild. in Bull. Jard. Bot. État **6**: 20 (1919). Type from Dem. Rep. Congo.

Hyparrhenia scabrimarginata (De Wild.) Robyns, Fl. Agrost. Congo Belge **1**: 184 (1929).

Laxly caespitose perennial arising from a short rhizome clad in white cataphylls; culms up to 130 cm high and 1–3 mm in diameter at the base, erect or ascending, wiry and slender, without stilt roots. Leaf sheaths glabrous (rarely the inner, protected, ones pubescent); leaf laminas up to 30 cm × 2–5 mm, firm but not rigid, scabrid on the margins and scaberulous on the nerves. False panicle 15–40 cm long, narrow, scanty; spatheoles 2–4 cm long, narrowly lanceolate, glabrous or thinly hirsute, turning reddish-brown; peduncles 1–2.5 cm long, a little over half as long as the spatheole, pilose above with yellowish hairs; racemes 1–2 cm long, 4–7-awned per pair, laterally exserted, deflexed; raceme-bases subequal, the superior 1–1.5 mm long, flattened, stiffly barbate, with an inconspicuous scarious rim. Homogamous spikelets 5–6 mm long, a single pair at the base of the inferior raceme only, villous with white hairs. Sessile spikelets 4.5–5 mm long; callus c. 0.5 mm long, cuneate, narrowly obtuse at the apex; inferior glume oblong-lanceolate, usually dark purple beneath a covering of white hairs; awn 1.5–2.5 cm long, the column puberulous with pallid hairs. Pedicelled spikelets 4–7 mm long, villous with white hairs, terminating in a short awn-point 1–3 mm long; callus scarcely developed; pedicel-tooth obscure.

Zambia. N: Mpika Distr., North Luangwa National Park, Muchinga Escarpment, 11°40'S, 32°04'E, 950 m, 6.v.1994, *P.P. Smith* 995 (K). C: Mkushi Distr., Munshiwemba, 10.xi.1941, *Stohr* 653 (K; SRGH). **Zimbabwe.** N: Hurungwe Distr., Dakwa Stream, 910 m, 19.xi.1953, *Wild* 4207 (K; PRE). W: Hwange Distr., Victoria Falls at Gregory's Pump House, 1.5 km north of Rain Forest Office, 13.vi.1977, *Simpathu* 45 (K). C: Shurugwi Distr., access road to Selukwe Mine, 1300 m, 30.iv.1973, *Simon & Biegel* 2401 (K). **Malawi.** N: Mzimba Distr., Viphya (Vipya), c. 35 km SW of Mzuzu, 1680 m, 10.iv.1971, *Pawek* 4621 (K). S: Maperere Mission, Chankalama Dambo, 2.xi.1950, *Jackson* 253 (K).

Mainly in eastern Africa but extending north to Sudan, west to Cameroon and south to South Africa (KwaZulu-Natal). Growing in upland grassland, deciduous bushland and wooded grassland, usually on damp soils, especially on riverbanks and in dambos; 910–1680 m.

The species is rather indistinctly separated from *H. dregeana*, *H. tamba* and *H. rudis*, but amalgamation of these taxa would undoubtedly be a retrograde step involving loss of useful information. Some specimens of *H. collina* (e.g. *Wild* 4207 and *Simon* 406, both from Zimbabwe) have pubescent inner basal sheaths (hairs may have been present on the outer sheaths but these would almost certainly have been lost through wear and tear) but can be distinguished from *H. tamba* by their slender culms.

25. **Hyparrhenia madaropoda** Clayton in Kew Bull., Addit. Ser. 2: 134 (1969). —Clayton & Renvoize in F.T.E.A., Gramineae: 812 (1982). TAB. **39**. Type from Uganda.

Annual; culms up to 300 cm high, stout, erect or the inferior internodes geniculately ascending, supported by stilt roots. Leaf sheaths glabrous, produced at the mouth into auricles up to 10 mm long; leaf laminas up to 45 cm × 13 mm, glabrous, the longer ones narrowed at the base into a short false petiole. False

Tab. 39. **HYPARRHENIA MADAROPODA**. 1, base of plant, showing stilt roots and upright habit of stem (× 2/3), from *A.S. Thomas* 4060; 2, inflorescence (× 2/3); 3, ligule and auricles (× 2); 4, raceme pair (× 2); 5 & 6, sessile spikelet: 5, inferior glume and callus (× 5); 6, superior glume (× 5), 2–6 from *Wilson* 1146. Drawn by Mary Grierson. From Kew Bull.

panicle c. 30 cm long, copious but becoming scanty in poorly grown specimens; spatheoles 2.5–5 cm long, lanceolate, glabrous; peduncles 3–5 mm long, much shorter than the spatheoles, densely barbate on one side near the apex with white bulbous-based hairs, deflexed; racemes 1.5–2 cm long, 2-awned per pair, not deflexed (instead the racemes exserted by flexion of the peduncle); raceme-bases unequal, the superior 2–2.5 mm long, subterete, thinly embellished with stiff bristles, the apex very oblique and produced into a scarious extension 0.5–0.7 mm long. Homogamous spikelets 8–9 mm long, a single pair at the base of the inferior raceme only, glabrous, awnless. Sessile spikelets 7–9.5 mm long; callus 1.5–3 mm long, pungent; inferior glume lanceolate, pallid, glabrous or hispidulous; awn 4.5–8 cm long, the column pubescent with pallid hairs c. 0.2 mm long. Pedicelled spikelets 8–9 mm long, linear-lanceolate, glabrous, with or without a terminal bristle up to 6 mm long; callus scarcely developed; pedicel-tooth obscure.

Zambia. N: Kaputa Distr., west shore of Lake Mweru Wantipa, 12.vi.1933, *Michelmore* 417 (K). **Mozambique**. N: Montepuez Distr., andados 43 km de Nairoto (Nantulo) para Mueda, c. 300 m, 10.iv.1964, *Torre & Paiva* 11858 (LISC).
Also in Uganda, Tanzania, Kenya, Sudan and Dem. Rep. Congo. Growing in deciduous bushland; 300–900 m.
A difficult species to place in a section of the genus with any certainty. The unequal raceme-bases with the superior subterete, the lack of significant appendage to the apex of the raceme-bases and the latter being almost glabrous and not deflexed are somewhat at odds with the circumscription of Sect. *Hyparrhenia*. However, it is also clear that there is nothing in any other section to which it is at all closely related. The lack of deflexion of the raceme-bases is compensated for by the flexure of the unilaterally barbate peduncle. The general aspect of the racemes, particularly the fact that they are only 2-awned, points to a probable relationship to other members of Sect. *Hyparrhenia*.

26. **Hyparrhenia niariensis** (Franch.) Clayton in Kew Bull., Addit. Ser. 2: 140 (1969). —Simon in Kirkia **8**: 51 (1971). —Clayton in F.W.T.A., ed. 2, **3**: 494 (1972). —Clayton & Renvoize in F.T.E.A., Gramineae: 813 (1982). Type from Dem. Rep. Congo.
    Andropogon niariensis Franch. in Bull. Soc. Hist. Nat. Autun **8**: 330 (1895).
    Andropogon viancinii Franch. in Bull. Soc. Hist. Nat. Autun **8**: 331 (1895). Type from Central African Republic.
    Cymbopogon welwitschii Rendle var. *minor* Rendle in Hiern et al., Cat. Afr. Pl. Welw. **2**: 159 (1899). Type from Angola.

Annual; culms up to 200 cm high, glabrous, supported by stilt roots. Leaf sheaths glabrous, or pubescent along the margins; leaf laminas up to 60 cm × 15 mm, glabrous or sometimes pubescent beneath, narrowed to the midrib at the base and sometimes falsely petiolate. False panicle 30–50 cm long, lax, leafy; spatheoles 3–5 cm long, narrowly lanceolate, glabrous or sparsely pilose; peduncles about as long as the spatheoles, barbate all round with yellow bulbous-based hairs, ± straight; racemes 1.5–1.8 cm long, 2-awned per pair (rarely 3-awned but not in the Flora Zambesiaca area), terminally or subterminally exserted, deflexed; raceme-bases subequal, the superior 1–1.5 mm long, flattened, stiffly setose with fulvous bulbous-based bristles, the apex produced into an oblong or narrowly oblong appendage 0.5–4 mm long. Homogamous spikelets 9–11 mm long, a single pair at the base of the inferior raceme only, glabrous, awnless. Sessile spikelets 7.5–11 mm long; callus 2–3 mm long, pungent; inferior glume linear-oblong, often shallowly hollowed on the back, ± pubescent; awn 6.5–8.5(10.5) cm long, the column pubescent with pale or rufous hairs 0.2–0.6 mm long. Pedicelled spikelets 8–13 mm long, linear-lanceolate, glabrous, terminating in a slender bristle 2–10 mm long; callus scarcely developed; pedicel-tooth obscure.

Zambia. N: Mpika Distr., North Luangwa National Park, upper escarpment, 11°40'S, 31°57'E, 1200 m, 28.iv.1994, *P.P. Smith* 991 (K). W: Mongu Distr., Mulundu, 960 m, 1.iv.1961, *Symoens* 8460 (K).
Around the perimeter of the Congo Basin in Uganda, Tanzania and western Africa from Cameroon to Angola. Growing in wooded grassland; 960–1200 m.
The species merges somewhat with *H. welwitschii* and is best distinguished by the larger spikelets, the raceme pairs with 2 rather than 3 awns, and the better developed appendage on the raceme-base.

27. **Hyparrhenia welwitschii** (Rendle) Stapf in Prain, F.T.A. **9**: 356 (1919). —Vesey-FitzGerald in Kirkia **3**: 107 (1963). —Clayton in Kew Bull., Addit. Ser. 2: 142 (1969). —Simon in Kirkia **8**: 18, 51 (1971). —Clayton in F.W.T.A., ed. 2, **3**: 494 (1972). —Clayton & Renvoize in F.T.E.A., Gramineae: 813 (1982). Types from Angola.

   Cymbopogon welwitschii Rendle in Hiern et al., Cat. Afr. Pl. Welw. **2**: 157 (1899).
   Andropogon welwitschii (Rendle) K. Schum. in Just's Bot. Jahresber. **27**(1): 454 (1901).
   Hyparrhenia gracilescens Stapf in Prain, F.T.A. **9**: 357 (1919). —Jackson & Wiehe, Annot. Check List Nyasal. Grass: 44 (1958). Types from Guinea and Nigeria.

   Annual; culms up to 300 cm high, glabrous or with a ring of white or brown hairs at each node, often supported by stilt roots. Leaf sheaths glabrous; leaf laminas up to 60 cm × 12 mm, glabrous, narrowed at the base. False panicle 30–60 cm long, lax, leafy; spatheoles 3–5 cm long, narrowly lanceolate, glabrous; peduncles a little shorter than the spatheoles, barbate all round above with white or yellow bulbous-based hairs, ± straight; racemes 1.2–1.7 cm long, 3-awned per pair, laterally exserted, deflexed; raceme-bases subequal, the superior c. 1 mm long, flattened, stiffly setose with pale rufous bulbous-based bristles, the apex produced into an oblong, reddish, irregularly lobed appendage 0.5–1 mm long. Homogamous spikelets 7–10 mm long, a single pair at the base of the inferior raceme only, glabrous, awnless. Sessile spikelet 5–7 mm long; callus c. 1.5 mm long, pungent; inferior glume linear-oblong, greyish, glabrescent to sparsely pubescent with white hairs; awn 4–7 cm long, the column pubescent with fulvous hairs 0.2–0.6 mm long. Pedicelled spikelets 6–8 mm long, linear-lanceolate, glabrous, terminating in a slender awn 2–11(14) mm long; callus scarcely developed; pedicel-tooth short.

   Zambia. N: 27 km east of Kawambwa on Mushota road, 7.iv.1961, *Angus* 2741 (K). W: Ndola Forest Reserve, 29.iv.1950, *Jackson* 6 (K). C: Mpika Distr., near Kapamba River, South Luangwa National Park, (Luangwa Valley Game Reserve), c. 760 m, 1.iv.1966, *Astle* 4760 (K). **Zimbabwe**. E: Mutasa Distr., Nyamkwarara, *Gilliland* 2022 (K; SRGH). **Malawi**. N: Mzimba Distr., c. 30 km west of Mzuzu, Lunyangwa River Bridge, c. 1180 m, 18.iv.1974, *Pawek* 8347 (K; PRE). C: Mchinji (Fort Manning), Nyoka, 21.iv.1952, *Jackson* 767 (K). S: Zomba, 950 m, 14.v.1949, *Wiehe* 103 (K). **Mozambique**. N: Malema, encosta oriental da serra Inago, 850 m, 20.iii.1964, *Torre & Paiva* 11273 (K; LISC). Z: Alto Molócuè Distr., between Alto Molócuè and Gilé, vii.1943, *Torre* 5706 (K; LISC). T: Cahora Bassa Distr., between Marueira and Songo, 3–4 km from Songo, 24.iii.1972, *Macêdo* 5080 (LISC).

   West tropical Africa from Guinea to Angola, eastwards to Sudan and the Comoro Islands. Growing in light or deep shade in wooded grassland, usually on deep moist soils, but occasionally on lighter sandy soils; 700–1220 m.

28. **Hyparrhenia bracteata** (Humb. & Bonpl. ex Willd.) Stapf in Prain, F.T.A. **9**: 360 (1919). —Stent & Rattray in Proc. & Trans. Rhodesia Sci. Assoc. **32**: 15 (1933). —Sturgeon in Rhodesia Agric. J. **51** 137 (1954). —Jackson & Wiehe, Annot. Check List Nyasal. Grass.: 43 (1958). —Clayton in Kew Bull., Addit. Ser. 2: 144 (1969). —Simon in Kirkia **8**: 17, 50 (1971). —Clayton in F.W.T.A., ed. 2, **3**: 494 (1972). —Clayton & Renvoize in F.T.E.A., Gramineae: 814 (1982). Type from Venezuela.

   Andropogon bracteatus Humb. & Bonpl. ex Willd., Sp. Pl. **4**: 914 (1806).
   Andropogon setifer Pilg. in Mildbraed, Wiss. Ergebn. Deutsch. Zentr.-Afrika Exped., Bot. **2**: 44 (1910). Type from Tanzania.
   Cymbopogon setifer (Pilg.) Pilg. in Bot. Jahrb. Syst. **54**: 287 (1917).
   Cymbopogon pilosovaginatus De Wild. in Bull. Jard. Bot. État **6**: 17 (1919). Types from Dem. Rep. Congo.
   Hyparrhenia contracta Robyns, Fl. Agrost. Congo Belge **1**: 189 (1929). Type from Dem. Rep. Congo.
   Cymbopogon giganteus sensu Stent & Rattray in Proc. & Trans. Rhodesia Sci. Assoc. **32**: 12 (1933), non Chiov.

   Densely caespitose perennial; culms up to 250 cm high, erect, glabrous or sometimes villous for a short distance below the nodes. Leaf sheaths usually sparsely pilose, sometimes densely villous, rarely glabrous, the basal nearly always tomentose with pale brown hairs below; leaf laminas up to 60 cm × 4 mm, rigid, scabrid along the margins, usually pubescent below. False panicle 20–60 cm long, narrow, dense; spatheoles 2–3 cm long, narrowly lanceolate, usually appressed-hirsute but sometimes only along the margins; peduncles a little shorter than the spatheoles, copiously barbate all round above with stiff bulbous-based yellow hairs; racemes 0.5–1.5 cm long, 2–4-awned per pair, dark purple, laterally exserted and deflexed;

raceme-bases slightly unequal, the superior 1.5–2 mm long, flattened or subterete, barbate with stiff yellow hairs, produced at the apex into a narrowly oblong, purple, irregularly bifid or lobed appendage 1–2.5 mm long, this glabrous, pubescent or even villous. Homogamous spikelets 4–7 mm long, a single pair at the base of the inferior raceme only, glabrous with scabrid margins, awnless. Sessile spikelets 4–6 mm long; callus c. 1 mm long, acute; inferior glume narrowly linear-oblong, coriaceous, glabrous or rarely hirsute, the midnerve raised between two fine longitudinal grooves; awn 1–2.5 cm long, the column shortly pubescent with white hairs. Pedicelled spikelets 4–6(7) mm long, glabrous, muticous or with a short mucro seldom over 1 mm long; callus scarcely developed; pedicel-tooth obtusely triangular, very obscure.

Zambia. B: Kalabo, vii.1933, *Trapnell* 1348 (K). N: Mbala Distr., Lake Chila, 1620 m, 10.iv.1958, *Vesey-FitzGerald* 1605 (K). W: Chingola Distr., between Chingola and Kitwe, 31.x.1965, *van Rensburg* 3046 (K). C: Serenje Distr., Bolelo River, 1370 m, 23.xii.1963, *Symoens* 10665 (K). E: Chipata Distr., between Katete and Kazimuli, 10.v.1963, *van Rensburg* 2116 (K). S: Kalomo Distr., Ihulu, 1220 m, iii.1934, *Trapnell* 1492 (K). **Zimbabwe**. N: Guruve Distr., Nyamunyeche, near Gurungwe Gap, 31.v.1985, *Martin* s.n. (K). C: Makoni Distr., Nyazura (Inyazura), 1250 m, iii.1920, *Lloyd* 2399 (K). S: Zaka Distr., Ndanga, v.1926, *Roberts in Dept. Agric.* 1231 (K). **Malawi**. N: Mzimba Distr., Mzuzu, 1370 m, 15.vi.1974, *Pawek* 8711 (K). C: Dedza Mt., 1490 m, 27.iii.1950, *Wiehe* 464 (K). S: Kirk Range, M'Vai, 14°52'S, 34°35'E, 13.vi.1950, *Wiehe* 589 (K). **Mozambique**. T: Angónia, monte Dzenza, 1500 m, 10.xii.1980, *Stefanesco & Nyongani* 585 (K).

Tropical Africa, mainly in the west, and tropical America. Boggy and marshy places in open and wooded grasslands; 1000–1650 m.

Very closely allied to *H. newtonii* with which it shares the characteristic purple racemes framed in yellow hairs. It is distinguished by its denser panicle, its smaller spikelets and awns, its preference for a marshy habitat and, most reliably, by its virtual absence of a pedicel tooth.

29. **Hyparrhenia newtonii** (Hack.) Stapf in Prain, F.T.A. **9**: 363 (1919). —Sturgeon in Rhodesia Agric. J. **51**: 137 (1954). —Chippindall in Meredith, Grasses & Pastures S. Africa: 513 (1955). —Clayton in Kew Bull., Addit. Ser. 2: 148 (1969). —Simon in Kirkia **8**: 18, 51 (1971). —Clayton in F.W.T.A., ed. 2, **3**: 494 (1972). —Clayton & Renvoize in F.T.E.A., Gramineae: 816, fig. 186 (1982). —Gibbs Russell et al., Grasses South. Africa [Mem. Bot. Surv. S. Africa No. 58]: 186 (1990). Type from Angola.

Andropogon newtonii Hack. in Bol. Soc. Brot. **3**: 137 (1885).

Andropogon lecomtei Franch. in Bull. Soc. Hist. Nat. Autun **8**: 329 (1895). Types from Congo-Brazzaville.

Cymbopogon lecomtei (Franch.) Rendle in J. Linn. Soc., Bot. **40**: 227 (1911).

Hyparrhenia lecomtei (Franch.) Stapf in Prain, F.T.A. **9**: 361 (1919). —Jackson & Wiehe, Annot. Check List Nyasal. Grass.: 44 (1958).

Hyparrhenia stolzii Stapf. in Prain, F.T.A. **9**: 364 (1919). —Vesey-FitzGerald in Kirkia **3**: 107 (1963). Type from Tanzania.

Hyparrhenia cirrosula Stapf, tom. cit.: 365. Type from Dem. Rep. Congo.

Hyparrhenia bisulcata Chiov. in Nuovo Giorn. Bot. Ital., n.s. **26**: 60 (1919). Types from Dem. Rep. Congo.

Hyparrhenia lecomtei var. *bisulcata* (Chiov.) Robyns, Fl. Agrost. Congo Belge **1**: 192 (1929).

Hyparrhenia squarrulosa Peter in Repert. Spec. Nov. Regni Veg. **40**(1): 375; Anh.: 118 (1936). Type from Tanzania.

Densely caespitose perennial; culms up to 210 cm high, erect, glabrous. Leaf sheaths glabrous to thinly pilose, the basal tomentose or glabrous below; leaf laminas up to 30 cm × 3 mm, rigid, scabrid along the margins, glabrous or pubescent below. False panicle 15–30 cm long, diffuse or scanty; spatheoles 2.5–5 cm long, narrowly lanceolate, glabrous or villous along the margins, rarely villous all over; peduncles a little shorter than the spatheoles, barbate all round above with stiff bulbous-based yellow hairs; racemes 1.5–2 cm long, 2–4-awned per pair, dark purple, laterally exserted and deflexed; raceme-bases unequal, the superior 1.5–3(4) mm long, flattened below, subterete above, barbate with stiff yellow or occasionally pallid hairs, produced at the apex into a linear, purple, entire or bidentate appendage 1–3 mm long. Homogamous spikelets 5–10 mm long, a single pair at the base of the inferior raceme only, glabrous, awnless. Sessile spikelets 6–10 mm long; callus 1.5–2 mm long, acute to pungent; inferior glume linear-oblong, coriaceous, glabrous or sometimes pilose towards the apex, often pubescent or pilose all over, the midnerve raised between two fine longitudinal grooves; awn 2.2–5.5 cm long, the column pubescent. Pedicelled spikelets 5–10 mm long, glabrous or sometimes pilose, terminating in a bristle 1–5 mm long; callus scarcely developed; pedicel-tooth 0.5–1.2 mm long, narrowly triangular or linear-oblong to subulate.

Var. **newtonii** TAB. **40**.

Inferior glume of sessile spikelet glabrous, sometimes thinly pilose towards the apex; pedicelled spikelets glabrous.

Zambia. B: Kaoma Distr., 56 km along Kafue Watershed, 2.iv.1964, *Verboom* 1367 (K). N: Mbala Distr., path to Sansia Falls on Kalambo River, 7.v.1961, *Vesey-FitzGerald* 3380 (K). W: Mufulira, Copperbelt Exp. Farm, 21.iv.1966, *Lawton* 1394 (K). S: Kalomo, 1220 m, v.1934, *Trapnell* 1446 (K). **Zimbabwe**. N: Bindura Distr., Kingstone Hill, 7.v.1969, *Wild* 7761 (K). C: Mazowe Distr., Spelonken Farm, c. 1280 m, 15.iii.1981, *H.H. Burrows* 1707 (K). E: Nyanga, 1700 m, 11.i.1931, *Norlindh & Weimarck* 4183a (K; PRE). **Malawi**. N: Mzimba Distr., Mzuzu, Marymount, 1370 m, 31.v.1973, *Pawek* 6773 (K). C: Mchinji (Fort Manning), Nyoka, 21.v.1952, *Jackson* 769 (K). S: Zomba Distr., Chongosi area, near Met. Station, 7.vi.1961, *Chapman* 1364 (K; PRE). **Mozambique**. T: entre Furancungo e a serra de Pandalanjala, 15.v.1948, *Mendonça* 4254 (K; LISC). MS: Sussundenga Distr., Zuira mountains, 15 km on road to Chimoio (Vila Pery), c. 1400 m, 5.iv.1966, *Torre & Correia* 15730 (LISC).

Tropical Africa, mainly in the west, South Africa and Madagascar; also in Thailand, Indo-China and Indonesia. Growing in grassland on stony hillsides, dambos and wooded grassland; 1070–1700 m.

One specimen (*Jackson* 819 from Mposa, Malawi) is discordant in *H. newtonii*. Its raceme-bases are only sparsely pilose, they are not deflexed, and their appendage is 4 mm long. The spikelet dimensions and colouring are otherwise typical of the species, but the grooves in the inferior glume of the sessile spikelet are coalescent into a single deep median furrow. The specimen is either an exceptional form of *H. newtonii*, which does not seem very likely, or it is a product of introgression from *Hyperthelia dissoluta* (Nees) Clayton.

Var. **macra** Stapf in Prain, F.T.A. **9**: 364 (1919). —Jackson & Wiehe, Annot. Check List Nyasal. Grass.: 45 (1958). —Simon in Kirkia **8**: 18, 51 (1971). —Clayton & Renvoize in F.T.E.A., Gramineae: 816 (1982). —Gibbs Russell et al., Grasses South. Africa [Mem. Bot. Surv. S. Africa No. 58]: 185 (1990). Type: Zimbabwe C: Harare (Salisbury), *Craster* 59 (K, lectotype).

Inferior glume of sessile spikelet pubescent to villous; pedicelled spikelets also sometimes villous.

Zambia. N: Mbala Distr., path to Sansia Falls on Kalambo River, 7.v.1961, *Vesey-FitzGerald* 3381 (K). W: Ndola, 1220 m, v.1961, *Wilberforce* A/87 (K). C: Serenje Distr., Kundalila Falls, 13 km SE of Kanona, 1370 m, 5.iv.1961, *Phipps & Vesey-FitzGerald* 2963 (K). S: Namwala Distr., upper Mulela floodplain, 19.iv.1963, *van Rensburg* 2062 (K). **Zimbabwe**. N: Hurungwe Distr., Mwami (Miami), K.34 Exp. Farm, 1370 m, 6.iii.1947, *Wild* 1676 (K). C: Harare (Salisbury), 1460 m, ii.1920, *Rowland* 2480 (K). E: Nyanga Distr., Susurumba, 23.v.1968, *Wild* 7719 (K; PRE). **Malawi**. N: Chitipa Distr., Nyika Plateau, 2130 m, 11.ii.1968, *Simon, Williamson & Ball* 1750 (K). C: Lilongwe Distr., Chitedze, vi.1950, *Palmer* in *Wiehe* 619 (K). S: Mulanje Mt., Chambe Plateau, 2130 m, 17.xi.1949, *Wiehe* 342 (K). **Mozambique**. Z: Gurué Distr., picos Namuli, 1500 m, 9.iv.1943, *Torre* 5131 (K; LISC). T: Mutarara Distr., NW of Dôa, 13.iv.1972, *Bond* 50 (LISC).

Mainly in the southern part of the range of var. *newtonii*, in similar habitats but reaching to 2350 m in Malawi.

The pilose sessile spikelets, southern bias in distribution and higher altitudes reached in parts of its range are all that justify maintaining var. *macra* as a distinct taxon.

30. **Hyparrhenia anemopaegma** Clayton in Kew Bull. Addit. Ser. 2: 154 (1969). —Simon in Kirkia **8**: 50 (1971). Type: Zambia C: Luangwa Valley, South Luangwa National Park, Kapamba River, *Astle* 4759 (K, holotype).

Slender annual; culms up to 90 cm high, erect. Leaf sheaths glabrous; leaf laminas up to 25 cm × 9 mm, dull green. False panicle scanty, comprising 3–10 raceme-pairs in 3 or 4 distant tiers; spatheoles 9–14 cm long, linear, green or tinged with red; peduncles $^1/_3$–$^3/_4$ as long as the spatheoles, densely barbate with pale yellow hairs near the apex; racemes c. 2 cm long, 4–6-awned per pair, deflexed; racemes-bases subequal, the superior not much longer than the inferior, flattened, glabrous, with a shallowly lobed scarious rim at the apex. Homogamous spikelets 11–16 mm long, a single pair at the base of each raceme, shortly pilose with white hairs on the back,

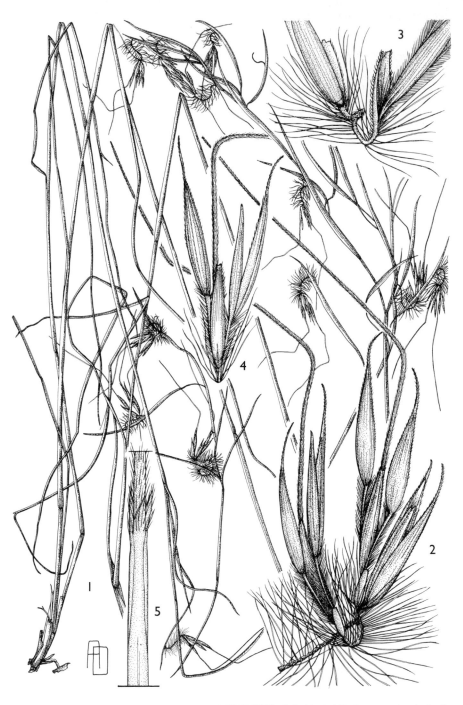

Tab. 40. HYPARRHENIA NEWTONII var. NEWTONII. 1, habit (× ¹/₂); 2, raceme pair (× 4); 3, raceme bases (× 6); 4, lower raceme (× 4); 5, tip of superior lemma (× 10), 1–5 from *Procter* 3313. Drawn by Ann Davies. From F.T.E.A.

pectinate-ciliate on the margins. Sessile spikelets c. 9 mm long; callus 2.5–3 mm long, pungent; inferior glume lanceolate, pubescent; awns 7.5–11 cm long, the column fulvously pilose with hairs up to 1.5 mm long. Pedicelled spikelets 10–11 mm long, narrowly lanceolate, pilose with white hairs on the back, pectinate-ciliate on the margins, terminating in a slender bristle 11–22 mm long; callus c. 0.5 mm long, oblong; pedicel-tooth short, triangular.

Zambia. N: Mpika Distr., North Luangwa National Park, Lubanga River, 11°46'S, 32°10'E, 690 m, 30.iii.1994, *P.P. Smith* 974 (K). C: Mpika Distr., South Luangwa National Park, Kapamba River, 760 m, 1.iv.1966, *Astle* 4759 (K).

Not known outside of the Luangwa National Park. Growing in riverine grassland and savanna woodland; 610–760 m.

31. **Hyparrhenia subplumosa** Stapf in Prain, F.T.A. **9**: 366 (1919). —Clayton in Kew Bull., Addit. Ser. 2: 164 (1969). —Simon in Kirkia **8**: 51 (1971). —Clayton in F.W.T.A., ed. 2, **3**: 496 (1972). —Clayton & Renvoize in F.T.E.A., Gramineae: 818 (1982). Type from Nigeria.

Robust perennial; culms up to 300 cm high, erect. Leaf sheaths glabrous or rarely pilose towards the summit; leaf laminas 20–60 cm × 3–10 mm, glaucous, glabrous or sparsely pilose below, usually with greyish hairs near the base. False panicle 20–50 cm long, loose; spatheoles 3–7 cm long, narrowly lanceolate, glaucous to purplish, glabrous; peduncles 1–3.5 cm long, glabrous or shortly and softly hirsute above, often recurved; racemes 1.5–2.5 cm long, 3–6-awned per pair, deflexed; raceme-bases slightly unequal, the superior 1.5–2 mm long, flattened, glabrous (except in the fork), truncate at the apex and lacking appendages. Homogamous spikelets 8–10 mm long, two pairs at the base of each raceme forming a kind of involucre, glabrous or rarely sparsely pubescent on the back, scabrid on the margins, initially green but becoming purplish, awnless. Sessile spikelets 6.5–7.5 mm long; callus c. 1.5 mm long, sharply acute; inferior glume narrowly lanceolate, glabrous or less commonly pubescent to tomentose; awns 4.5–7.5 cm long, the column subplumose with white or fulvous hairs 0.5–1.3(1.7) mm long. Pedicelled spikelets 7–8 mm long, narrowly lanceolate, glabrous, terminating in a bristle 2–7 mm long; callus scarcely developed; pedicel-tooth obscure.

Zambia. N: Mansa Distr., Mote Area, road to Kapalala from Kasanka, Lumanya Dambo, 1.iii.1996, *Renvoize* 5743 (K). C: Muchinga Escarpment, 11°25'S, 32°01'E, 1200 m, 29.iv.1994, *P.P. Smith* 994 (K). **Malawi**. N: Mzimba Distr., Marymount, Mzuzu, 1370 m, 30.v.1971, *Pawek* 4879 (K).

Widespread in West tropical Africa west of the Cameroon Mountains, but with scattered records to the south and east in Dem. Rep. Congo, Angola, Zambia and Malawi. Growing in *Brachystegia* woodland, along riverbanks and in dambos; 1200–1710 m.

The species forms a distinct tetraploid (2n = 40) population related to *H. diplandra* (2n = 20) to the west of the Cameroon Mountains but with a subsidiary population to the east centred on Zambia. According to Clayton (1969, op. cit.) variation in the material from Zambia and Malawi conforms more to that of *H. diplandra*, with respect to peduncle length and awn length, than it does with *H. subplumosa*, but the point is equivocal and in awn-hair length these specimens very much agree with *H. subplumosa*. In terms of the immense difficulties that the genus presents to the taxonomist it is probably better for now to retain the distinction than to begin a process of amalgamating taxa that may be difficult to stop.

32. **Hyparrhenia diplandra** (Hack.) Stapf in Prain, F.T.A. **9**: 368 (1919). —Sturgeon in Rhodesia Agric. J. **51**: 137 (1954). —Jackson & Wiehe, Annot. Check List Nyasal. Grass.: 43 (1958), in part (excl. *Jackson* 556 cited in error). —Clayton in Kew Bull., Addit. Ser. 2: 166 (1969). —Simon in Kirkia **8**: 18, 51 (1971). —Clayton in F.W.T.A., ed. 2, **3**: 496 (1972). —Clayton & Renvoize in F.T.E.A., Gramineae: 818 (1982). Types from Sudan.

Andropogon diplandrus Hack. in Flora **68**: 123 (1885); in A. & C. de Candolle, Monogr. Phan. **6**: 627 (1889).

Sorghum diplandrum (Hack.) Kuntze, Revis. Gen. Pl. **2**: 791 (1891).

Cymbopogon phoenix Rendle in Hiern et al., Cat. Afr. Pl. Welw. **2**: 156 (1899). Types from Angola.

Andropogon phoenix (Rendle) K. Schum. in Just's Bot. Jahresber. **27**(1): 454 (1901).

Hyparrhenia pachystachya Stapf in Prain, F.T.A. **9**: 370 (1919). —Sturgeon, loc. cit. (1954). Syntypes: Zambia C: *Macauley* s.n. (K, syntype) and Zimbabwe E: *Swynnerton* 993 (K, syntype).

Cymbopogon diplandrus (Hack.) De Wild. in Bull. Jard. Bot. État **6**: 11 (1919).

Coarse, caespitose perennial; culms up to 300 cm high, erect. Leaf sheaths glabrous, or pilose towards the summit, rarely pilose throughout; leaf laminas 20–60 cm × 3–10 mm, glabrous or sometimes sparsely hirsute below, usually with long grey hairs at the base. False panicle 20–40 cm long, narrow, usually purplish; spatheoles 2–4.5 cm long, narrowly lanceolate, brownish-red, glabrous, barbate or not at the base; peduncles 0.3–1.5 cm long, glabrous or shortly hirsute above; racemes 1.5–2(2.5) cm long, (3)4–6(9)-awned per pair, deflexed; raceme-bases subequal, the superior 1–2 mm long, flattened, glabrous (except in the fork), truncate at the apex and lacking appendages. Homogamous spikelets 7–9 mm long, 2 pairs at the base of each raceme forming a kind of involucre, scabrid on the margins, muticous. Sessile spikelets 6–8 mm long; callus c. 1.5 mm long, sharply acute; inferior glume narrowly lanceolate to lanceolate-oblong, glabrous to pubescent; awns 2–5.5 mm long, the column pubescent with white or fulvous hairs 0.2–0.4(0.5) mm long, rarely quite awnless. Pedicelled spikelets 5–7.5 mm long, narrowly lanceolate, muticous or with a terminal bristle up to 5 mm long; callus scarcely developed; pedicel-tooth obscure.

Throughout tropical Africa but with a limited distribution west of the Cameroon Mountains; also in Indo-China and Indonesia. Habitat very variable, favouring damp places in deciduous bushland and wooded grassland, and extending onto stony hillsides and into areas of cultivation; 0–1830 m.

Var. **diplandra**

Superior lemma of sessile spikelet awned.

Zambia. B: Kaoma Distr., 113 km on Mongu road, 17.iv.1964, *Verboom* 1384 (K). N: Samfya Distr., Kalasa Mukoso Flats, 1200 m, 23.ii.1996, *Renvoize* 5681 (K). W: Solwezi Distr., Solwezi Dambo on road to Chifubwa Gorge, c. 5 km south of Solwezi, 1310 m, 20.xii.1969, *Simon & Williamson* 1860 (K). C: Mkushi Distr., Munshiwemba, x.1941, *Stohr* 630 (K). S: Mumbwa, 20.vi.1932, *Trapnell* 2083 (K). **Zimbabwe**. E: Chimanimani (Melsetter), 1830 m, iii.1934, *Brain* 10586 (K; SRGH). **Malawi**. N: Chitipa Distr., Misuku Hills, Mugesse (Mughesse), 1590 m, 7.vii.1973, *Pawek* 7068 (K). C: Lilongwe Distr., Dzalanyama Forest Reserve, 27.iv.1958, *Jackson* 2222 (K). **Mozambique**. Z: Morrumbala Distr., Metolola, 23.v.1943, *Torre* 5377 (K; LISC). T: Angónia Distr., Posto Zootécnico, 12.v.1948, *Mendonça* 4186 (K; LISC). MS: Mossurize Distr., Espungabera Mission, 8.vi.1942, *Torre* 4263 (K; LISC).

Var. **mutica** (Clayton) Cope, comb. et stat. nov. TAB. **41**. Type: Liberia, Nimba, *Adames* 746 (K, holotype).
  Hyparrhenia mutica Clayton in Kew Bull., Addit. Ser. 2: 161 (1969). —Simon in Kirkia **8**: 51 (1971). —Clayton in F.W.T.A., ed. 2, **3**: 494 (1972). —Clayton & Renvoize in F.T.E.A., Gramineae: 818 (1982).

Superior lemma of sessile spikelet quite awnless.

Zambia. N: Mpika Distr., Mpika to Serenje 48 km, Mufubushi School, Ngulube Dambo, 12°12'S, 31°13'E, 1600 m, 27.v.1995, *Bingham* 10565 (K). W: Copperbelt Distr., Mufulira, 23.v.1934, *Eyles* 8392 (K; SRGH). **Malawi**. S: Machinga Distr., Chikweo (Chikwewo), 16.v.1952, *Jackson* 828 (K). **Mozambique**. N: Lichinga (Vila Cabral), 26.v.1934, *Torre* 103 (LISC). Z: R. Licungo, serra do Gurué, 20.xi.1944, *Mendonça* 2129 (K; LISC).
  The awnless variant of the species was hitherto regarded as a separate species on account of the distinctive morphology and apparent difference in chromosome number (var. *mutica* 2n = 60; var. *diplandra* 2n = 20) based on a single count for each. The latter character has not been thoroughly explored and it is not known for sure whether the two varieties are consistently different in chromosome number. They often occur in mixed populations more or less throughout the range of the species with var. *mutica* much the scarcer and this would seem to indicate that a rank no higher than variety is appropriate.

33. **Hyparrhenia gossweileri** Stapf in Prain, F.T.A. **9**: 371 (1919). —Clayton in Kew Bull., Addit. Ser. 2: 172 (1969). —Simon in Kirkia **8**: 51 (1971). —Clayton & Renvoize in F.T.E.A., Gramineae: 819 (1982). Type from Angola.
  Cymbopogon bequaertii De Wild. in Bull. Jard. Bot. État **6**: 8 (1919). Type from Dem. Rep. Congo.
  Hyparrhenia bequaertii (De Wild.) Robyns, Fl. Agrost. Congo Belge **1**: 197 (1929).

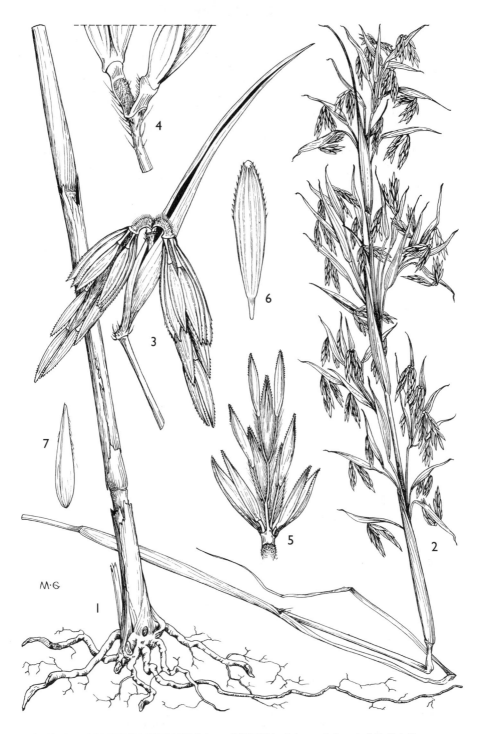

Tab. 41. HYPARRHENIA DIPLANDRA var. MUTICA. 1, base of plant ($\times \frac{2}{3}$); 2, inflorescence ($\times \frac{2}{3}$); 3, raceme pair ($\times$ 4); 4, raceme bases ($\times$ 12); 5, single raceme, spread open ($\times$ 4); 6, sessile spikelet, showing inferior glume and callus ($\times$ 6); 7, superior lemma of sessile spikelet ($\times$ 6), 1–7 from *Adames* 746. Drawn by Mary Grierson. From F.T.E.A.

Caespitose perennial; culms up to 200 cm high, erect. Leaf sheaths glabrous or shortly pubescent towards the summit and with greyish hairs at the mouth; leaf laminas up to 30 cm × 3–6 mm, glabrous. False panicle 30–60 cm long, scanty, lax; spatheoles 4–7 cm long, narrowly lanceolate, reddish-purple, glabrous, the peduncles 1.5–3 cm long, glabrous or shortly hirsute above; racemes 2.5–3 cm long, 6–12-awned per pair, deflexed; raceme-bases unequal, the inferior c. 1 mm long, the superior 3–4 mm long, flattened, pubescent below but not barbate, truncate at the apex and lacking appendages. Homogamous spikelets 8–10 mm long, 1 pair at the base of each raceme, scabrid on the margins near the apex, awnless. Sessile spikelets c. 8 mm long; callus c. 2 mm long, sharply acute; inferior glume narrowly lanceolate, glabrous, faintly 11–13-nerved; awns 3.5–5 cm long, the column pubescent with fulvous hairs 0.2–0.3 mm long. Pedicelled spikelets 7–8 mm long, narrowly lanceolate, glabrous, muticous; callus scarcely developed; pedicel-tooth obscure.

Zambia. N: Kawambwa Distr., Chishinga Ranch, 1430 m, 14.v.1964, *Astle* 3023 (K). W: Ndola, 1933, *Duff* s.n. (K).
Distribution centred on Dem. Rep. Congo, but also in Tanzania and Angola. Growing in savanna woodland; 1430–1590 m.
A somewhat anomalous species in Sect. *Apogonia* because of the single homogamous spikelet pair at the base of each raceme. It otherwise resembles *H. diplandra* which also on occasion has only 1 homogamous pair (at least in other parts of Africa: in the Flora Zambesiaca area all material seems to conform to the description given above). Other distinguishing characters are provided by the unusually long superior raceme-base and by the longer racemes with more awns per pair.

27. EXOTHECA Andersson

By T.A. Cope

Exotheca Andersson in Nova Acta Regiae Soc. Sci. Upsal., ser. 3, **2**: 253 (1857). — Clayton in Kew Bull. **20**: 447 (1966).

Perennial. Ligule scarious, adnate to the sheath-auricles; leaf laminas linear. Inflorescence composed of paired racemes, the pairs each subtended by a spatheole and gathered into a scanty false panicle; racemes short, the superior borne on a very long raceme-base, its basal spikelets at ± the same level as the apex of the inferior raceme, the two thus appearing to be arranged end-to-end, each raceme with 2 homogamous spikelet pairs; pedicels linear. Sessile spikelet terete; callus pungent, applied obliquely; inferior glume subcoriaceous with a herbaceous beak 2–3 mm long, rounded on the back and sides; superior glume awnless; inferior floret reduced to a hyaline lemma; superior lemma hyaline, bidentate, passing between the teeth into an awn with pubescent column. Caryopsis narrowly oblong. Pedicelled spikelet male, without a distinct callus.

A genus of one species, occurring in Indo-China and tropical Africa.

Exotheca abyssinica (Hochst. ex A. Rich.) Andersson, in Nova Acta Regiae Soc. Sci. Upsal., ser. 3, **2**: 253 (1857). —Stapf in Prain, F.T.A. **9**: 384 (1919). —Jackson & Wiehe, Annot. Check List Nyasal. Grass.: 42 (1958), in part (excl. *Jackson* 89, which is *Themeda triandra*). —Hood, A Guide to the Grasses of Zambia: 57 (1967). —Simon in Kirkia **8**: 50 (1971). —Clayton & Renvoize in F.T.E.A., Gramineae: 821 (1982). TAB. **42**. Types from Ethiopia.
　　Anthistiria abyssinica Hochst. ex A. Rich., Tent. Fl. Abyss. **2**: 449 (1851).
　　Andropogon monatheros A. Rich., Tent. Fl. Abyss. **2**: 462 (1851). Type from Ethiopia.
　　Hyparrhenia monathera (A. Rich.) Schweinf., Beitr. Fl. Aethiop.: 300, 310 (1867), *comb. inval., ante publ. gen.*
　　Andropogon exothecus Hack. in A. & C. de Candolle, Monogr. Phan. **6**: 632 (1889), based on *Anthistiria abyssinica*, non *Andropogon abyssinicus* Fresen. (1837). Types as for *Anthistiria abyssinica*.
　　Hyparrhenia abyssinica (Hochst. ex A. Rich.) Roberty in Boissiera **9**: 108 (1960).

Densely caespitose; culms up to 200 cm high. Leaf laminas 5–45 cm × 1–4 mm, the sheath produced at the mouth into narrow auricles 3–18 mm long. False panicle lax, narrow, of 1–10 distant raceme-pairs, these long-exserted from linear spatheoles 8–12

Tab. 42. EXOTHECA ABYSSINICA. 1, habit (× ²/₃), from *Mwangangi* 355; 2, raceme pair (×
2); 3, lower raceme (× 2); 4, superior lemma (× 4), 2–4 from *Grassl* 46–13. Drawn by Ann
Davies. From F.T.E.A.

cm long; racemes 1.3–3 cm long, 2-awned per pair, green with purple tinge; raceme-bases unequal, the inferior up to 5 mm long (but usually much shorter), the superior filiform and 15–25 mm long. Homogamous spikelets narrowly oblong, forming an involucre at the base of each raceme. Sessile spikelet 12–15 mm long (including the callus), narrowly lanceolate; callus slender, 3–4 mm long; inferior glume becoming brownish, pubescent with fulvous or pallid hairs; superior lemma with a stout awn 6–10 cm long, its column fulvously pubescent. Pedicelled spikelet 14–16 mm long, narrowly lanceolate, glabrous, with or without a terminal bristle up to 10 cm long.

Zambia. N: Mbala Distr., 1.6 km east of Ndundu, 1680 m, 24.xii.1967, *Simon, Williamson, Richards & Vesey-FitzGerald* 1539 (K; PRE). **Malawi.** N: Nkhata Bay Distr., Viphya (Vipya) Plateau, 37 km SW of Mzuzu, Lusangadzi Forest, 1700 m, 23.vi.1974, *Pawek* 8740 (K; PRE). S: Mulanje Distr., Chambe Plateau, Mt. Mulanje, 2130 m, 16.xi.1949, *Wiehe* 334 (K). **Mozambique.** N: Lago Distr., serra Jéci, prox. de Malulo, a c. 60 km de Lichinga (Vila Cabral), 1500 m, 3.iii.1964, *Torre & Paiva* 10984 (K; LISC; PRE). Z: Gurué Distr., Pico Namuli, 1500 m, 9.iv.1943, *Torre* 5134 (K; LISC; PRE). T: Angónia Distr., between Vila Coutinho and the frontier, 11.v.1948, *Mendonça* 4149 (LISC).

Tropical Africa and Vietnam. Growing in montane grassland and savanna; 1200–2250 m.

The species has an oddly disjunct distribution which is perhaps best explained by coastal trading dating back to antiquity.

28. HYPERTHELIA Clayton

By T.A. Cope

Hyperthelia Clayton in Kew Bull. **20**: 438 (1966).

Tall annuals or perennials. Ligule scarious, sometimes adnate to the sheath-auricles; leaf laminas linear. Inflorescence composed of paired racemes subtended by spatheoles and crowded into a large leafy false panicle; racemes short, with (1)2(10) sessile spikelets per pair, deflexed or not, with a single homogamous spikelet pair at the base of the inferior raceme; raceme-bases unequal, terete, the apex oblique and produced into a scarious appendage, this flat or rolled into a funnel about the base of the raceme; internodes and pedicels linear. Sessile spikelet terete; callus pungent, applied obliquely to the apex of the internode; inferior glume coriaceous, with a median longitudinal groove, otherwise rounded on the back and sides, herbaceous or membranous at the apex; superior glume with or without an awn; inferior floret reduced to a hyaline lemma; superior lemma hyaline, bidentate, passing between the teeth into a stout awn with hirsute column. Caryopsis narrowly elliptic. Pedicelled spikelet male, narrowed at the base to an indistinct conical callus, muticous or aristulate.

A genus of 6 species; one distributed throughout tropical Africa and introduced in tropical America, the remainder localized in southern Sudan and the Central African Republic.

Hyperthelia dissoluta (Nees ex Steud.) Clayton, in Kew Bull. **20**: 441 (1966). —Simon in Kirkia **8**: 18, 51 (1971). —Clayton in F.W.T.A., ed. 2, **3**: 496 (1972). —Clayton & Renvoize in F.T.E.A., Gramineae: 786, fig. 183 (1982). —Gibbs Russell et al., Grasses South. Africa [Mem. Bot. Surv. S. Africa No. 58]: 189 (1990). TAB. **43**. Type from tropical Africa (probably Ghana).

  Anthistiria dissoluta Nees ex Steud., Syn. Pl. Glumac. **1**: 400 (1854).
  Andropogon ruprechtii Hack. in Flora **68**: 126 (1885). Type from Mexico.
  Hyparrhenia ruprechtii (Hack.) E. Fourn., Mexic. Pl. Gram.: 67 (1886). —Stapf in Prain, F.T.Á. **9**: 326 (1919). —Stent & Rattray in Proc. & Trans. Rhodesia Sci. Assoc. **32**: 14 (1933).
  Cymbopogon ruprechtii (Hack.) Rendle in Hiern et al., Cat. Afr. Pl. Welw. **2**: 160 (1899).
  Hyparrhenia dissoluta (Nees ex Steud.) C.E. Hubb. in F.W.T.A. **2**: 591 (1936). —Sturgeon in Rhodesia Agric. J. **51**: 135 (1954). —Chippindall in Meredith, Grasses & Pastures S. Africa: 512 (1955). —Jackson & Wiehe, Annot. Check List Nyasal. Grass.: 44 (1958).

Caespitose perennial; culms 100–300 cm high. Leaf laminas up to 30 cm × 6 mm. False panicle erect, rather stiff, composed of 4–6 fastigiate tiers; spatheole 5–7 cm long, narrowly lanceolate, glabrous or sometimes hirsute, yellowish-green eventually becoming reddish, the peduncle $^1/_2$–$^2/_3$ as long; racemes 2–3 cm long, 2-awned per

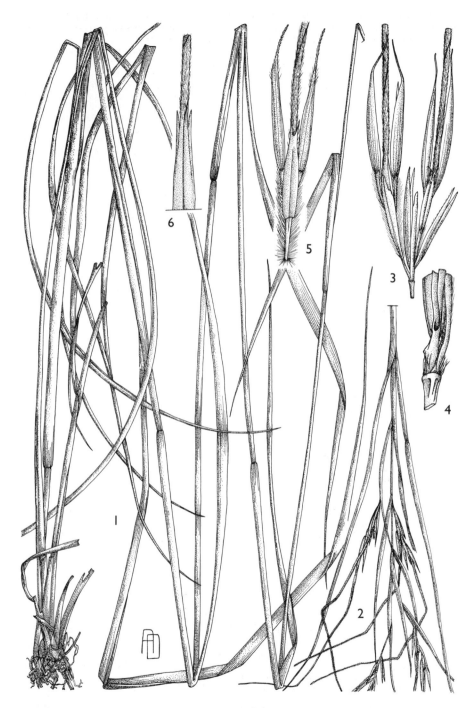

Tab. 43. **HYPERTHELIA DISSOLUTA.** 1, habit (× ½), from *Milne-Redhead & Taylor* 1647; 2, part of inflorescence (× ½); 3, raceme pair (× 2), 2 & 3 from *Rees* 35; 4, raceme bases (× 7); 5, upper raceme (× 2); 6, tip of superior lemma (× 6), 4–6 from *Milne-Redhead & Taylor* 1647. Drawn by Ann Davies. From F.T.E.A.

pair, not deflexed; superior raceme-base 2–3(5) mm long, the appendage 4–11 mm long, linear to narrowly lanceolate. Sessile spikelet 10–14 mm long (including the callus); callus 3–6 mm long; superior glume with or without a mucro up to 2 mm long; superior lemma with a yellowish awn 5–10 cm long. Pedicelled spikelet 9–14 mm long, with or without a terminal bristle up to 6 mm long.

Caprivi Strip. c. 32 km from Katima Mulilo on road to Linyanti, 910 m, 26.xii.1958, *Killick & Leistner* 3120 (K; PRE). **Botswana**. N: Ngamiland Distr., Kwando R., eastern edge of airfield, 18°22'S, 23°32'E, 1000 m, 19.iv.1977, *D.T. Williamson* 125 (K; PRE). SE: South East Distr., near Otse, north of Lobatse, 1000 m, 4.iii.1987, *Long & Rae* 33 (K). **Zambia**. B: Mongu Airport, 22.iii.1964, *Verboom* 1324 (K). N: Chambeshi Pontoon, 17.iv.1958, *Vesey-FitzGerald* 1668 (K). C: Lusaka Distr., Chilanga, Quien Sabi Farm, 1100 m, ix–x.1929, *Mrs C.I. Sandwith* s.n. (K). E: Katete Distr., St. Francis Hospital, 1070 m, 22.ii.1957, *Wright* 159 (K). S: Mazabuka Distr., Burdett's Farm, Monze to Magoye, km 9.6, 19.ii.1960, *White* 7234 (K). **Zimbabwe**. N: Gokwe Distr., Sengwa R. off southern border of Charama Plateau, 1220 m, 3.v.1965, *Simon* 422A (K). W: Bubi Distr., Lonely Mine, Dromoland, 14.vi.1968, *Wild* 7735 (K; PRE). C: Chegutu (Hartley) to Norton, 45 km west of Harare (Salisbury), 31.iii.1946, *Weber* 241 (K). E: Nyanga Distr., Cheshire, 1300 m, 15.i.1931, *Norlindh & Weimarck* 4429 (K). S: Chiredzi Distr., Chitsa's Kraal, 240 m, 4.vi.1950, *Wild* 3359 (K; PRE). **Malawi**. N: Nkhata Bay Distr., Ngani, Likoma (Lukoma) Is., 25.viii.1984, *Salubeni & Nachamba* 3886 (K). C: Lilongwe Agric. Research Station, 20.iv.1951, *Jackson* 463A (K). S: Zomba, 940 m, 14.v.1949, *Wiehe* 101 (K). **Mozambique**. N: Mogincual Distr., andados 30 km na picada nova do Namialo para Quixaxe, 120 m, 27.xi.1963, *Torre & Paiva* 9343 (K; LISC; PRE). Z: Mocuba, 5.vi.1943, *Torre* 5394 (K; LISC). T: Angónia Distr., Posto Zootécnico, 12.v.1948, *Mendonça* 4189 (K; LISC). MS: Parque Nacional da Gorongosa (Gorongosa National Park), iv.1973, *Tinley* 2802 (K). GI: Zavala Distr., between Chidenguele (Chidenguel) and Quissico (Zavala), 11.xii.1944, *Mendonça* 3381 (K). M: entre Moamba e Ressano Garcia, 25.ii.1948, *Torre* 7420 (K; LISC).

Throughout tropical Africa and introduced in tropical America. Common in deciduous bushland and wooded grassland, in fixed dunes, along riverbanks and by roads and in disturbed places, including mining areas contaminated with nickel and molybdenum; 0–1350 m.

29. ELYMANDRA Stapf

By T.A. Cope

Elymandra Stapf in Prain, F.T.A. **9**: 407 (1919). —Clayton in Kew Bull. **20**: 287 (1966).

Tall annuals or perennials. Ligule scarious; leaf laminas linear. Inflorescence composed of paired (rarely solitary) racemes long-exserted from narrow spatheoles and gathered into a false panicle; racemes slender, with up to 10 homogamous spikelet-pairs (only 1 in the Flora Zambesiaca area); raceme-bases filiform, unequal, rarely deflexed; internodes and pedicels linear. Sessile spikelet ± terete; callus narrowly obtuse to pungent, applied obliquely to the apex of the internode with its tip free; inferior glume coriaceous, rounded on the back and sides, becoming dark brown; superior glume nearly always awned; inferior floret reduced to a hyaline lemma; superior lemma stipitiform, bilobed, passing between the lobes into an awn with ± hirsute column. Caryopsis narrowly ellipsoid. Pedicelled spikelet male or barren, narrowly lanceolate, olive-green tinged with brown, the callus well developed.

A genus of 6 species; occurring in tropical Africa, with one also in Brazil.

The facies of the genus is quite distinctive, with olive-green homogamous spikelets and dark sessile spikelets. The well developed callus of the pedicelled spikelet is also characteristic of this and a few other closely related genera.

Awn of fertile lemma 3–5 cm long; callus of sessile spikelet 2.5–3.5 mm long · · · · · 1. *grallata*
Awn of fertile lemma 1.2–2.1 cm long; callus of sessile spikelet 1–1.5 mm long · · · 2. *lithophila*

1. **Elymandra grallata** (Stapf) Clayton in Kew Bull. **20**: 292 (1966). —Simon in Kirkia **8**: 17, 50 (1971). —Clayton in F.W.T.A., ed. 2, **3**: 498 (1972). —Clayton & Renvoize in F.T.E.A., Gramineae: 823, fig. 189 (1982). Gibbs Russell et al., Grasses South. Africa [Mem. Bot. Surv. S. Africa No. 58]: 132 (1990). TAB. **44**. Types from Angola.
    Hyparrhenia grallata Stapf in Prain, F.T.A. **9**: 320 (1919).

Tab. 44. ELYMANDRA GRALLATA. 1, habit (× ½); 2, raceme (× 3); 3, spikelet pair (× 4); 4, superior lemma (× 5), 1–4 from *Milne-Redhead & Taylor* 9439. Drawn by Ann Davies. From F.T.E.A.

Hyparrhenia eylesii C.E. Hubb. in Bull. Misc. Inform., Kew **1928**: 37 (1928). —Stent & Rattray in Proc. & Trans. Rhodesia Sci. Assoc. **32**: 13 (1933). —Sturgeon in Rhodesia Agric. J. **51**: 134 (1954). —Jackson & Wiehe, Annot. Check List Nyasal. Grass.: 44 (1958). Type: Zimbabwe, Goromonzi, *Eyles* 4880 (K, holotype, isotypes).

Densely caespitose perennial; culms 100–200 cm high, up to 4 mm in diameter. Leaf sheaths glabrous or sparsely pilose; ligules 1–1.5 mm long, ciliate; leaf laminas up to 30 cm × 6 mm, rigid, glaucous, glabrescent, those on the tillers much shorter (c. 5 cm) and silvery-villous. Panicle scanty, narrow, rarely reduced to a single pair of racemes; spatheoles linear, 7.5–10 cm long, green or purplish, glabrous or rarely sparsely hirsute; peduncle glabrous or rarely pubescent. Racemes 4–5.5 cm long, c. 12(18)-awned per pair, slender, loose; raceme-bases pubescent where they meet, the inferior very short, the superior 3–4 mm long. Homogamous spikelet-pairs 1 at the base of the inferior raceme (rarely of both racemes), c. 7 mm long, similar to the pedicelled spikelets. Sessile spikelets (7)9–10(12) mm long (including the callus); callus 2.5–3.5(4.5) mm long, slender, pungent, white-pubescent; superior glume with a slender awn 8–18 mm long; superior lemma 5–6 mm long, bilobed almost to the middle; awn (2.5)3–5 cm long, the column thinly and shortly pubescent with white hairs. Pedicelled spikelets 6–10 mm long (excluding the callus), white-villous, acute; callus 2–3 mm long.

Botswana. N: Kwando–Movombe road, 18°09'S, 23°14'E, 9.iv.1982, *P.A. Smith* 3823 (K; PRE). **Zambia**. B: Mongu Distr., 2.iv.1964, *Verboom* 1356 (K; PRE). N: Mpika Distr., North Luangwa National Park, 11°47'S, 32°21'E, 630 m, 22.iii.1994, *P.P. Smith* 875 (K). W: Mwinilunga Distr., Luakela R., 26 km north of Mwinilunga, 1370 m, 23.xii.1969, *Simon & Williamson* 1958 (K). C: Old Mkushi, Maunda R. tributary, 14°20'S, 29°16'E, 1190 m, 2.iv.1963, *Vesey-FitzGerald* 4091 (K). E: Lundazi, 1100 m, 1.vi.1954, *E.A. Robinson* 801 (K). S: Namwala Distr., Nkola R. at Ngoma, Kafue National Park, 4.iii.1964, *B.L. Mitchell* 24/93 (K; PRE). **Zimbabwe**. N: Guruve Distr., Nyamunyeche Estate, 6.iii.1979, *Nyariri* 744 (K; PRE). W: Hwange Distr., Kazuma Range, c. 1000 m, 9.v.1972, *Simon* 2169 (K). C: Chirumanzu Distr., Mtao Forest Reserve, 1460 m, 4.v.1955, *Mullin* 25/55 (K; PRE; SRGH). **Malawi**. N: Karonga Distr., Vinthukutu (Vintukhutu) Forest, 3 km north of Chilumba, 550 m, 26.iv.1975, *Pawek* 9587 (K). C: Lilongwe, Kasiya road, 29.iv.1966, *Salubeni* 443 (K). S: Zomba Distr., Lake Chilwa (Shirwa), c. 610 m, 12.iii.1950, *Wiehe* 434 (K). **Mozambique**. N: Malema Distr., andados 16 km de Entre Rios (Malema) para Ribáuè, c. 650 m, 3.ii.1964, *Torre & Paiva* 10415 (LISC). Z: between Mocuba and Milange, 11.iii.1943, *Torre* 4915 (K; LISC; PRE). Tropical Africa. Growing in savanna woodland; 550–1460 m.

2. **Elymandra lithophila** (Trin.) Clayton in Kew Bull. **20**: 292 (1966). —Simon in Kirkia **8**: 17 (1971). Type from Brazil.
   Andropogon lithophilus Trin. in Mém. Acad. Imp. Sci. St.-Pétersbourg, Sér. 6, Sci. Math. **2**: 277 (1832).
   Sorghum lithophilum (Trin.) Kuntze, Revis. Gen. Pl. **2**: 792 (1891).
   Andropogon bovonei Chiov. in Nuovo Giorn. Bot. Ital., n.s. **26**: 59 (1919). Type from Dem. Rep. Congo.
   Hyparrhenia lithophila (Trin.) Pilg. in Engler & Prantl, Nat. Pflanzenfam., ed. 2, **14e**: 174 (1940).

Caespitose perennial; culms 30–100 cm high, slender, erect. Leaf sheaths glabrous or puberulous, with a tuft of long hairs at the mouth; ligules c. 0.5 mm long, glabrous or ciliolate; leaf laminas 3–20 cm × 2–6 mm, stiff, glabrous or puberulent. Panicle scanty, narrow, of 1–3 pairs of racemes; spatheoles linear, 5–15 cm long, glabrous or sparsely hirsute; peduncle glabrous or softly villous. Racemes 2.5–7 cm long, 6- to 16-awned per pair, slender, loose; raceme-bases glabrous or villous, the inferior very short, the superior 5–10 mm long. Homogamous spikelet-pairs 1 at the base of the inferior raceme, c. 6 mm long, similar to the pedicelled spikelets. Sessile spikelets c. 7 mm long (including the callus); callus 1–1.5 mm long, narrowly obtuse at the base, white-pubescent; superior glume with a slender awn 3–5 mm long; superior lemma 4–4.5 mm long, bilobed to the middle; awn 1.2–2.1 cm long, the column hispidulous with white hairs. Pedicelled spikelets 6–8 mm long (excluding the callus), white-villous, acute; callus 1–1.5 mm long.

Zimbabwe. E: Chimanimani Distr., Bundi Valley, below mountain hut, Chimanimani Mts., 1680 m, 6.iv.1969, *Simon & Kelly* 1845 (K; PRE; SRGH).
Also in Dem. Rep. Congo and Brazil. Growing among rocks in mountains between 1600 and 1700 m.

The species is very closely related to *E. grallata*, differing only in such minor characters as lengths of awns, spikelets and callus. However, there seems to be a clear disjunction between the species and until further collections indicate otherwise they should be regarded as distinct.

Specimens from the Dem. Rep. Congo are much the more robust, up to about 1 m high, and grow in savanna woodland. On the other hand, specimens from Zimbabwe are dwarfed wiry plants, up to 40 cm high, found in rock crevices at high altitude. These differences, while remarkable, are confined to the vegetative parts and are presumably induced by the different habitats.

30. ANADELPHIA Hack.

By T.A. Cope

Anadelphia Hack. in Bot. Jahrb. Syst. **6**: 240 (1885). —Clayton in Kew Bull. **20**: 275–285 (1966).

Monium Stapf in Prain, F.T.A. **9**: 399 (1919).

Pobeguinea (Stapf) Jacq.-Fél. in Rev. Int. Bot. Appl. Agric. Trop. **30**: 172 (1950).

Annuals or perennials. Ligule a short ciliolate membrane; leaf laminas linear. Inflorescence a solitary raceme, exserted from or enclosed by a narrow spatheole and gathered into a scanty or copious false panicle; raceme loose, with few spikelets (sometimes only 1), without homogamous spikelet pairs; internodes and pedicels long and slender. Sessile spikelet slightly dorsally compressed to subterete; callus usually pungent, applied obliquely to the apex of the internode with its tip free; inferior glume usually not grooved on the back (grooved in *A. scyphofera*); superior glume usually awned; inferior floret reduced to a hyaline lemma; superior lemma bilobed, passing between the lobes into a glabrous to pubescent awn. Caryopsis narrowly ellipsoid to subcylindrical. Pedicelled spikelets as long as the sessile, linear-lanceolate, acuminate, usually glabrous, with a narrowly oblong to linear callus, sometimes the spikelet much reduced or quite absent and its pedicel likewise much reduced.

A genus of 14 species, occurring from Senegal to Mozambique, but mostly in the Fouta Djallon plateau of Guinea.

Annual; peduncle produced into a trumpet-like appendage embracing the callus of the sessile
    spikelet; inferior glume of sessile spikelet grooved on the back · · · · · · · · · · 1. *scyphofera*
Perennial; peduncle without an appendage; inferior glume of sessile spikelet not grooved on
    the back · 2. *trispiculata*

1. **Anadelphia scyphofera** Clayton in Kew Bull. **20**: 278 (1966). TAB. **45**. Type: Zambia N: Luwingu Distr., *Astle* 647 (K, holotype).

Annual; culms up to 40 cm high, slender, branched near the base. Leaf sheaths glabrous, produced into 2 narrowly lanceolate hyaline auricles 2–3 mm long; ligule adnate to the auricles; leaf laminas up to 20 cm × 0.5 mm, glabrous. False panicle up to 30 cm long, linear; spatheoles 2.5–3.5 cm long, narrowly linear; peduncles filiform, glabrous, expanded into a 2 mm long trumpet-like appendage embracing the callus of the sessile spikelet. Raceme exserted from the side of the spatheole, bearing a solitary sessile spikelet and two reduced pedicelled spikelets. Sessile spikelet 6–7 mm long (including the callus); callus c. 1.5 mm long, shortly barbate; inferior glume narrowly oblong, firmly chartaceous, 6-nerved, rounded on the back with a median groove, shortly bidentate at the apex; superior glume similar to the inferior but 3-nerved and drawn out at the apex into a slender awn 12–14 mm long; superior lemma shortly bilobed; awn 4.5–5.5 cm long, the column dark brown, scaberulous. Pedicels filiform, c. 2 mm long, ciliate. Pedicelled spikelet 2–4 mm long, linear, reduced to the glumes, the callus c. 0.25 mm long, oblong, truncate.

Zambia. N: Mporokoso Distr., Kasanshi Dambo, 55 km ESE of Mporokoso, 13.v.1962, *E.A. Robinson* 5171 (K). W: Mwinilunga Distr., Kalenda Plain, Matonchi, 1400 m, 16.iv.1960, *E.A. Robinson* 3598a (K).

Not known elsewhere. Growing in shallow soil over sandstone or laterite; at about 1400 m.

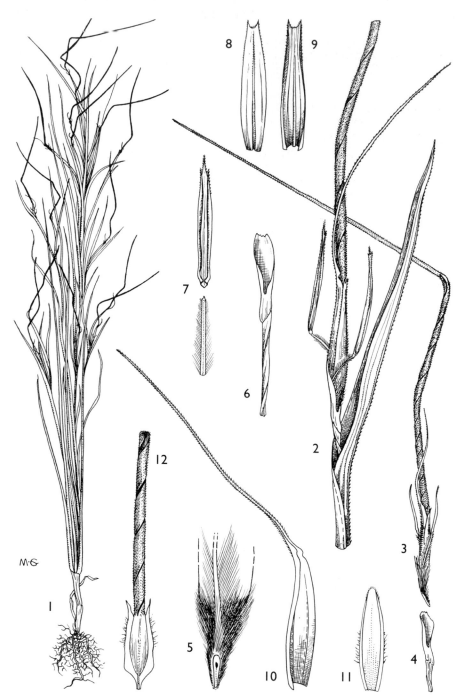

Tab. 45. ANADELPHIA SCYPHOFERA. 1, habit (× ²⁄₃); 2, raceme (× 6); 3, spikelet, separated from peduncle (× 3); 4, peduncle (× 3); 5, callus and base of pedicel (× 18); 6, tip of peduncle (× 8); 7, pedicelled spikelet, with pedicel separated from spikelet (× 6); 8–12, sessile spikelet: 8, inferior glume, back view (× 6); 9, inferior glume, face view (× 6); 10, superior glume (× 6); 11, interior lemma (× 6); 12, superior lemma (× 6), 1–12 from *Astle* 647. Drawn by Mary Grierson. From Kew Bull.

2. **Anadelphia trispiculata** Stapf in Prain, F.T.A. **9**: 398 (1919). Type from Guinea.
   Pobeguinea trispiculata (Stapf) Jacq.-Fél. in Rev. Int. Bot. Appl. Agric. Trop. **30**: 174 (1950).
   Hypogynium trispiculatum (Stapf) Roberty in Boissiera **9**: 183 (1960).
   Pobeguinea gabonensis Koechlin in Bull. Soc. Bot. France **108**: 243 (1961). Type from Gabon.

Caespitose perennial; culms up to 120 cm high, stout, unbranched. Leaf sheaths glabrous, produced into 2 inconspicuous rounded auricles; ligule adnate to the auricles; leaf laminas up to 25 cm × 5 mm, glabrous except towards the base. False panicle 30–40 cm long, linear; spatheoles 2–3 cm long, linear-lanceolate; peduncles filiform, glabrous, without an appendage. Raceme exserted from the side of the spatheole, bearing a solitary sessile spikelet and two well developed pedicelled spikelets. Sessile spikelet 8–10 mm long (including the callus); callus 3–3.5 mm long, fulvously barbate; inferior glume narrowly oblong, coriaceous, 7-nerved, rounded on the back without a median groove, narrowly truncate at the apex; superior glume similar to the inferior but 3-nerved, usually awnless; superior lemma shortly bilobed; awn 3.5–7 cm long, the column yellowish to brown, scaberulous, twice as long as the limb. Pedicels filiform, ³⁄₄ the length of the sessile spikelet, fulvously ciliate. Pedicelled spikelet c. 4 mm long, linear, male, the callus c. 1.5 mm long, oblong, truncate.

Mozambique. MS: Cheringoma coastal area, Zuni drainage, 5 km west of Nyamaruza Camp, v.1973, *Tinley* 2843 (K).
Also in Sierra Leone, Guinea, Ghana and Gabon. Growing in swampy places; sea level.

31. MONOCYMBIUM Stapf

By T.A. Cope

Monocymbium Stapf in Prain, F.T.A. **9**: 386 (1919). —C.E. Hubbard in Kew Bull. **4**: 375 (1949). —Jacques-Félix in Rev. Int. Bot. Appl. Agric. Trop. **30**: 175 (1950).

Perennials. Ligule scarious, very short; leaf laminas linear. Inflorescence composed of solitary racemes, these densely spiculate, shorter than and enclosed by the cymbiform spatheole, loosely gathered into a false panicle; racemes without homogamous spikelet pairs; internodes and pedicels short, filiform, ciliate. Sessile spikelet dorsally compressed; callus obtuse, applied obliquely to the apex of the internode with its tip free; inferior glume coriaceous, broadly convex with rounded sides; superior glume awned; inferior floret reduced to a hyaline lemma; superior lemma hyaline, bilobed, passing between the lobes into an awn with glabrous column. Caryopsis oblong, slightly dorsally compressed. Pedicelled spikelet male, resembling the sessile spikelet but awnless, with a distinct callus.

A genus of 3 species, occurring in tropical and South Africa.

Monocymbium ceresiiforme (Nees) Stapf in Prain, F.T.A. **9**: 387 (1919). —Stent & Rattray in Proc. & Trans. Rhodesia Sci. Assoc. **32**: 15 (1933). —Sturgeon in Rhodesia Agric. J. **51**: 138 (1954). —Chippindall in Meredith, Grasses & Pastures S. Africa: 515 (1955). —Jackson & Wiehe, Annot. Check List Nyasal. Grass.: 49 (1958). —Simon in Kirkia **8**: 18, 51 (1971). —Clayton in F.W.T.A., ed. 2, **3**: 489 (1972). —Clayton & Renvoize in F.T.E.A., Gramineae: 825, fig. 190 (1982). —Gibbs Russell et al., Grasses South. Africa [Mem. Bot. Surv. S. Africa No. 58]: 223 (1990). TAB. **46**. Type from South Africa.
   Andropogon ceresiiformis Nees, Fl. Afr. Austral. Ill.: 109 (1841).
   Andropogon ceresiiformis var. *breviaristatus* Hack. in A. & C. de Candolle, Monogr. Phan. **6**: 399 (1889). Type from Sudan.
   Andropogon ceresiiformis var. *submuticus* Hack. in A. & C. de Candolle, Monogr. Phan. **6**: 399 (1889). Type from Sudan.
   Andropogon ceresiiformis var. *hirtellus* Franch. in Bull. Soc. Hist. Nat. Autun **8**: 328 (1895). Types from Congo-Brazzaville.
   Monocymbium ceresiiforme subvar. *hirtulum* Chiov. in Nuovo Giorn. Bot. Ital., n.s. **26**: 74 (1919). Type from Dem. Rep. Congo.
   Monocymbium nimbanum Jacq.-Fél. in Rev. Int. Bot. Appl. Agric. Trop. **30**: 176 (1950). Type from Guinea.
   Hypogynium ceresiiforme (Nees) Roberty in Boissiera **9**: 192 (1960).

Tab. 46. MONOCYMBIUM CERESIIFORME. 1, habit (× ¹/₂), from *Bullock* 2755; 2, spatheole and raceme (× 2); 3, spikelet pair (× 7); 4, superior lemma (× 8), 2–4 from *Milne-Redhead & Taylor* 9416. Drawn by Ann Davies. From F.T.E.A.

Perennial of variable habit ranging from plants up to 130 cm high with weak culms easily lodged by rain and sometimes straggling, to plants scarcely exceeding 30 cm with wiry culms arising from a basal tussock of leaves mixed with the fibrous remains of old sheaths; leaf laminas 5–25 cm × 2–5 mm, glabrous or pilose. False panicle linear, open; spatheoles 2–4 cm long, narrowly lanceolate, reddish-brown. Sessile spikelet 3–4 mm long, elliptic; callus longer than wide, pilose on both sides and at the base; inferior glume softly pilose to villous; superior glume with an awn 2–6 mm long; superior lemma with an awn 6–20 mm long. Pedicelled spikelet similar to the sessile spikelet, but awnless and with a callus 0.5–1 mm long.

Caprivi Strip. c. 80 km from Singalamwe on road to Katima Mulilo, c. 900 m, 3.i.1959, *Killick & Leistner* 3289 (K; PRE). **Botswana**. N: east of Shishikola, 18°06'S, 23°00'E, 28.i.1978, *P.A. Smith* 2289 (K; PRE). **Zambia**. B: Kaoma Distr., Kafue Watershed, 2.iv.1964, *Verboom* 1363 (K; PRE). N: Chinsali Distr., Mbwingimfumu (Bwingimfumu), 40 km north of Mpika, ii.1969, *G. Williamson* 1463 (K; PRE). W: Mwinilunga Distr., 42 km from Mwinilunga along road to Solwezi, 22.xi.1972, *Strid* 2602 (K). C: Mkushi Distr., Mkushi River, dambo, 1430 m, 12.xii.1967, *Simon & Williamson* 1400 (K). **Zimbabwe**. N: Makonde Distr., Silverside Mine, 28.iv.1965, *Wild* 7361 (K; PRE). W: Matobo Distr., Besna Kobila Farm, 1460 m, iv.1960, *Miller* 7271 (K). C: Harare Distr., Cleveland Dam Park, c. 10 km west of Harare, 1500 m, 15.xii.1990, *Laegaard* 15802 (K). E: Chimanimani Distr., Melsetter Pasture Research Station, 1615 m, 20.iii.1949, *F.R. Williams* 13 (K). S: Mberengwa Distr., near summit of Mt. Buhwa, 1640 m, 3.v.1973, *Simon, Pope & Biegel* 2437 (K; PRE). **Malawi**. N: Rumphi Distr., Nyika Plateau, 2130 m, 6.ii.1968, *Simon, Williamson & Ball* 1656 (K; PRE). C: Ntchisi Distr., on Ntchisi Forest Reserve, road near Chitembwene, 13.iv.1991, *Radcliffe-Smith* 5963 (K). S: Zomba Plateau, 1830 m, 25.iv.1949, *Wiehe* 90 (K). **Mozambique**. Z: Maganja da Costa Distr., c. 59 km from Maganja da Costa (Vila da Maganja) on road to Mocuba, c. 100 m, 9.i.1968, *Torre & Correia* 16992 (LISC). MS: entre Beira e Dondo, 31.xii.1943, *Torre* 6338 (K; LISC; PRE). GI: Bilene Distr., between R. Incomáti and Magul, 5.ii.1948, *Torre* 7249 (LISC). M: Matutuíne Distr., Ponta do Ouro, 22.v.1952, *Myre & Carvalho* 1161 (LISC).

Throughout tropical Africa and in South Africa. Grasslands on a wide variety of soils including those contaminated by heavy metals around copper mines; 50–2150 m.

The variation in habit is less pronounced in the Flora Zambesiaca area, where plants are mostly small and caespitose with wiry culms. Plants with a taller looser habit are rather less common.

32. HETEROPOGON Pers.

By T.A. Cope

Heteropogon Pers., Syn. Pl. **2**: 533 (1807).

Annuals or perennials. Ligule very short, membranous, ciliate on the upper edge; leaf laminas linear. Inflorescence a single raceme, these terminal and axillary, sometimes loosely aggregated into a false panicle, partially enclosed by a narrow spatheole; racemes with homogamous spikelet pairs for the inferior $^{1}/_{4}$–$^{2}/_{3}$ of their length; internodes linear, the pedicels reduced to a minute stump. Sessile spikelet terete; callus pungent; inferior glume coriaceous, convex without a median groove, obtuse at the apex; inferior floret reduced to a hyaline lemma; superior lemma stipitiform, entire, passing directly into a stout pilose awn. Caryopsis lanceolate, channelled on one side. Pedicelled spikelet male or barren, lanceolate, larger than the sessile spikelet, awnless, with a long slender callus functioning as a pedicel.

A genus of 6 species; occurring in tropical and warm temperate regions.

Plant annual; pedicelled spikelet with a row of depressed glands down the middle of the
  inferior glume · 1. *melanocarpus*
Plant perennial; pedicelled spikelet without glands · 2. *contortus*

1. **Heteropogon melanocarpus** (Elliott) Benth. in J. Linn. Soc., Bot. **19**: 71 (1881). —Stapf in Prain, F.T.A. **9**: 413 (1919). —Stent & Rattray in Proc. & Trans. Rhodesia Sci. Assoc. **32**: 16 (1933). —Sturgeon in Rhodesia Agric. J. **51**: 139 (1954). —Chippindall in Meredith, Grasses & Pastures S. Africa: 494 (1955). —Jackson & Wiehe, Annot. Check List Nyasal. Grass. 43 (1958). —Simon in Kirkia **8**: 17, 50 (1971). —Clayton in F.W.T.A., ed. 2, **3**: 473 (1972). —Hall-Martin & Drummond in Kirkia **12**: 158 (1980). —Clayton & Renvoize in F.T.E.A., Gramineae: 827 (1982). —Müller, Grasses of South West Africa/Namibia: 180

(1984). —Gibbs Russell et al., Grasses South. Africa [Mem. Bot. Surv. S. Africa No. 58]: 179 (1990). Type from U.S.A.

Andropogon melanocarpus Elliott, Sketch Bot. S. Carolina **1**: 146 (1816).

Andropogon polystictus Hochst. ex Steud., Syn. Pl. Glumac. **1**: 369 (1854) as '*polystichus.*' Type from Ethiopia.

Heteropogon polystictus (Hochst. ex Steud.) Hochst. in Flora **39**: 28 (1856).

Tall annual; culms up to 250 cm high, robust, erect, supported below by stilt roots. Leaf laminas flat, up to 50 cm × 10 mm, gradually acuminate at the apex. Racemes 3–6 cm long, emerging from the side of the spatheole, aggregated into a copious false panicle; homogamous spikelet pairs 1–3, resembling the pedicelled spikelets. Sessile spikelet 10–11 mm long (including a sharply pungent and rufously barbate callus c. 4 mm long); inferior glume elliptic-oblong, dark brown, densely hispidulous; awn 7–12 cm long, pubescent. Pedicelled spikelet 15–25 mm long; inferior glume light green, herbaceous, glabrous, with a median line of depressed glands running along its length, long-acuminate at the apex; callus 2–3 mm long.

Botswana. N: Aha Hills (northern range), near Qabi (Xabe) at c. 19°39'S, 21°01'E, 29.iv.1980, *P.A. Smith* 3506 (K). **Zambia**. B: Mongu Airport, 29.iii.1964, *Verboom* 1353 (K). N: Samfya Distr., Kalasa Mukoso Flats, Masenga Causeway, south of Katumba, 1200 m, 22.ii.1996, *Renvoize* 5668 (K). C: South Luangwa National Park (Luangwa Valley Game Reserve South), 5 km south of Lubi R., 610 m, 10.iv.1967, *Prince* 455 (K). E: Petauke Distr., Luangwa Valley, South Luangwa National Park, Chilongozi Game Camp, 14.iv.1963, *Verboom* 951 (K). S: Choma Distr., Mapanza west, 1070 m, 21.iii.1954, *E.A. Robinson* 631 (K). **Zimbabwe**. N: Guruve Distr., Nyamunyeche Estate, 24.iv.1979, *Nyariri* 805 (K). W: Matobo Distr., Matopos Research Station, 8.iv.1965, *de Winter* 8335 (K; PRE). C: Charter Distr., Wiltshire Purchase Area, 12.ii.1964, *Cleghorn* 853 (K). E: Mudzi Distr., Lawleys Concession, 21.ii.1954, *West* 3407 (K). S: Chiredzi Distr., west of Sabi R, western end of Chionja Hills, 370 m, 28.iii.1961, *Phipps* 2891 (K). **Malawi**. N: Nkhata Bay Distr., Bandawe Point, 470 m, 8.vi.1974, *Pawek* 8700 (K). C: Mchinji (Fort Manning), Nyoka, 21.iv.1952, *Jackson* 778 (K). S: Zomba, 940 m, 11.vi.1949, *Wiehe* 128 (K). **Mozambique**. N: Montepuez Distr., encosta do monte Matuta, c. 5 km a sul do rio M'salo, próximo de Nantulo, c. 400 m, 9.iv.1964, *Torre & Paiva* 11802 (K; LISC). Z: Morrumbala Distr., Região de M'bobo, 20.iv.1943, *Torre* 5209 (K; LISC). T: Cahora Bassa Distr., between Inhacapirire and Chicoa, 15 km from Inhacapirire, 19.ii.1972, *Macêdo* 4864 (K; LISC). MS: Parque Nacional da Gorongosa (Gorongosa National Park), c. 120 m, 8.vii.1972, *Ward* 7789 (K). GI: Inhambane Distr., Inhampossa, 10 km south of Inhambane town, iii.1938, *Gomes e Sousa* 2112 (K).

Tropical Africa, extending through Arabia to India; also in tropical America. Growing in patches in deciduous bushland and wooded grassland; also in abandoned cultivation and along tracks and roads; 120–1200 m.

2. **Heteropogon contortus** (L.) P. Beauv. ex Roem. & Schult., Syst. Veg. **2**: 836 (1817). —Stapf in Prain, F.T.A. **9**: 411 (1919). —Stent & Rattray in Proc. & Trans. Rhodesia Sci. Assoc. **32**: 15 (1933). —Sturgeon in Rhodesia Agric. J. **51**: 139 (1954). —Chippindall in Meredith, Grasses & Pastures S. Africa: 492 (1955). —Jackson & Wiehe, Annot. Check List Nyasal. Grass.: 43 (1958). —Simon in Kirkia **8**: 17, 50 (1971). —Clayton in F.W.T.A., ed. 2, **3**: 473 (1972). —Jacobsen in Kirkia **9**: 148 (1973). —Clayton & Renvoize in F.T.E.A., Gramineae: 827, fig. 191 (1982). —Müller, Grasses of South West Africa/Namibia: 178 (1984). —Gibbs Russell et al., Grasses South. Africa [Mem. Bot. Surv. S. Africa No. 58]: 179 (1990). TAB. **47**. Type from India.

Andropogon contortus L., Sp. Pl. **2**: 1045 (1753).

Heteropogon hispidissimus Hochst. ex A. Rich., Tent. Fl. Abyss. **2**: 453 (1851), in synon.

Andropogon hispidissimus Hochst. ex Steud., Syn. Pl. Glumac. **1**: 367 (1854). Type from Ethiopia.

Tab. 47. HETEROPOGON CONTORTUS. 1, habit (× ⅔); 2, ligule (× 3); 3, pedicelled spikelet of homogamous pairs (× 3); 4, inferior glume of sessile homogamous spikelet (× 3); 5, superior glume of sessile homogamous spikelet (× 3); 6 & 7, empty lemmas of sessile homogamous spikelet (× 3); 8, flower of male pedicelled spikelet (× 3); 9, two pairs of heterogamous spikelets (× 3); 10 & 11, inferior glume of sessile fertile spikelet (× 3); 12 & 13, superior glume of sessile fertile spikelet (× 3); 14, palea of fertile floret (× 3); 15, lemma and portion of awn of superior floret (× 3); 16, female flower (× 3); 17, caryopsis, showing hilum and embryo (× 3), 1–17 from *Vesey-FitzGerald* 1232/5. Drawn by Derrick Erasmus. From Fl. Iraq.

1

2

16

12

13

17

14 15

10 11

9

8

6 7

4 5

3

D.E.

Caespitose perennial with laterally compressed leaf sheaths; culms up to 100 cm high, erect. Leaf laminas 3–30 cm × 2–8 mm, abruptly narrowed at the apex. Racemes 3–10 cm long, mostly long-exserted, solitary or aggregated into a scanty false panicle, the awns eventually twisted together into a terminal spire; homogamous spikelet pairs 3–17, resembling the pedicelled spikelets. Sessile spikelet 5.5–10 mm long (including the ferociously pungent and rufously barbate callus 2–3 mm long); inferior glume elliptic-oblong, brown, hispidulous; awn 5–8 cm long, hirtellous. Pedicelled spikelet 5–15 mm long; inferior glume green with yellowish membranous margins, without glands, the indumentum ranging from absent to tuberculate-villous; callus 2–3 mm long.

Caprivi Strip. c. 72 km from Katima Mulilo on road to Singalamwe, 910 m, 30.xii.1958, *Killick & Leistner* 3196 (K; PRE). **Botswana.** N: Ngamiland Distr., Kwando, James Camp, c. 18°22'S, 23°32'E, 16.iv.1975, *D.T. Williamson* 32 (K; PRE). SW: Ghanzi Farm, 6.i.1970, *Brown* s.n. (K). SE: Lobatse, near the railway, 20.x.1929, *Sandwith* 98 (K; PRE). **Zambia.** B: Sichinga Forest near Sesheke, 29.xii.1952, *Angus* 1068 (K; PRE). N: Mpika Distr., North Luangwa National Park, Luangwa Valley, 11°46'S, 32°10'E, 690 m, 31.iii.1993, *P.P. Smith* 854 (K). C: Lusaka Distr., Chilanga, Quien Sabe, 1100 m, 20.viii.1929, *Sandwith* 47 (K). F: Chipata Distr., Lugomo, 13°15'S, 32°12'E, 610 m, 4.ii.1958, *D.R.M. Stewart* 116 (K). S: Mazabuka Distr., 40 km north of Pemba, near Kanchale Village, 11.ii.1960, *White* 6941 (K). **Zimbabwe.** N: Binga Distr., Chizarira Game Reserve, 1020 m, 16.i.1968, *W.R. Thomson* 54 (K). W: Bulawayo, Malindela, Albermarle Road, 1370 m, 20.ii.1965, *Simon* 142 (K). C: Chegutu (Hartley) to Norton, 45 km west of Harare (Salisbury), 10.ii.1946, *Weber* 203 (K). E: Mudzi Distr., Nyanga North, Lawleys Concession, 21.ii.1954, *West* 3405 (K). S: Chiredzi Distr., Save/Runde (Sabi/Lundi) Junction, Chitsa's Kraal, 240 m, 4.vi.1950, *Wild* 3360 (K; PRE). **Malawi.** N: St. Patrick's Mission, 2 km east of Rumphi, 1070 m, 2.iv.1977, *Pawek* 12574 (K). C: Dedza Distr., Chongoni Forest, near Kangoli (Kanjoli), 12.i.1967, *Salubeni* 485 (K; PRE). S: Mulanje Distr., Lukulezi Swamps, 650 m, 20.vi.1962, *E.A. Robinson* 5396 (K). **Mozambique.** N: Mossuril Distr., Lumbo, near aerodrome, 5.v.1948, *Pedro & Pedrogão* 3112 (K). Z: Morrumbala Distr., Região de M'Bobo, 27.iv.1943, *Torre* 5211 (K; LISC). T: Angónia Distr., Ulónguè (Ulongwe), R. Livelange, 17.xii.1980, *Macuácua* 1468 (K; LISC; PRE). MS: Chemba Distr., Chiou, Estação Experimental do C.I.C.A., 13.iv.1960, *Lemos & Macuácua* 91 (K; LISC; PRE). GI: Inhassoro Distr., Santa Carolina Is., 21°27'S, 35°20'E, 1–15 m, 3–5.xi.1958, *Mogg* 28799 (K; LISC). M: between Boane and Porto Henrique, 19.ii.1981, *de Koning & Boane* 8668 (K; LISC).

Tropical and warm temperate regions generally. Growing in deciduous bushland, wooded grassland and sand dunes, and often dominating open grassy places on dry soils; 0–1400 m.

33. THEMEDA Forssk.

By T.A. Cope

Themeda Forssk., Fl. Aegypt.-Arab.: 178 (1775).

Annuals or perennials. Ligule very short, membranous; leaf laminas linear. Inflorescence composed of solitary racemes embraced by colourful sheathing spatheoles, the racemes single or more often densely packed in fan-shaped clusters on a flexuous peduncle and gathered into a leafy false compound panicle; racemes comprising 2 homogamous spikelet pairs forming a sort of involucre, and 1–4 sessile spikelets with their pedicelled attendants; internodes linear. Homogamous spikelets all sessile, the pairs separated by a short internode, persistent. Sessile spikelet terete or dorsally compressed; callus obtuse to pungent; inferior glume coriaceous, not grooved, obtuse at the apex; inferior floret reduced to a hyaline lemma; superior lemma stipitiform and passing directly into a puberulous or pubescent awn, or hyaline and awnless. Caryopsis lanceolate, channelled on one side. Pedicelled spikelet male or barren, narrowly lanceolate, awnless, with a long slender callus as long as or longer than the true pedicel (this often reduced to a little stump).

A genus of 18 species; occurring in tropical and subtropical regions of the Old World, but mostly in Asia.

Themeda triandra Forssk., Fl. Aegypt.-Arab.: 178 (1775). —Stapf in Prain, F.T.A. **9**: 416 (1919). —Stent & Rattray in Proc. & Trans. Rhodesia Sci. Assoc. **32**: 16 (1933). —Sturgeon in Rhodesia Agric. J. **51**: 140 (1954). —Chippindall in Meredith, Grasses & Pastures S. Africa: 490 (1955). —Simon in Kirkia **8**: 19, 52 (1971). —Clayton in F.W.T.A., ed. 2, **3**: 471 (1972). —Clayton & Renvoize in F.T.E.A., Gramineae: 829, fig. 192 (1982). —Müller, Grasses of

Tab. 48. THEMEDA TRIANDRA. 1, habit (× ¹/₂); 2, ligule (× 2); 3, raceme, showing homogamous spikelets (× 2¹/₂); 4, heterogamous spikelet pair (× 8). Drawn by W.E. Trevithick. From F.T.E.A.

South West Africa/Namibia: 262 (1984). —Gibbs Russell et al., Grasses South. Africa [Mem. Bot. Surv. S. Africa No. 58]: 335 (1990). TAB. 48. Type from Yemen.

Anthistiria imberbis Retz., Observ. Bot. 3: 11 (1783). Type from South Africa.

Themeda polygama J.F. Gmel., Syst. Nat. 2: 149 (1791), *nom. superfl.* based on *T. triandra* Forssk.

Anthistiria australis R. Br., Prodr. 1: 200 (1810). Type from Australia.

Anthistiria forsskalii Kunth, Révis. Gramin. 1: 162 (1829), *nom. superfl.* based on *Themeda polygama* J.F. Gmel.

Anthistiria punctata Hochst. ex A. Rich., Tent. Fl. Abyss. 2: 448 (1851). Types from Ethiopia.

Themeda forsskalii Hack. in A. & C. de Candolle, Monogr. Phan. 6: 659 (1889), *nom. superfl.* based on *Anthistiria imberbis* Retz.

Themeda forsskalii var. *burchellii* Hack., op. cit.: 661. Types from South Africa.

Themeda forsskalii var. *glauca* Hack., op. cit.: 663. Type from Algeria.

Themeda imberbis (Retz.) T. Cooke, Fl. Bombay 2: 993 (1908).

Themeda triandra var. *imberbis* (Retz.) Hack. in Proc. & Trans. Rhodesia Sci. Assoc. 7: 63 (1908). —Chippindall in Meredith, Grasses & Pastures S. Africa: 491 (1955).

Themeda triandra var. *burchellii* (Hack.) Domin in Biblioth. Bot. 85: 280 (1915). —Stapf in Prain, F.T.A. 9: 419 (1919). —Sturgeon in Rhodesia Agric. J. 51: 140 (1954). —Chippindall in Meredith, op. cit.: 492.

Themeda triandra var. *hispida* Stapf in Prain, tom. cit.: 418. —Sturgeon, loc. cit. —Chippindall in Meredith, op. cit.: 492. —Jackson & Wiehe, Annot. Check List Nyasal. Grass.: 62 (1958). Type from South Africa.

Themeda australis (R. Br.) Stapf in Prain, tom. cit.: 420.

Themeda triandra var. *punctata* (Hochst. ex A. Rich.) Stapf in Prain, tom. cit.: 419. —Jackson & Wiehe, loc. cit.

Themeda triandra var. *trachyspathea* Gooss. in Bull. Misc. Inform., Kew 1934: 195 (1934). —Chippindall in Meredith, op. cit.: 491 (1955). Type from Swaziland.

Themeda triandra var. *sublaevigata* Chiov. in Webbia 8: 61 (1951). Type from Ethiopia.

Caespitose perennial; culms up to 200 cm high, erect. Leaf laminas up to, or exceeding, 30 cm × 1–8 mm, flat. False panicle up to c. 30 cm long, composed of wedge-shaped clusters of 2–8 racemes enfolded by spatheoles and spathes; spatheole 1.5–3.5 cm long, russet-coloured, glabrous to tuberculate-pilose; raceme containing 1 fertile spikelet. Homogamous spikelet pairs arising at about the same level, the internode barely perceptible; inferior glume 6–14 mm long, narrowly elliptic, glabrous to tuberculate-pilose. Sessile spikelet 6–11 mm long (including the pungent rufously barbate callus 2–4 mm long); inferior glume brown, smooth except for the appressed-pubescent apex; awn 2.5–7 cm long, puberulous. Pedicelled spikelet 6–14 mm long, glabrous or tuberculate-pilose; callus 2–3 mm long.

Caprivi Strip. Mpalela (Mpilila) Is., 910 m, 15.i.1959, *Killick & Leistner* 3385 (K; PRE). **Botswana**. N: Ngamiland Distr., floodplain of Kwando R., 18°03'S, 23°18'E, 10.iv.1982, *P.A. Smith* 3826 (K; PRE). SE: 47.6 km from Shoshong on Shoshong–Serowe road, 22°40'S, 16°25'E, 1240 m, 1.iv.1989, *Terry et al.* 141 (K). **Zambia**. B: Mongu, 6.i.1966, *E.A. Robinson* 6777 (K). N: Mbala Distr., Saisi R., near Jericho Range, 27.ii.1958, *Vesey-FitzGerald* 1545 (K). W: Mwinilunga Distr., Kalenda Dambo, 10.xii.1937, *Milne-Redhead* 3595 (K; PRE). C: Serenje Distr., Kundalila Falls, c. 53 km ENE of Serenje, 4.ii.1973, *Strid* 2904 (K). E: Chipata Distr., Kamkulo Dambo, 31.xii.1963, *V.J. Wilson* 30 (K). S: Mazabuka Distr., Tara Protected Forest Area, 26.i.1960, *White* 6387 (K). **Zimbabwe**. N: Gokwe Distr., Sengwa R. off southern border of Charama Plateau, 1220 m, 3.v.1965, *Simon* 414 (K). W: Matobo Distr., Rhodes Matopos Estate, c. 1370 m, 1946/7, *West* 2238 (K; PRE). C: Chegutu (Hartley)–Norton, c. 45 km west of Harare, 24.i.1946, *Weber* 174 (K). E: Chimanimani Mts., Dead Cow Camp, 1650 m, 5.ii.1965, *Simon* 114 (K). S: Mberengwa Distr., western slopes of Mt. Buhwa, c. 1300 m, 3.v.1973, *Simon & Pope* 2433 (K; PRE). **Malawi**. N: Rumphi Distr., Nyika Plateau, Mbuzinandi, 1980 m, 28.xii.1975, *E. Phillips* 770 (K). C: Mchinji (Fort Manning), Bua R., 1250 m, 7.i.1959, *Robson* 1082 (K; PRE). S: Blantyre Distr., Limbe, Maone Hill, 23.iii.1969, *C.H. Williams* 60 (K). **Mozambique**. N: Lago Distr., serra Jéci, prox. de Malulo, a c. 60 km de Lichinga (Vila Cabral), 1700 m, 3.iii.1964, *Torre & Paiva* 11006 (K; LISC; PRE). Z: entre Quelimane e Mocuba, 26.iii.1943, *Torre* 4996 (K; LISC). T: Moatize Distr., encosta ocidental do monte Mt. Zóbuè, c. 1000 m, 11.iii.1964, *Torre & Paiva* 11142 (LISC). MS: Parque Nacional da Gorongosa (Gorongosa National Park), entrance area, Bue Maria, iii.1971, *Tinley* 2057 (K; LISC). GI: Massingir, 14.iv.1972, *Myre, Lousã & Rosa* 5772 (K). M: between Boane and Namaacha, 20.xii.1944, *Torre* 6889 (K; LISC).

Tropical and subtropical regions of the Old World. Widespread and very common in grassland and open woodland, along rivers and in dambos, particularly on sandy and alluvial soils; often the dominant species; 0–2200 m.

Highly polymorphic and sometimes formally divided according to the degree and manner of hairiness of the spatheoles and spikelets. There is only poor correlation between morphology and geographical distribution, habitat or chromosome number, so the traditional varieties are of extremely limited value.

34. URELYTRUM Hack.

By T.A. Cope

Urelytrum Hack. in Engler & Prantl, Nat. Pflanzenfam. II, **2**: 25 (1887).

Perennials, rarely annual (not in the Flora Zambesiaca area). Ligule membranous; leaf laminas flat or folded. Inflorescence terminal, composed of 1–many subdigitate (rarely paniculate) racemes; racemes cylindrical or slightly dorsally flattened, very fragile, moderately slender; rhachis internodes clavate, glabrous to pubescent, the summit crateriform with a lobed scarious rim. Sessile spikelet narrowly ovate to narrowly oblong; callus oblique, broadly obtuse to cuneate, pubescent, inserted into the summit of the internode and mostly concealed by the rim; inferior glume coriaceous, broadly convex, the margins sharply inflexed and becoming keeled towards the apex, wingless or very narrowly winged, without a prominent ciliate fringe, smooth on the back, especially below, with the nerves raised only towards the apex or not at all, often muricate on the sides, usually entire, rarely biaristulate; superior glume awnless; inferior floret male with palea; superior lemma entire and awnless. Caryopsis ellipsoid. Pedicelled spikelet almost as long as the sessile spikelet or much reduced, its inferior glume drawn out into a long curved awn, rarely awnless; pedicel free, resembling the internode.

A genus of 7 species; occurring in tropical Africa.
A difficult genus in the Flora Zambesiaca area with species only moderately distinct; some intermediate specimens can be hard to place.

1. Racemes 3–10, rarely fewer but then the awns 2–5 cm long and the leaf sheaths densely pilose or pubescent above; ligule short, 1–3(5) mm long, blunt and inconspicuous · · · 1. *digitatum*
– Racemes solitary, rarely 2(3) together; ligule blunt or acute · · · · · · · · · · · · · · · · · · 2
2. Awn 3–12 cm long, rarely less but then the leaf sheaths usually glabrous above; ligule usually pointed and conspicuous, (2)3–6 mm long · 2. *agropyroides*
– Awn not more than 2 cm long, often absent; leaf sheaths usually pubescent or pilose above; ligule blunt and inconspicuous, 0.8–1.3 mm long · 3. *henrardii*

1. **Urelytrum digitatum** K. Schum. in Engler, Pflanzenw. Ost-Afrikas **C**: 97 (1895). —Stapf in Prain, F.T.A. **9**: 48 (1917). —Simon in Kirkia **8**: 53 (1971). —Clayton & Renvoize in F.T.E.A., Gramineae: 833, fig. 194 (1982). TAB. **49**. Type from Tanzania.
     Urelytrum stapfianum C.E. Hubb. in Kew Bull. **4**: 366 (1949). —Jackson & Wiehe, Annot. Check List Nyasal. Grass.: 64 (1958) (as 'sp. aff. *stapfianum*'). Type from Angola.
     Urelytrum fasciculatum C.E. Hubb. in Kew Bull. **4**: 370 (1949). —Clayton in F.W.T.A., ed. 2, **3**: 504 (1972). Type from Cameroon.

Caespitose perennial; culms up to 200 cm high. Leaf sheaths commonly pubescent to pilose above; ligule usually short, 1–3(5) mm long, truncate, inconspicuous; leaf laminas 10–40 cm × 4–10 mm, fairly broad. Inflorescence of (1)3–4(10) racemes each 15–30 cm long, subdigitate or borne upon a central axis 2–7(12) cm long; rhachis internodes and pedicels glabrous. Sessile spikelet narrowly oblong; inferior glume 6–10 mm long, glabrous except for the spinulose to ciliolate keels, not muricate, acute to bidenticulate (very rarely biaristulate) at the apex. Pedicelled spikelet 2–7 mm long, with an awn 2–3.5(5) cm long.

Zambia. N: Chinsali Distr., near Mbesuma Ranch, 1280 m, 20.vii.1961, *Astle* 818 (K). **Malawi**. N: Karonga Distr., Igembe (Yembe), 27.vi.1951, *Jackson* 556 (K). C: Nkhotakota (Nkhota Kota), 2.v.1963, *Verboom* 984 (K).
     Also in Uganda, Kenya, Tanzania, Cameroon and Angola. Growing in *Brachystegia* woodland and along roadsides in ditches; 470–1280 m.

2. **Urelytrum agropyroides** (Hack.) Hack. in A. & C. de Candolle, Monogr. Phan. **6**: 272 (1889). —Stapf in Prain, F.T.A. **9**: 45 (1917). —Clayton & Renvoize in F.T.E.A., Gramineae: 833 (1982). —Gibbs Russell et al., Grasses South. Africa [Mem. Bot. Surv. S. Africa No. 58]: 349 (1990). Type from Angola.

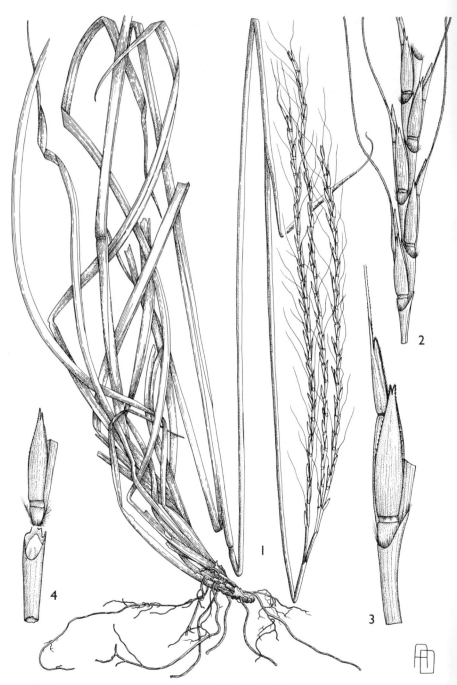

Tab. 49. **URELYTRUM DIGITATUM.** 1, habit (× ½), from *Kabuye* 299; 2, portion of raceme
(× 2); 3, spikelet pair (× 4); 4, raceme joint, disarticulated (× 3), 2–4 from *Grassl* 46–2.
Drawn by Ann Davies. From F.T.E.A.

Rottboellia agropyroides Hack. in Bol. Soc. Brot. **3**: 135 (1885).
Urelytrum squarrosum Hack. in A. & C. de Candolle, Monogr. Phan. **6**: 272 (1889). —Stapf in Prain, F.T.Å. **9**: 43 (1917). —Sturgeon in Rhodesia Agric. J. **51** 142 (1954). —Chippindall in Meredith, Grasses & Pastures S. Africa: 516 (1955). —Simon in Kirkia **8**: 20, 53 (1971). Type from South Africa.
Urelytrum squarrosum Hack. var. *robustum* Stapf in Dyer, F.C. **7**: 331 (1898). —Chippindall, loc. cit. Type from South Africa.
Urelytrum spp. sensu Simon in Kirkia **8**: 53 (1971).

Caespitose perennial forming hard tussocks; culms up to 200 cm high. Leaf sheaths usually glabrous above; ligule (2)3–6 mm long, usually conspicuous, acute, lacerate, often adnate to lateral auricles; leaf laminas 10–40 cm × 1–6 mm, narrow, harsh. Inflorescence of 1(2) racemes each 8–25 cm long; rhachis internodes and pedicels often pubescent. Sessile spikelet narrowly elliptic; inferior glume 6–9 mm long, glabrous or pubescent, smooth, spinulose or muricate on the keels, acute at the apex. Pedicelled spikelet 2–7 mm long, with an awn (1)3–7(12) cm long.

Caprivi Strip. near Andara Mission Station, 23.ii.1956, *de Winter & Marais* 4819 (K; PRE). **Botswana**. N: near Movombi (Movombe) Village, 18°07'S, 23°11'E, 14.ii.1983, *P.A. Smith* 4036 (K). SE: Morapedi Ranch, 25°38'S, 25°03'E, 1100 m, 29.iii.1978, *Hansen* 3394 (K; PRE). **Zambia**. B: Mongu Boma, 26.iii.1964, *Verboom* 1345 (K; SRGH). N: Samfya Distr., Kalasa Mukoso Flats, Masenga Causeway, south of Katumba, 1200 m, 22.ii.1996, *Renvoize* 5657 (K). C: c. 32 km north of Lusaka on the Ngwerere–Chisamba road, 14.iii.1956, *Hinds* 310 (K; SRGH). **Zimbabwe**. W: Insiza Distr., Shangani, Main Road A5, near turnoff to Shangani Station, 1400 m, 28.xi.1974, *Crook* 2041 (K; PRE). C: Seke Distr., Mexico road at 83 km post south of Harare, 1 km east of main highway, 1370 m, 15.ii.1974, *Davidse, Simon & Pope* 6680 (K; PRE). E: Nyanga Distr., Juliasdale, c. 1950 m, 23.i.1973, *Biegel* 4159 (K). S: Masvingo Distr., Makaholi Experimental Station, 10.iii.1978, *Senderayi* 150 (K; PRE). **Malawi**. N: Mzimba Distr., Kasitu Valley, 1100 m, 30.i.1938, *Fenner* 239 (K). **Mozambique**. N: Muecate Distr., andados 30 km de Imala para Mecubúri (Mocuburi), c. 450 m, 16.i.1964, *Torre & Paiva* 10009 (K; LISC). MS: Cheringoma Distr., 10 km south of Sengo, 20 m, 16.xii.1971, *Müller & Pope* 2057 (K; LISC). M: Matutuíne Distr., Matutuíne (Bela Vista), para Porto Henrique, 30.x.1960, *de Koning & Nuvunga* 8555 (K).
Also in Tanzania southwards to South Africa, and in Ghana and Madagascar. Growing in open and wooded grassland, and in regenerating *Brachystegia* woodland; 20–1400 m.
A variable species with little constancy in many characters. The long pointed ligule and long awns on solitary or paired racemes are the most reliable characters. Ornamentation of the inferior glume of the sessile spikelet, and indumentum of spikelets and leaf sheaths are less so.

3. **Urelytrum henrardii** Chippind. in Blumea, Suppl. **3**: 25 (1946). —Simon in Kirkia **8**: 53 (1971). Type: Zambia, Munshiwemba, *Stohr* 759 (PRE, holotype; K; SRGH).

Caespitose perennial; culms up to 100 cm high. Leaf sheaths usually pilose or pubescent above; ligule short, 0.8–1.3 mm long, truncate, inconspicuous; leaf laminas up to 40 cm × 3–4 mm, fairly narrow. Inflorescence a solitary raceme 15–30 cm long; rhachis internodes and pedicels scaberulous. Sessile spikelet narrowly oblong-lanceolate; inferior glume 7.5–8.5 mm long, glabrous below, spinulose above and on the keels, subacute at the apex. Pedicelled spikelet 2–3 mm long, with an awn up to 2 cm long, but often absent.

Caprivi Strip. c. 80 km from Singalamwe on road to Katima Mulilo, 910 m, 3.i.1959, *Killick & Leistner* 3287 (K). **Zambia**. N: near the Chambeshi R., Kasama–Kayumbi, 1220 m, 30.i.1962, *Astle* 1335 (K; SRGH). W: Ndola, ii.1957, *Fanshawe* 3083 (K; PRE). C: Serenje Distr., Kundalila Falls, c. 14 km east of Kanona, 1300 m, 17.iii.1974, *Davidse & Handlos* 7259 (K; PRE). S: Namwala Distr., Kafue National Park, Shakalonga Plain, 17.iii.1961, *B.L. Mitchell* 7/28 (K; PRE).
Unknown outside Zambia and the Caprivi Strip. Growing in open woodland and savanna on sandy soils, on dry open hillsides and on river terraces; 900–1500 m.
The complete absence of an awn on the glume of the pedicelled spikelet can be misleading because the plant can then superficially resemble a species of *Rhytachne*. Examination of the articulation of the joint should resolve any problem since in *Rhytachne* the callus of the sessile spikelet is horizontal with a central peg, not oblique.

35. LOXODERA Launert

By T.A. Cope

Loxodera Launert in Bol. Soc. Brot., sér. 2, **37**: 80 (1963); in Senckenberg. Biol. **46**: 121 (1965). —Clayton in Kew Bull. **20**: 258 (1966).

Perennials. Ligule membranous; leaf laminas flat or folded. Inflorescence a solitary raceme; raceme ± dorsally flattened, tardily disarticulating, stout; internodes narrowly oblong to clavate, pubescent to pilose, the summit crateriform with a lobed scarious rim. Sessile spikelet narrowly ovate to narrowly oblong; callus oblique, broadly obtuse, conspicuously barbate, inserted into the summit of the internode, the hairs exserted to form a ring around each node; inferior glume coriaceous, broadly convex, the margins sharply inflexed and becoming keeled towards the apex, usually with the nerves raised into longitudinal ridges running the whole length of the glume, sometimes muricate on the sides, entire at the apex; superior glume awnless; inferior floret male; superior lemma entire and awnless. Caryopsis ellipsoid. Pedicelled spikelet about as long as the sessile spikelet or much reduced, awnless or rarely awned from the inferior glume.

A genus of 5 species; occurring in tropical Africa.
Rather similar to *Urelytrum* and linked to it by *U. henrardii* (spikelets usually awnless, q.v.) and *L. strigosa* (spikelets awned; not in the Flora Zambesiaca area). In *Loxodera* the hairs on the callus of the sessile spikelet are very pronounced, forming a conspicuous hairy ring around the summit of the internode. In *Urelytrum* the much shorter callus hairs are more or less concealed by the scarious rim of the internode.

Inferior glume of the sessile spikelet glabrous, or if pubescent then only towards the apex; scarious rim of the internode incomplete, exposing much of the callus · · · · · 1. *caespitosa*
Inferior glume of the sessile spikelet conspicuously pilose on the back; scarious rim of the internode complete, concealing much of the callus · · · · · · · · · · · · · · · · · · 2. *bovonei*

1. **Loxodera caespitosa** (C.E. Hubb.) Simon in Kirkia **8**: 8 (1971). —Simon in Kirkia **8**: 19, 53 (1971). —Clayton & Renvoize in F.T.E.A., Gramineae: 840, fig. 196 (1982). TAB. **50**. Type: Zimbabwe, Harare (Salisbury), *Eyles* 1940 (K, holotype; SRGH).
    Phacelurus caespitosus C.E. Hubb. in Bull. Misc. Inform., Kew **1928**: 35 (1928). —Stent & Rattray in Proc. & Trans. Rhodesia Sci. Assoc. **32**: 5 (1933). —Sturgeon in Rhodesia Agric. J. **51**: 144 (1954).
    Lasiurus epectinatus Napper in Kirkia **3**: 121 (1963). Type from Tanzania.
    Loxodera epectinata (Napper) Launert in Bol. Soc. Brot., sér. 2, **37**: 82 (1963).

Densely caespitose perennial; culms up to 100 cm high; leaf laminas 5–30 cm × 2–5 mm. Raceme 6–10 cm long, stiff, grey; internodes clavate, pubescent, the scarious rim at the summit incomplete and exposing much of the callus. Sessile spikelet narrowly oblong-elliptic; inferior glume 5–7 mm long, glabrous or pubescent only towards the apex, the nerves prominently raised and rib-like, the margins not muricate. Pedicelled spikelet almost as long as the sessile spikelet, sometimes reduced.

Zambia. N: Kasama Distr., Mungwi, 8.x.1960, *E.A. Robinson* 3916 (K; SRGH). C: Mkushi Distr., Mkushi River Dambo, 1430 m, 14.x.1967, *Simon & Williamson* 979 (K; SRGH). **Zimbabwe**. C: Harare, 1520 m, xi.1919, *Eyles* 1940 (K; SRGH).
Also in Tanzania. Growing in upland dambos and along riverbanks; 1400–1550 m.
The almost glabrous, strongly ribbed inferior glumes are instantly recognizable. Only *L. ledermanii* (Pilg.) Launert from Nigeria, Cameroon and Uganda is at all similar, but this has muricate margins to the glume.

2. **Loxodera bovonei** (Chiov.) Launert in Senckenberg. Biol. **46**: 122 (1965). Type from Dem. Rep. Congo.
    Rottboellia bovonei Chiov. in Ann. Bot. (Rome) **13**: 36 (1914).
    Rhytachne bovonei (Chiov.) Chiov. in Nuovo Giorn. Bot. Ital., n.s. **26**: 73 (1919).
    Loxodera rigidiuscula Launert in Bol. Soc. Brot., sér. 2, **37**: 81 (1963). Type: Zambia N: Mbala Distr., *Vesey-FitzGerald* 2936 (BM, holotype; SRGH).
    Rhytachne pilosa F. Ballard & C.E. Hubb. in Bull. Misc. Inform., Kew **1934**: 108 (1934). Type: Zambia W: Mwinilunga Distr., *Milne-Redhead* 987 (K, holotype).

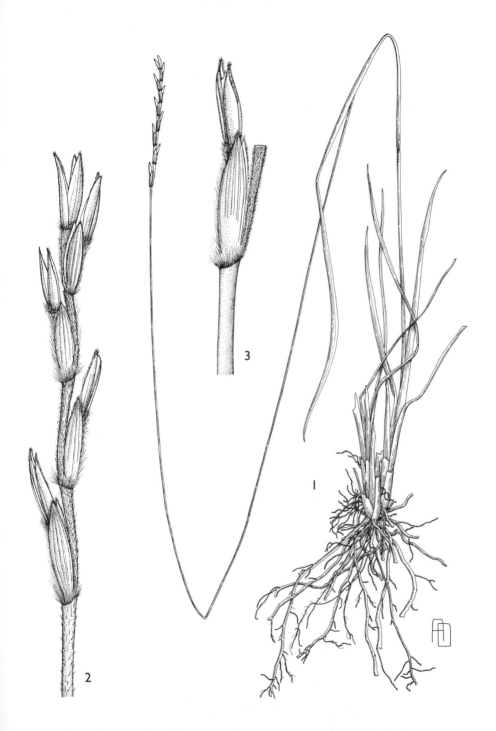

Tab. 50. LOXODERA CAESPITOSA. 1, habit (× ½), from *Robinson* 3916; 2, raceme (× 3); 3, spikelet pair (× 4), 2 & 3 from *McCallum Webster* A169. Drawn by Ann Davies. From F.T.E.A.

Densely caespitose perennial; culms up to 95 cm high; leaf laminas 7–20 cm × 2–5 mm. Raceme 6–9 cm long, stiff, grey; internodes narrowly clavate, densely pubescent, the scarious rim at the summit complete and concealing much of the callus. Sessile spikelet lanceolate-oblong; inferior glume 6.5–10 mm long, densely long-pilose to appressed tuberculate-hispid, the nerves raised but not prominently rib-like, the margins not muricate. Pedicelled spikelet much reduced, up to 6 mm long including the occasional awn of the inferior glume.

Zambia. N: Mbala Distr., Safu Dambo, north of Chilwa School, 1680 m, 23.x.1967, *Simon, Williamson & Richards* 1165 (K; PRE). W: Mwinilunga Distr., Chinkobolo (Sinkabolo) Dambo, 20.x.1937, *Milne-Redhead* 2866 (K; PRE).

Also in Dem. Rep. Congo. Growing in upland dambos; 1370–1680 m.

A variable species, especially in the indumentum of the inferior glume. This ranges from long-pilose to stiffly hispid, but always covers the glume from base to apex. The nerves of the glume are not so prominently rib-like as in other members of the genus, and the glume of the pedicelled spikelet is sometimes awned.

36. ELIONURUS Willd.

By T.A. Cope

Elionurus Willd., Sp. Pl. **4**: 941 (1806). —Kunth in Humboldt Bonpland & Kunth, Nov. Gen. Sp. **1**: 192 (1817). —Renvoize in Kew Bull. **32**: 665–675 (1978).

Perennials, rarely annuals. Ligule a very short densely ciliate membrane; leaf laminas flat or rolled. Inflorescence a single raceme, these terminal or sometimes axillary and gathered into a spathate false panicle; racemes flexuous, dorsally compressed; rhachis internodes columnar to subclavate, strongly oblique at the summit, neither crateriform nor scariously rimmed. Sessile spikelet lanceolate to narrowly ovate; callus often large, applied obliquely to the summit of the internode; inferior glume subcoriaceous to herbaceous, broadly convex on the back, smooth or sometimes toothed on the keels (not in the Flora Zambesiaca area), laterally 2-keeled, the keels ciliately fringed and often with an oil-streak on the inner side, mostly cuspidate to a bifid apex; superior glume awnless; inferior floret reduced to a hyaline lemma; superior lemma entire and awnless. Caryopsis ellipsoid, dorsally compressed. Pedicelled spikelet well developed, muticous or aristulate.

A genus of 15 species; distributed through tropical Africa, SW Asia and tropical America, with a single species in Australia.

The species in the Flora Zambesiaca area are not all that distinct, with reliance being placed on number of racemes, hairiness of glumes and rhachis internodes, and flatness of leaves for distinguishing them.

1. Culms mostly unbranched and bearing a single raceme (occasionally branched and bearing 2, rarely 3 racemes); leaves mostly confined to a basal tuft (but some cauline when the culms are branched); rhachis internodes spreading-pilose; leaf laminas tightly inrolled; inferior glume of sessile spikelet pilose on the back · · · · · · · · · · · · · · · · · · 1. *muticus*
   – Culms branched and bearing several to many racemes gathered into a spathate false panicle; leaves mostly cauline · 2
2. Rhachis internodes glabrous or appressed-pilose; leaf laminas mostly tightly inrolled; inferior glume of sessile spikelet mostly glabrous or only sparsely pilose on the back · · 2. *tripsacoides*
   – Rhachis internodes spreading-pilose; leaf laminas mostly flat; inferior glume of sessile spikelet densely pilose on the back · 3. *platypus*

1. **Elionurus muticus** (Spreng.) Kuntze, Revis. Gen. Pl. **3**: 350 (1898). —Clayton & Renvoize in F.T.E.A., Gramineae: 837, fig. 195 (1982). —Gibbs Russell et al., Grasses South. Africa [Mem. Bot. Surv. S. Africa No. 58]: 131 (1990). TAB. **51**. Type from Uruguay.
   Lycurus muticus Spreng., Syst. Veg. **4**, cur. post.: 32 (1827).
   Elionurus argenteus Nees, Fl. Afr. Austral. Ill. **1**: 95 (1841). —Stapf in Prain, F.T.A. **9**: 70 (1917). —Stent & Rattray in Proc. & Trans. Rhodesia Sci. Assoc. **32**: 6 (1933). —Jackson & Wiehe, Annot. Check List Nyasal. Grass.: 38 (1958). —Simon in Kirkia **8**: 19, 52 (1971). —Clayton in F.W.T.A., ed. 2, **3**: 505 (1972). Type from South Africa.

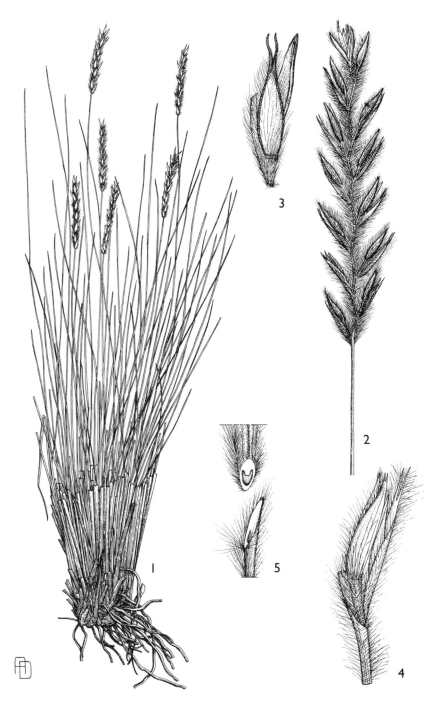

Tab. 51. ELIONURUS MUTICUS. 1, habit (× ¹/₂); 2, raceme (× 2), 1 & 2 from *Vesey-FitzGerald* 1884; 3, spikelet pair, showing sessile spikelet (× 5); 4, spikelet pair, from rear, showing pedicelled spikelet (× 5); 5, raceme joint, disarticulated (× 10), 3–5 from *Greenway* 7839. Drawn by Ann Davies. From F.T.E.A.

Elionurus thimiodorus Nees, Fl. Afr. Austral. Ill. **1**: 95 (1841). Type from South Africa.
Andropogon caespitosus A. Rich., Tent. Fl. Abyss. **2**: 451 (1851). Type from Ethiopia.
Andropogon thimiodorus (Nees) Steud., Syn. Pl. Glum. **1**: 365 (1854).
Elionurus argenteus var. *caespitosus* (A. Rich.) Hack. in A. & C. de Candolle, Monogr.
Phan. **6**: 340 (1889). —Stent & Rattray in Proc. Rhodesia Sci. Assoc. **32**: 6 (1933). —
Sturgeon in Rhodesia Agric. J. **51**: 142 (1954). —Chippindall in Meredith, Grasses &
Pastures S. Africa: 518 (1955).
Elionurus argenteus var. *thimiodorus* (Nees) Stapf in Dyer, F.C. **7**: 333 (1898).
Elionurus chevalieri Stapf in Mém. Soc. Bot. France **8**: 100 (1908). —Stapf in Prain, tom.
cit.: 70. Type from Guinea.
Elionurus gobariensis Vanderyst in Bull. Agric. Congo Belge **13**: 326 (1922). Type from
Dem. Rep. Congo.
Elionurus glaber E.P. Phillips in Bothalia **3**: 261 (1937). Type from South Africa.
Elionurus glaber var. *villosus* E.P. Phillips in Bothalia **3**: 261 (1937). Type from South Africa.
Elionurus pretoriensis E.P. Phillips in Bothalia **3**: 261 (1937). Type from South Africa.
Elionurus brazzae sensu Simon in Kirkia **8**: 19, 52 (1971), non Franch.

Densely caespitose perennial; culms up to 100 cm high, mostly unbranched.
Leaves mostly confined to the base and compacted into a dense tussock, ± aromatic;
leaf sheaths papery, sometimes becoming fibrous or horny, occasionally tomentose;
leaf laminas 5–30 cm × 1–2 mm, involute and filiform. Racemes 4–14 cm long,
terminal on the culm or sometimes with 1–2 axillary racemes, silvery to grey or
purple; rhachis internodes densely villous, the hairs conspicuously spreading. Sessile
spikelet lanceolate; callus 1–1.5 mm long, broadly cuneate; inferior glume 4–8 mm
long, villous on the back, rarely thinly pilose or very rarely quite glabrous, entire or
divided at the apex into 2 teeth up to 7 mm long. Pedicelled spikelet 4–7 mm long,
lanceolate, pubescent to villous, acuminate or with a short awn-point.

Botswana. N: Kwando River floodplain, 18°05'S, 23°20'E, 24.i.1978, *P.A. Smith* 2231 (K; PRE).
SE: Kweneng Distr., Lephephe Pasture Station, i.1965, *Yalala* 485 (K; PRE). **Zambia.** N: Mbala
(Abercorn), 1680 m, 9.ix.1954, *Siame* 510 (SRGH). W: Mwinilunga Distr., 16 km from Matonchi
Farm, 1200 m, 17.xi.1962, *Richards* 17261 (K). C: Lusaka Distr., Kanyanja, 25 km SE of Lusaka,
1300 m, 4.ii.1996, *Bingham & Nefdt* 10872 (K). E: Chama Distr., Nyika Plateau, ix.1968, *G.
Williamson* 1021 (K). S: near Choma, c. 1220 m, 21.xii.1962, *Astle* 1854 (K). **Zimbabwe.** N:
Gokwe North Distr., Chirisa, Sengwa Research Area, 12.ii.1981, *Mahlangu* 427 (K; PRE). W:
Matobo Distr., Matopos Research Station, 1370 m, 4.iii.1954, *Rattray* 1680 (K; SRGH). C:
Chegutu Distr., Poole Farm, 13.ix.1954, *R.M. Hornby* 3349 (K; PRE; SRGH). E: Chimanimani
Distr., Chimanimani Mts., summit of Mt. Peza, 2180 m, 30.xii.1957, *R. Goodier* 491 (K; PRE;
SRGH). **Malawi.** N: Nyika Plateau, 22.ii.1949, *Hawksworth* in *GHS* 22680 (K; PRE; SRGH). C:
Dedza Distr., Dedza Mt., 13.xi.1960, *Chapman* 1042 (SRGH). **Mozambique.** Z: Maganja da
Costa Distr., c. 64 km from Maganja da Costa (Vila da Maganja) on road to Mocuba, c. 100 m,
9.i.1968, *Torre & Correia* 17002 (LISC). T: Chifunde Distr., entre Furancungo e Mualadze (Vila
Gamito), 21.x.1943, *Torre* 6074 (LISC). MS: Chimanimani Mts., between Skeleton Pass and The
Plateau, 1580 m, 27.ix.1966, *Simon* 875 (K; LISC; PRE). GI: Govuro Distr., c. 50 km from Save
(Vila Franca do Save) on road to Búzi, c. 100 m, 22.xi.1967, *Torre & Correia* 15815 (LISC). M:
Matutuíne Distr., between Santaca and Catuane, 13.iv.1949, *Myre & Balsinhas* 583 (SRGH).
    Throughout tropical Africa, South Africa and Yemen; also in tropical and subtropical
America. Growing in montane and lowland grassland, dambos, deciduous bushland and open
woodland on dry sandy and stony soils; 100–2180 m.
    Extremely variable, with many synonyms from Africa and many more from America. Typical
plants, with a basal cushion of leaves and solitary racemes, are no problem to recognize, but the
species intergrades with several others, in particular the West African (and South American) *E.
ciliaris* Kunth (*E. pobeguinii* Stapf) which is taller and has cauline leaves and axillary racemes.
The latter are more positively gathered into a spathate false panicle than the 2 or 3 of *E. muticus*.
The distinction is rather vague and species boundaries in this part of the genus would repay
further investigation.

2. **Elionurus tripsacoides** Willd., Sp. Pl. **4**: 941 (1806). —Kunth in Humboldt, Bonpland &
    Kunth, Nov. Gen. Sp. **1**: 192, t. 62 (1816). —Clayton & Renvoize in F.T.E.A., Gramineae:
    838 (1982). —Gibbs Russell et al., Grasses South. Africa [Mem. Bot. Surv. S. Africa No. 58]:
    131 (1990). Type from Venezuela.
    Elionurus welwitschii Rendle in Hiern et al., Cat. Afr. Pl. Welw. **2**: 137 (1899). —Stapf in
    Prain, F.T.A. **9**: 68 (1917). Type from Angola.
    Elionurus trapnellii C.E. Hubb. in Bull. Misc. Inform., Kew **1933**: 498 (1933). —Sturgeon
    in Rhodesia Agric. J. **51**: 113 (1954). —Jackson & Wiehe, Annot. Check List Nyasal. Grass.:
    38 (1958). —Vesey-FitzGerald in Kirkia **3**: 108 (1963). —Simon in Kirkia **8**: 19, 52 (1971).
    Type: Zambia S, Mazabuka, *Trapnell* 846 (K, holotype; SRGH); 874 (K; SRGH, isoparatypes).

Loosely caespitose perennial; culms up to 150 cm high. Leaf laminas 20–30 cm × 1–3 mm, narrow, often involute, mainly cauline, filiform to subulate at the apex. Racemes 3–12 cm long, the branched culms bearing 3–20 axillary racemes gathered into a lax spathate false panicle; rhachis internodes glabrous or appressed-pubescent, rarely with spreading hairs up to 2 mm long. Sessile spikelet lanceolate; callus 1–2 mm long, broadly cuneate; inferior glume 4–6 mm long, glabrous on the back or almost so, with a conspicuous brown oil-streak bordering each keel, entire or bidentate at the apex, the teeth up to 2 mm long. Pedicelled spikelet 4–5 mm long, lanceolate, glabrous, entire.

Caprivi Strip. 16 km from Katima Mulilo on road to Singalamwe, c. 900 m, 30.xii.1958, *Killick & Leistner* 3189 (K; PRE). **Botswana**. N: Ngamiland Distr., Kwando, 18°22'S, 23°32'E, James Camp Airfield, 30.xi.1975, *D.T. Williamson* 55 (K; PRE). **Zambia**. B: Sesheke Distr., Mashi R. fringe, 30.x.1964, *Verboom* 1139 (K). N: Mbala Distr., Chisungu Estate, 1460 m, 20.iv.1959, *McCallum Webster* A335 (K; SRGH). C: Lusaka Distr., Mt. Makulu Agric. Research Station, 8.iii.1968, *van Rensburg* 3117 (K). E: Chipata (Fort Jameson), *Verboom* 586 (K). S: Mazabuka Distr., v.1932, *Trapnell* 2021 (K; SRGH). **Zimbabwe**. N: Hurungwe Distr., Karambazunga Store, Hurungwe East, 21.i.1958, *Cleghorn* 388 (PRE; SRGH). C: Chegutu Distr., Poole Farm, 29.i.1946, *Hornby* 2303 (K; PRE; SRGH). **Malawi**. N: Mzimba Distr., Mbawa, iv.1954, *Jackson* 1297 (K; SRGH). C: Lilongwe Distr., Nsaru, 910 m, 10.iv.1950, *I. Sturgeon in Wiehe* 559 (K).

Tropical Africa from Uganda and Kenya southwards; also in tropical America. Growing in open places in wooded grassland; 900–1460 m.

The species intergrades with *E. muticus*, and plants with sparingly branched culms and slightly hairy spikelets are a particular problem. Specimens from Caprivi Strip are like this, but they retain the glabrous inferior spikelet; they are best regarded as forms of *E. tripsacoides*. The species is also sometimes hard to distinguish from *E. platypus* which has broader, flat leaf laminas and villous rhachis internodes.

3. **Elionurus platypus** (Trin.) Hack. in Bol. Soc. Brot. **3**: 135 (1885). —Stapf in Prain, F.T.A. **9**: 66 (1917). —Clayton in F.W.T.A., ed. 2, **3**: 505 (1972). Type from Sierra Leone.

Andropogon platypus Trin. in Mém. Acad. Imp. Sci. St.-Pétersbourg, Sér. 6, Sci. Math. **2**: 261 (1832).

Elionurus pallidus K. Schum. in Bot. Jahrb. Syst. **24**: 326 (1897). Type from Sierra Leone.

Densely caespitose perennial; culms up to 180 cm high. Leaf laminas 5–20 cm × 3–6 mm, usually flat, mainly cauline. Racemes 6–8 cm long, the branched culms bearing 3–20 axillary racemes gathered into a lax spathate false panicle; rhachis internodes spreading-pilose, the hairs usually more than 2 mm long. Sessile spikelet lanceolate; callus c. 1.5–2 mm long, cuneate; inferior glume 5–6 mm long, glabrous to thinly pilose on the back, with a conspicuous brown oil-streak bordering each keel, bidentate at the apex, the teeth up to 2 mm long. Pedicelled spikelet 4–5 mm long, narrowly lanceolate, glabrous, entire.

Zambia. B: Sesheke Distr., 6.5 km south of Masese, xi.1972, *G. Williamson* 2269 (K; PRE).

Tropical Africa mainly on the western side. Collected from a clearing in teak forest; 1050 m.

The specimen cited is doubtfully *E. platypus* but it does not convincingly correspond with any other species. It has a sparse panicle with fewer racemes than is normal but it does have the flat leaves 3–4 mm wide that are typical of the species. It is close to *E. ciliaris* Kunth, from West Africa and tropical America, but this has a shorter callus (c. 1 mm) and inferior glume villous on the back with considerably longer marginal cilia.

37. PHACELURUS Griseb.

By T.A. Cope

Phacelurus Griseb., Spic. Fl. Rumel. **2**: 423 (1846). —Clayton in Kew Bull. **33**: 175–179 (1978).

Jardinea Steud., Syn. Pl. Glumac. **1**: 360 (1854).

Thyrsia Stapf in Prain, F.T.A. **9**: 48 (1917).

Pseudovossia A. Camus in Bull. Mus. Hist. Nat. (Paris) **26**: 665 (1920).

Pseudophacelurus A. Camus, op. cit. **27**: 370 (1921).

Perennial. Ligule membranous; leaf laminas setaceous or flat. Inflorescence terminal, usually of ± flattened digitate racemes, these rarely single or paniculate, often tardily disarticulating; rhachis internodes columnar to inflated. Sessile spikelet flat or slightly convex (rarely concave) across the back; callus truncate, flat or with a central peg; inferior glume membranous to coriaceous, smooth or muricate, the keels winged or not; inferior floret male, or barren and then with or without a palea; superior lemma entire and awnless. Caryopsis oblong, dorsally compressed. Pedicelled spikelet usually resembling the sessile spikelet but mostly smaller and rarely bisexual, occasionally vestigial, rarely with an elongated callus (not in the Flora Zambesiaca area); pedicel free, resembling the internode.

A genus of 9 species; occurring in the Old World tropics and subtropics, and extending northwards into the Mediterranean region.

1. Leaf laminas setaceous · 1. *franksiae*
– Leaf laminas linear, flat · 2
2. Inferior glume muricate on the back · 2. *gabonensis*
– Inferior glume smooth on the back · 3
3. Inferior glume glabrous, indistinctly nerved, clearly winged · · · · · · · · · · · · 3. *huillensis*
– Inferior glume ciliate, prominently nerved, obscurely winged · · · · · · · · · · · 4. *schliebenii*

1. **Phacelurus franksiae** (J.M. Wood) Clayton in Kew Bull. **33**: 178 (1978). —Gibbs Russell et al., Grasses South. Africa [Mem. Bot. Surv. S. Africa No. 58]: 267 (1990). Type from South Africa.
   Ischaemum franksiae J.M. Wood in Bull. Misc. Inform., Kew **1908**: 226 (1908). —Chippindall in Meredith, Grasses & Pastures S. Africa: 488 (1955).
   Ischaemum juncifolium F. Ballard & C.E. Hubb. in Bull. Misc. Inform., Kew **1934**: 107 (1934). —Simon in Kirkia **8**: 52 (1971). Type: Zambia W: Mwinilunga, *Milne-Redhead* 1004 (K, holotype; SRGH).

Densely caespitose perennial, the basal sheaths broad and papery, sometimes flabellate; culms up to 80 cm high, erect. Leaf laminas up to 40 cm × 1 mm, setaceous. Inflorescence 7–9 cm long, with a short central axis up to 2 cm long; racemes 2–4, each pedunculate and 4–6 cm long; internodes and pedicels linear, subclavate to inflated upwards, smooth and glabrous to shortly pilose. Sessile spikelet oblong-ovate, purplish; callus with a conspicuous central peg; inferior glume 6–8.5 mm long, thinly chartaceous with prominent nerves to firmly chartaceous with fainter nerves, ± flat across the back, rarely somewhat concave, scabrid on the sides below and smooth on the back to muricate on back and sides, the protuberances bearing a sharp prickle, ciliolate to rigidly ciliate on the margins above, these not winged, acute to acuminate at the apex. Pedicelled spikelet similar to the sessile spikelet but usually smaller, with a well developed callus, this bearing a central peg; pedicels half as long to as long as the internodes.

Zambia. W: Mwinilunga Distr., Chinkobolo (Sinkabolo) Dambo, 20.x.1937, *Milne-Redhead* 2867 (K; PRE). **Zimbabwe**. E: Chimanimani Distr., Bundi Valley, 2 km north of Southern Lakes, Chimanimani Mts., c. 1400 m, 7.xi.1972, *Simon* 2279 (K; PRE).
   Also known from Dem. Rep. Congo and South Africa (KwaZulu-Natal); growing in boggy grassland at c. 1400 m.
   Rather variable with respect to ornamentation of the inferior glume of the sessile spikelet and indumentum of the rhachis internodes and pedicels, but quite uniform in habit with the distinctive setaceous leaves often overtopping the inflorescence. There is not much material available, but what there is forms a continuum from one extreme to the other.

2. **Phacelurus gabonensis** (Steud.) Clayton in Kew Bull. **35**: 817 (1981). TAB. **52**. Type from Gabon.
   Jardinea gabonensis Steud., Syn. Pl. Glumac. **1**: 360 (1854). —Stapf in Prain, F.T.A. **9**: 51 (1917). —Simon in Kirkia **8**: 53 (1971).
      Rhytachne gabonensis (Steud.) Hack. in A. & C. de Candolle, Monogr. Phan. **6**: 276 (1889).
      Rhytachne congoensis Hack. in A. & C. de Candolle, Monogr. Phan. **6**: 277 (1889). Type from Dem. Rep. Congo.
      Jardinea congoensis (Hack.) Franch. in Bull. Soc. Hist. Nat. Autun **8**: 321 (1895). —Stapf in Prain, F.T.A. **9**: 53 (1917). —Simon in Kirkia **8**: 53 (1971).
      Rottboellia angolensis Rendle in Hiern et al., Cat. Afr. Pl. Welw. **2**: 139 (1899). Type from Angola.
      Jardinea angolensis (Rendle) Stapf in Prain, F.T.A. **9**: 52 (1917).

Tab. 52. PHACELURUS GABONENSIS. 1, habit ($\times \frac{1}{2}$); 2, ligule ($\times 4\frac{1}{2}$); 3, portion of raceme ($\times 4\frac{1}{2}$). Drawn by Victoria Friis. From Ghana Gr.

Robust, caespitose perennial, the basal sheaths broad and papery but not flabellate; culms up to 300 cm high, erect. Leaf laminas up to 45 cm × 12 mm, flat or loosely inrolled. Inflorescence up to 35 cm long, with a central axis up to 15 cm long; racemes 6–12, each 15–25 cm long, occasionally some of them branched; internodes and pedicels linear, subclavate upwards, smooth or scabrid. Sessile spikelet narrowly lanceolate, yellowish-green or purplish; callus with a short central peg; inferior glume 4–8 mm long, firmly chartaceous, flat across the back, faintly to prominently nerved, muricate on the back and sides below, or smaller spikelets on the sides only, the protuberances often bearing a sharp prickle, rigidly ciliate on the margins above, these scarcely winged, acuminate to apiculate or mucronate at the apex. Pedicelled spikelet sometimes similar to the sessile spikelet, but often smaller or even almost suppressed, rigidly apiculate at the apex, with a short callus, this with a short central peg; pedicels from half as long to almost as long as the internode.

Zambia. N: Lushiba Marsh, 930 m, 27.v.1961, *Astle* 728 (K). S: Namwala Distr., Kafue National Park, Musa–Kafue R confluence, 12.i.1964, *B.L. Mitchell* 24/51 (K; PRE; SRGH).
Tropical Africa, on the western side. Growing in seasonally wet grassland and on river banks; 930–1060 m.

3. **Phacelurus huillensis** (Rendle) Clayton in Kew Bull. **33**: 177 (1978). —Clayton & Renvoize in F.T.E.A., Gramineae: 847, fig. 199 (1982). TAB. **53**. Type from Angola.
    Rottboellia huillensis Rendle in Hiern et al., Cat. Afr. Pl. Welw. **2**: 140 (1899).
    Rottboellia undulatifolia Chiov. in Ann. Bot. (Rome) **13**: 36 (1914). Type from Dem. Rep. Congo.
    Thyrsia inflata Stapf in Prain, F.T.A. **9**: 49 (1917). Type from Dem. Rep. Congo.
    Thyrsia huillensis (Rendle) Stapf in Prain, F.T.A. **9**: 50 (1917).
    Coelorachis undulatifolia (Chiov.) Chiov. in Nuovo Giorn. Bot. Ital, n.s. **26**: 73 (1919).
    Thyrsia undulatifolia (Chiov.) Robyns, Fl. Agrost. Congo Belge **1**: 53 (1929). —Jackson & Wiehe, Annot. Check List Nyasal. Grass.: 63 (1958). —Simon in Kirkia **8**: 53 (1971).
    Rottboellia sulcata Peter in Repert. Spec. Nov. Regni Veg., Beih. **40**: 133; Anh.: 14, t. 15/2 (1929). Type from Tanzania.
    Coelorachis fasciculata Peter, op. cit.: 357 (1936), *nom. superfl.*, based on the preceding.

Caespitose perennial; culms up to 180 cm high, erect. Leaf laminas 15–40 cm × 2–7 mm, flat. Inflorescence 10–25 cm long, with a short central axis 2–9 cm long; racemes 4–14, each 8–20 cm long; internodes and pedicels clavate, often inflated upwards, mostly glabrous. Sessile spikelet ovate, pallid, sometimes purplish; callus without a central peg; inferior glume 3.5–5 mm long, thinly coriaceous, flat across the back, indistinctly nerved, smooth and glabrous, bordered by a distinct but narrow coriaceous wing, obtuse to emarginate at the apex. Pedicelled spikelet occasionally resembling the sessile spikelet but usually smaller and sometimes vestigial, without a callus; pedicel from half as long to as long as the internode.

Zambia. N: Mbala Distr., 13 km NW of Mbala (Abercorn), c. 1520 m, 11.iv.1961, *Phipps & Vesey-FitzGerald* 3055 (K; PRE; SRGH). W: Kitwe, 6.ii.1958, *Fanshawe* 4249 (SRGH). C: Mkushi Distr., Mkushi River Dambo, c. 1430 m, 16.xii.1967, *Simon & Williamson* 1391 (K; PRE). **Malawi**. N: Rumphi Distr., Nyika Plateau foothills, c. 1830 m, 12.ii.1968, *Simon, Williamson & Ball* 1772 (K). C: Nkhotakota Distr., 19.xi.1953, *Jackson* 1065 (K; SRGH). **Mozambique**. N: between Lichinga (Vila Cabral) and Litunde, 37 km from Litunde, 6.iv.1961, *M.F. Carvalho* 487 (K).
Also in Tanzania and Dem. Rep. Congo. Growing in woodland and riverine grassland; 1000–1830 m.

4. **Phacelurus schliebenii** (Pilg.) Clayton in Kew Bull. **33**: 178 (1978). —Clayton & Renvoize in F.T.E.A., Gramineae: 847 (1982). Type from Tanzania.
    Thyrsia schliebenii Pilg. in Notizbl. Bot. Gart. Berlin-Dahlem **11**: 649 (1932). —Simon in Kirkia **8**: 53 (1971).

Densely caespitose perennial with flabellate leaf sheaths; culms up to 90 cm high. Leaf laminas 15–30 cm × 3–5 mm, flat. Inflorescence 9–11 cm long, with a short central axis 1–5 cm long; racemes 2–8, each 3–9 cm long; internodes and pedicels clavate, usually pubescent. Sessile spikelet narrowly ovate, dark purple; callus without a central peg; inferior glume 4–5 mm long, chartaceous with prominent raised intercarinal nerves, these ± ciliate, the keels ciliate above and scarcely winged,

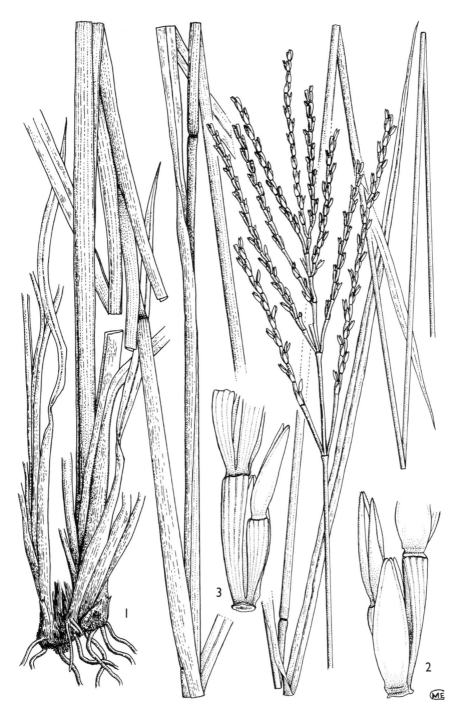

Tab. 53. PHACELURUS HUILLENSIS. 1, habit (× ²/₃); 2, spikelet pair, showing sessile spikelet (× 6); 3, spikelet pair, from rear, showing pedicelled spikelet (× 6), 1–3 from *Milne-Redhead & Taylor* 9358. Drawn by M.E. Church. From F.T.E.A.

subacute to emarginate at the apex. Pedicelled spikelet resembling the sessile spikelet or smaller, without a callus; pedicel about half as long as the internode.

Zambia. N: 113 km north of Mpika on Great North Road, 1370 m, 4.i.1962, *Astle* 1165 (K). **Malawi**. N: Rumphi Distr., Nyika Plateau, Lake Kaulime, 2340 m, 9.vii.1970, *Brummitt* 11913 (K; PRE).
Also in Tanzania. Growing in bogs and marshes; 1370–2340 m.

38. VOSSIA Wall. & Griff.

By T.A. Cope

Vossia Wall. & Griff. in J. Asiat. Soc. Bengal **5**: 572 (1836), *nom. conserv.*

Aquatic perennial. Ligule a short pilose membrane; leaf laminas flat. Inflorescence composed of digitate or subdigitate racemes, borne terminally; racemes ± flattened, tardily disarticulating; internodes and pedicels thickened, clavate, hollowed at the summit. Sessile spikelet flat or slightly convex across the back; callus truncate with an irregular central convexity and without a pronounced central peg; inferior glume coriaceous, 2-keeled, smooth except for the scabrid keels, narrowly winged above and drawn out into a long linear flattened tail; inferior floret male, with hyaline lemma and palea; superior lemma entire and awnless. Pedicelled spikelet resembling the sessile; pedicel free.

A genus of one species; distributed throughout tropical Africa up to the Nile Delta and in India.

Vossia cuspidata (Roxb.) Griff., Not. Pl. Asiat. **3**, Index: 12 (1851). —Stapf in Prain, F.T.A. **9**: 41 (1917). —Jackson & Wiehe, Annot. Check List Nyasal. Grass.: 65 (1958). —Simon in Kirkia **8**: 20, 53 (1971). —Clayton in F.W.T.A., ed. 2, **3**: 502 (1972). —Clayton & Renvoize in F.T.E.A., Gramineae: 832, fig. 193 (1982). —Gibbs Russell et al., Grasses South. Africa [Mem. Bot. Surv. S. Africa No. 58]: 353, t. 233 (1990). TAB. **54**. Type from Bangladesh.
    Ischaemum cuspidatum Roxb., Fl. Ind. **1**: 324 (1820).
    Vossia procera Wall. & Griff., op. cit.: 573, *nom. superfl.*

Perennial; culms submerged or floating, up to 7 m long and 1 cm in diameter, spongy, with fibrous roots from the nodes, standing 1–2 m out of the water. Leaf laminas 30–100 cm × 6–18 mm, linear. Racemes 1–12, each 10–30 cm long. Sessile spikelet narrowly ovate; inferior glume 2–4 cm long, the body 6–8 mm long, yellowish with green tail. Pedicelled spikelet a little smaller than the sessile.

Caprivi Strip. Katima Mulilo, Ngoma Bridge, 950 m, 1.v.1975, *D. Edwards* 4384 (K; PRE). **Botswana**. N: Ngamiland Distr., west bank of Okavango River, 18°46'S, 22°01'E, 28.i.1996, *Snow & Chatukuta* 6869 (K; PRE). **Zambia**. B: Barotse Floodplain, 19.iii.1964, *Verboom* 1150 (K; SRGH). N: Lake Mweru, near Nchelenge, xii.1968, *G. Williamson* 1239 (K). S: Kafue Flats, near Mazabuka, 1000 m, iv.1932, *Trapnell* 1094 (K). **Zimbabwe**. N: Binga Distr., between mouth of Binga inlet and Chimere Estuary, Lake Kariba, 7.iv.1960, *Phipps* 2802 (K; PRE). W: Hwange Distr., Victoria Falls, 890 m, 23.iv.1976, *Gonde* 57 (K; PRE). **Malawi**. N: Rukuru River, Feb/Mar 1903, *McClounie* 9 (K). C: Salima Distr., Grand Beach, iv.1972, *G. Williamson* 2188 (K). S: Mangochi (Fort Johnston), Lake Malawi (Nyasa) Dally's Hotel, 490 m, 20.iv.1949, *Wiehe* 75 (K). **Mozambique**. N: Sanga Distr., Unango, rio Malagui, c. 1100 m, 2.iii.1964, *Torre & Paiva* 10956 (K; LISC; PRE). MS: Gorongosa Distr., Urema Plains, Parque Nacional da Gorongosa (Gorongosa National Park), iii.1970, *Tinley* 1905A (K).
Distribution as for genus. Growing in or near water, often floating; 400–1100 m.

39. HEMARTHRIA R. Br.

By T.A. Cope

Hemarthria R. Br., Prodr. Fl. Nov. Holl.: 207 (1810).

Caespitose or stoloniferous, mostly rambling perennials. Ligule a very short ciliate membrane; leaf laminas flat. Inflorescence a single axillary raceme embraced below by the subtending leaf sheath; racemes tough, dorsally compressed; internodes

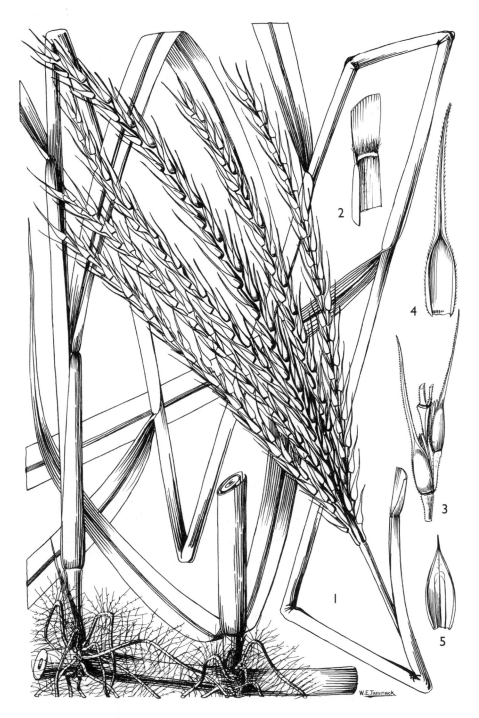

Tab. 54. **VOSSIA CUSPIDATA.** 1, habit ($\times \frac{1}{2}$); 2, ligule ($\times \frac{3}{4}$); 3, spikelet pair ($\times 1$); 4, sessile spikelet, inferior glume ($\times 2$); 5, superior glume ($\times 2$). Drawn by W.E. Trevithick. From F.T.E.A.

thickened, clavate, nearly always obliquely articulated, fused to the adjacent pedicel. Sessile spikelet dorsally compressed; callus obtuse to cuneate, rarely truncate; inferior glume narrowly elliptic, rigidly herbaceous, broadly convex, 2-keeled, smooth, indistinctly winged above, obtuse, caudate or bifid at the apex; superior glume obtuse to long-acuminate; inferior floret reduced to a hyaline lemma; superior lemma entire and awnless. Caryopsis narrowly obovoid, slightly dorsally compressed. Pedicelled spikelet similar to the sessile spikelet but truncate at the base and without a callus; pedicel flattened, broadly linear.

A genus of 12 species; occurring in the Old World tropics and subtropics; possibly also native in America.

Inferior glume of sessile spikelet obtuse to emarginate at the apex; racemes mostly borne singly and exserted from the subtending sheaths · 1. *altissima*
Inferior glume of sessile spikelet acute to acuminate at the apex; racemes borne in fascicles and not or scarcely exserted from the subtending sheaths · · · · · · · · · · · · · · · · · · 2. *natans*

1. **Hemarthria altissima** (Poir.) Stapf & C.E. Hubb. in Bull. Misc. Inform., Kew **1934**: 109 (1934). —Sturgeon in Rhodesia Agric. J. **51**: 144 (1954). —Chippindall in Meredith, Grasses & Pastures S. Africa: 519 (1955). —Jackson & Wiehe, Annot. Check List Nyasal. Grass.: 42 (1958). —Vesey-FitzGerald in Kirkia **3**: 108 (1963). —Simon in Kirkia **8**: 19, 53 (1971). —Clayton in F.W.T.A., ed. 2, **3**: 506 (1972). —Clayton & Renvoize in F.T.E.A., Gramineae: 851 (1982). —Gibbs Russell et al., Grasses South. Africa [Mem. Bot. Surv. S. Africa No. 58]: 178 (1990). Type from Algeria.
    Rottboellia altissima Poir., Voy. Barbarie **2**: 105 (1789).
    Rottboellia fasciculata Lam., Tab. Encycl. **1**: 204 (1791), *nom. superfl.*, based on the preceding.
    Hemarthria fasciculata (Lam.) Kunth, Révis. Gramin. **1**: 153 (1829). —Stent & Rattray in Proc. & Trans. Rhodesia Sci. Assoc. **32**: 5 (1933).
    Rottboellia heterochroa Gand. in Bull. Soc. Bot. France **66**: 302 (1920). Type from South Africa.
    Manisuris altissima (Poir.) Hitchc. in J. Wash. Acad. Sci. **24**: 292 (1934).
    Hemarthria compressa (L.f.) R. Br. subsp. *altissima* (Poir.) Maire, Fl. Afr. Nord **1**: 261 (1952).

Stoloniferous perennial; culms up to 250 cm long and 2–4 mm in diameter, prostrate and rooting at the nodes below. Leaf laminas 5–15 cm × 3–4 mm, flat. Racemes 4–10 cm long, mostly borne singly in the axils and exserted from the axillary sheath. Sessile spikelet elliptic-oblong, with a triangular callus; inferior glume 4–6 mm long, with or without a constriction near the apex, obtuse to emarginate at the apex; superior glume obtuse to acute at the apex. Pedicelled spikelet narrowly triangular, 4–6 mm long, truncate at the base and without a callus, subacute to acute at the apex.

Caprivi Strip. c. 13 km from Katima Mulilo on road to Ngoma, 910 m, 22.xii.1958, *Killick & Leistner* 3044 (K; PRE). **Botswana**. N: Kwara Bochai floodplain, 14.iv.1973, *P.A. Smith* 537 (K; PRE). **Zambia**. N: Chinsali Distr., Mbesuma Ranch, Chambeshi River, 1220 m, 28.i.1958, *Vesey-FitzGerald* 1414 (K; SRGH). C: Lusaka Distr., Mt. Makulu Agric. Research Station, 27.xi.1957, *Angus* 1794 (SRGH). E: Chipata Distr., Chinzombo, 5 km SE of Mfuwe, 610 m, 30.xii.1968, *Astle* 5401 (K). S: Mazabuka Distr., i.1932, *Trapnell* 733 (SRGH). **Zimbabwe**. N: Kariba Distr., Kariba Gorge, just below Dam wall, c. 460 m, v.1960, *Goldsmith* 79/60 (K; PRE; SRGH). W: Nkayi Distr., Gwampa Forest Reserve, c. 910 m, i.1956, *Goldsmith* 13/56 (K; PRE; SRGH). C: Harare Distr., 1460 m, iii.1920, *Eyles* 2143 (K; PRE; SRGH). E: Chimanimani (Melsetter), 8.iii.1931, *Otterson* in GHS 4116 (SRGH). S: Masvingo Distr., Makaholi Experimental Station (Farm), 10.iii.1948, *D.A. Robinson* 153 (K; SRGH). **Malawi**. N: Nkhata Bay Distr., Chombe Estate, Limphasa (Limpasa) Dambo, 490 m, 21.ii.1961, *Vesey-FitzGerald* 3028 (SRGH). C: Lilongwe Agric. Research Station, Canbandwe Dambo, 1.ii.1951, *Jackson* 392 (K). S: Nsanje Distr., Shire River at Chiromo, 80 m, 24.iii.1960, *Phipps* 2682 (K; PRE; SRGH). **Mozambique**. N: Mogincual Distr., andados 3 km de Liúpo para Namaponda, c. 100 m, 18.i.1964, *Torre & Paiva* 10065 (LISC). Z: Maganja da Costa Distr., Maganja da Costa (Vila da Maganja), km 42 on road to the coast, c. 15 m, 15.ii.1966, *Torre & Correia* 14649 (LISC). MS: Lower Búzi, 90–120 m, xii.1906, *Swynnerton* in *Eyles* 5861 (SRGH). M: between Santaca and Catuane, 13.iv.1949, *Myre & Balsinhas* 577 (K; LISC; SRGH).
    Mediterranean region southwards to Nigeria and Ethiopia, and then from the Flora Zambesiaca area to South Africa (Cape); also in Madagascar and with isolated records from Burma, Thailand and Borneo; introduced in America. Growing in damp ground in marshes and dambos, along riverbanks and on the shores of lakes; 15–1500 m.

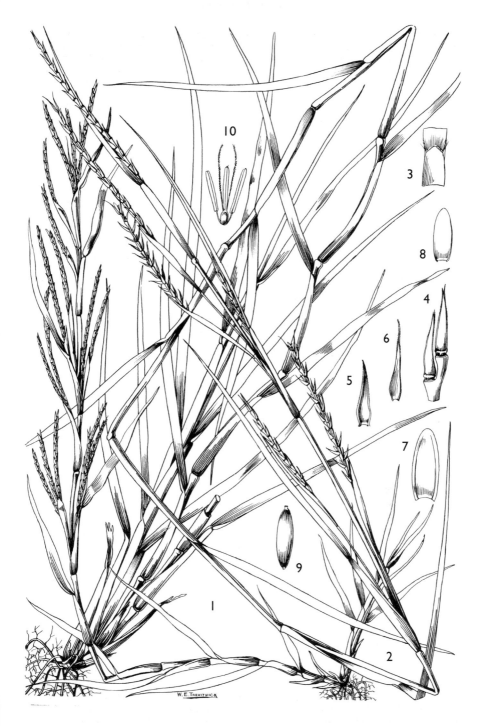

Tab. 55. HEMARTHRIA NATANS. 1 & 2, two plants showing range of habit (× $\frac{1}{2}$); 3, ligule (× 1); 4, pedicelled spikelets (× 4); 5, inferior glume (× 4); 6, superior glume (× 4); 7, inferior lemma (× 14); 8, superior lemma (× 14); 9, caryopsis (× 14); 10, flower (× 14). Drawn by W.E. Trevithick. From F.T.E.A.

2. **Hemarthria natans** Stapf in Prain, F.T.A. **9**: 56 (1917). —Jackson & Wiehe, Annot. Check List
Nyasal. Grass.: 43 (1958). —Simon in Kirkia **8**: 53 (1971). —Clayton & Renvoize in
F.T.E.A., Gramineae: 853, fig. 202 (1982). TAB. **55**. Syntypes: Malawi N: Umbaka River,
Scott s.n. (K); and without locality, *Buchanan* 1310 (K).
  Hemarthria fasciculata auct. non (Poir.) Stapf & C.E. Hubb.

Stoloniferous perennial; culms up to 250 cm long and 2–5 mm in diameter,
prostrate and rooting at the nodes below. Leaf laminas 4–15 cm × 2–5 mm, flat.
Racemes 3–7 cm long, fastigiate in the axils and not or scarcely exserted from the
axillary sheath. Sessile spikelet elliptic-oblong with a triangular callus; inferior glume
3.5 mm long, with or without a constriction towards the apex, sharply acute to
acuminate or caudate-acuminate at the apex, or the apex minutely bidentate;
superior glume acute to acuminate at the apex. Pedicelled spikelet narrowly
triangular, 4–7 mm long, truncate at the base and without a callus, acute to
acuminate at the apex.

Zambia. N: Mbala Distr., Sumbu National Park, Kasaba Camp, shore of Lake Tanganyika, 760
m, 16.ii.1959, *McCallum Webster* A64 (K). **Malawi**. N: Nkhata Bay, Lake Malawi, 490 m,
20.ii.1961, *Vesey-FitzGerald* 2994 (SRGH). C: Nkhotakota Distr., Dwangwa River, lagoon near
river mouth, 25.ii.1953, *Jackson* 1112 (K). S: Mangochi Distr., Monkey Bay, Thumbi (Tumbi)
Island, 480 m, 26.i.1969, *Eccles* 244 (K).
  Tropical Africa from Ethiopia and Dem. Rep. Congo to Angola. Growing on lake shores and
along riverbanks; 470–980 m.

40. RHYTACHNE Desv.

By T.A. Cope

Rhytachne Desv. in Hamilton, Prodr. Pl. Ind. Occid.: 11 (1825). —Clayton in Kew
Bull. **32**: 767–771 (1978).

Annuals or perennials. Basal leaf sheaths rarely compressed; ligule a short
membrane; leaf laminas narrow, often setaceous. Inflorescence a single terminal
cylindrical raceme; internodes clavate, as long as or longer than the sessile spikelet.
Sessile spikelet convex on the back; callus truncate, with prominent central peg;
inferior glume crustaceous, wingless (very obscurely winged in *latifolia*), sharply
inflexed in the inferior half, smooth, muricate or rugose on the back, usually
glabrous, sometimes awned; superior glume sometimes awned; inferior floret male;
superior lemma entire and awnless. Caryopsis oblong, dorsally compressed.
Pedicelled spikelet almost as long as the sessile spikelet or, more often, much
reduced, sometimes awned; pedicel curved and subfoliaceous.

A genus of 12 species; occurring in tropical Africa and tropical America, with one species
widespread in both.

1. Plant annual · 1. *triaristata*
 – Plants perennial · 2
2. Pedicels and internodes shortly pilose · 2. *robusta*
 – Pedicels and internodes glabrous · 3
3. Internodes conspicuously longer at the base of the raceme than at the summit · · · · · · ·
   · 3. *subgibbosa*
 – Internodes approximately equal in length throughout the raceme · · · · · · · · · · · · · · · · 4
4. Inferior glume smooth or granular, acute; leaf laminas flat · · · · · · · · · · · · · · 4. *latifolia*
 – Inferior glume rugose or muricate, at least on the keels, if smooth then leaf laminas
   setaceous · 5. *rottboellioides*

1. **Rhytachne triaristata** (Steud.) Stapf in Prain, F.T.A. **9**: 85 (1917). —Simon in Kirkia **8**: 53
(1971). —Clayton in F.W.T.A., ed. 2, **3**: 511 (1972). Type from Guinea.
  Lepturopsis triaristata Steud., Syn. Pl. Glumac. **1**: 358 (1854).
  Rhytachne triseta Hack. in A. & C. de Candolle, Monogr. Phan. **6**: 275 (1889). Type from
  Sudan.
  Rottboellia triaristata (Steud.) Roberty in Boissiera **9**: 71 (1960).

Annual up to 60 cm high. Leaf laminas up to 20 cm long, setaceous, involute. Racemes 9–15 cm long, the internodes equalling or slightly exceeding the spikelets. Sessile spikelet 3.5–5.5 mm long (excluding the awns), narrowly ovate-oblong to linear-oblong; inferior glume prominently rugose in the inferior $^2/_3$, with 5–7 laterally projecting transverse ridges, bifid at the apex into slightly unequal awns 1.5–4(6) mm long; superior glume with an awn 5–9.5 mm long. Pedicelled spikelet almost completely suppressed, represented by a curved pedicel exceeding the sessile spikelet by 1–2 mm and surmounted by 2 unequal awns, the longer 4–7 mm long.

Zambia. W: Mwinilunga Distr., 7 km from Kalene Hill, 16.iv.1965, *E.A. Robinson* 6582 (K; SRGH).
Also from Senegal to Cameroon and Sudan. Growing in laterite soils; c. 1200–1350 m.

2. **Rhytachne robusta** Stapf in Prain, F.T.A. **9**: 82 (1917). —Simon in Kirkia **8**: 53 (1971). —Gibbs Russell et al., Grasses South. Africa [Mem. Bot. Surv. S. Africa No. 58]: 282 (1990). Type from Angola.
   Rottboellia robusta (Stapf) Keng in Sinensia **10**: 306 (1939).

Densely caespitose perennial; culms up to 120 cm high. Leaf laminas up to 30 cm × 8–9 mm, flat with a long setaceous apex. Raceme 12–20 cm long, the internodes longer than the sessile spikelets, shortly pilose along the edges and around the summit. Sessile spikelet 8.5–11 mm long, narrowly ovate; inferior glume unornamented on the back, faintly scaberulous below, scabrid above, subacute to acute at the apex, awnless; superior glume awnless. Pedicelled spikelet much reduced, represented by a curved pedicel a little shorter than the sessile spikelet, surmounted by a pair of glumes 1.5–5 mm long.

Zambia. B: Kaoma Distr., 16 km east of Luampa River on verge of main road, 12.ii.1987, *Jeanes* 414 (K).
Also in Angola. Growing in woodland; altitude unknown.

3. **Rhytachne subgibbosa** (Winkl. ex Hack.) Clayton in Kew Bull. **20**: 261 (1966). —Renvoize, Grasses of Bahia: 287 (1984). Type from Brazil.
   Rottboellia loricata Trin. subsp. *subgibbosa* Winkl. ex Hack. in Martius, Fl. Bras. **2**(3): 311, t. 7 1/2 (1883).
   Rottboellia loricata subsp. *glaberrima* Hack. in Martius, Fl. Bras. **2**(3): 311, t. 7 1/3 (1883). Type from Brazil.

Caespitose perennial; culms up to 120 cm high. Leaf laminas 15–30 cm long, setaceous, involute. Racemes 16–30 cm long, the internodes exceeding the sessile spikelets, those towards the base of the raceme (5)9–18(23) mm long, those in the middle part 7–11 mm long, glabrous. Sessile spikelet 5–9 mm long, narrowly ovate; inferior glume variously rugose or muricate (in the New World), rarely almost smooth (in the Flora Zambesiaca area), acute at the apex, awnless; superior glume awnless. Pedicelled spikelet represented by a curved pedicel shorter than the sessile spikelet, surmounted by a vestigial glume up to 1 mm long (in the New World), sometimes the spikelet up to 2.5 mm long, well developed and containing a floret (also in the New World), rarely the spikelet up to 6.5 mm long, fully developed, bisexual, awnless (in the Flora Zambesiaca area).

Zambia. N: Samfya Distr., Lake Bangweulu, Chibambo Lagoon, SW of Ncheta Is., 1062 m, 11.ii.1996, *Renvoize* 5576 (K).
Also in South America (see below). Growing in flooded grassland at about 1050 m.
It is extraordinary, and currently quite inexplicable, that this widespread South American species (Brazil, Colombia, Venezuela, Paraguay, Argentina) should occur by Lake Bangweulu in northern Zambia. While extreme in almost all of its features for *R. subgibbosa* it has not proved possible to separate it as a distinct taxon. One possibility is that *R. subgibbosa* is not in fact a species distinct from *R. rottboellioides* but rather that the combination of characters that distinguish it have occurred by chance, from the large gene pool of the very variable *R. rottboellioides*, on both continents. The genus is badly in need of extensive revision to examine the relationships between many of its species, particularly in Africa. *Chabwela* 35 (K), also from Lake Bangweulu and mentioned by Clayton in his revision, belongs here.

4. **Rhytachne latifolia** Clayton in Kew Bull. **32**: 770 (1978). —Clayton & Renvoize in F.T.E.A., Gramineae: 845 (1982). —Gibbs Russell et al., Grasses South. Africa [Mem. Bot. Surv. S. Africa No. 58]: 282 (1990). Type from Tanzania.

Caespitose perennial; culms up to 100 cm high. Leaf laminas 15–50 cm × 4–8 mm, flat, with a brown collar at the junction with the sheath. Raceme 10 cm long or more, the internodes 5–10 mm long, glabrous. Sessile spikelet 5.5–8 mm long, narrowly ovate; inferior glume smooth or granular, very obscurely winged, acute at the apex; superior glume awnless. Pedicelled spikelet suppressed or much reduced, awnless, represented by a curved pedicel as long as the sessile spikelet.

Zambia. C: South Luangwa National Park, Mfuwe, c. 600 m, 12.iii.1969, *Astle* 5585 (K).
Also in Tanzania and South Africa (KwaZulu-Natal). Growing in brown clay soil under *Colophospermum mopane*, c. 600 m.

5. **Rhytachne rottboellioides** Desv. in Hamilton, Prodr. Pl. Ind. Occid.: 12 (1825). —Stapf in Prain, F.T.A. **9**: 83 (1917). —Stent & Rattray in Proc. & Trans. Rhodesia Sci. Assoc. **32**: 6 (1933). —Sturgeon in Rhodesia Agric. J. **51**: 143 (1954). —Chippindall in Meredith, Grasses & Pastures S. Africa: 519 (1955). —Jackson & Wiehe, Annot. Check List Nyasal. Grass.: 56 (1958). —Vesey-FitzGerald in Kirkia **3**: 109 (1963). —Clayton in Kew Bull. **20**: 261 (1966). —Hood, A Guide to the Grasses of Zambia: 73 (1967). —Simon in Kirkia **8**: 20, 53 (1971). —Clayton in F.W.T.A., ed. 2, **3**: 511 (1972). —Clayton & Renvoize in F.T.E.A., Gramineae: 843, fig. 198 (1982). —Gibbs Russell et al., Grasses South. Africa [Mem. Bot. Surv. S. Africa No. 58]: 283 (1990). TAB. **56**. Type from West Indies.
　　Rottboellia rhytachne Hack. in Bol. Soc. Brot. **3**: 136 (1885), *nom. superfl.*, based on *Rhytachne rottboellioides.*
　　Rottboellia caespitosa Baker in J. Linn. Soc., Bot. **22**: 533 (1887). Type from Madagascar.
　　Rottboellia setifolia K. Schum. in Engler, Pflanzenw. Ost-Afrikas **C**: 96 (1895). Type from Tanzania.
　　Rhytachne benguellensis Rendle in Hiern et al., Cat. Afr. Pl. Welw. **2**: 138 (1899). —Stapf in Prain, F.T.A. **9**: 84 (1917). Type from Angola.
　　Rhytachne mannii Stapf in Prain, F.T.A. **9**: 85 (1917). Type from Mbini (Rio Muni).
　　Rhytachne rottboellioides var. *guineensis* A. Camus & Schnell in Rev. Gén. Bot. **57**: 291 (1950). Type from Guinea.

Slender to robust caespitose perennial; culms up to 100 cm high. Leaf laminas 5–25 cm long, setaceous, involute. Racemes 2–20 cm long, the internodes all ± equalling the spikelets, glabrous. Sessile spikelet 2–5(6) mm long, narrowly ovate to oblong; inferior glume variously rugose or muricate, rarely almost smooth, obtuse, acuminate or bidenticulate at the apex, with or without an awn up to 5 mm long; superior glume awnless or with an awn up to 5 mm long. Pedicelled spikelet suppressed or almost so, represented by a curved pedicel as long as the sessile spikelet, sometimes surmounted by an awn up to 5 mm long.

Zambia. B: Kaoma Distr., Luampa R., Sikelenge, 19.xi.1959, *Drummond & Cookson* 6634 (SRGH). N: Kasama Distr., Chishimba Falls, 12.ii.1961, *E.A. Robinson* 4416 (K; PRE; SRGH). W: Mwinilunga Distr., 29 km from Mwinilunga on Matonchi road, 1370 m, 21.xii.1969, *Simon & Williamson* 1916 (K). C: Serenje Distr., Kundalila Falls, 13 km SE of Kanona, 1430 m, 17.xii.1967, *Simon & Williamson* 1409 (K). **Zimbabwe.** C: Makoni Distr., 8 km west of Rusape (Rusapi), 1400 m, 22.i.1955, *Chase* 5398 (K; SRGH). E: Chimanimani Distr., 10 km north of Chimanimani (Melsetter), 16.i.1955, *Crook* 554 (K; PRE; SRGH). S: Mberengwa Distr., Sikanjena, 26 km SE of Mberengwa (Belingwe), 1220 m, 4.v.1973, *Simon, Pope & Biegel* 2456 (K; PRE). **Malawi.** N: Nkhata Bay, Chombe Estate, Limphasa (Limpasa) Dambo, 490 m, 21.ii.1961, *Vesey-FitzGerald* 3024 (SRGH). C: Mchinji (Fort Manning), Bua River, 31.i.1952, *Jackson* 719 (K). S: Machinga Distr., Kasupe, 760 m, 16.vi.1949, *Wiehe* 140 (K). **Mozambique.** N: Montepuez Distr., andados 4 km de Montepuez para Namuno, 430 m, 25.xii.1963, *Torre & Paiva* 9678 (K; LISC). Z: Gurué Distr., rio Malema, near Picos Namuli, 1500 m, 29.vi.1943, *Torre* 5617 (K; LISC; PRE). MS: Cheringoma Distr., 4 km west of Safrique Hunting Camp, Roda, c. 18°57'S, 35°26'E, c. 55 m, 12.vii.1972, *Ward* 7851 (K; PRE).
Tropical Africa, South Africa and Madagascar; also in the West Indies and South America. Growing in swamps and seasonally wet grassland; 55–1500 m.
An extremely variable species with respect to habit, spikelet size, ornamentation of inferior glume and in the presence or absence of awns, but difficult to partition into meaningful subordinate taxa.

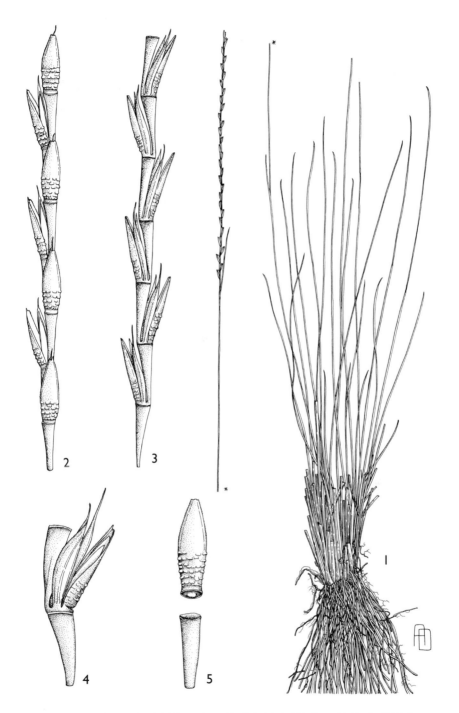

Tab. 56. RHYTACHNE ROTTBOELLIOIDES. 1, habit (× ¹/₂), from *Robinson* 4883; 2, raceme, showing sessile spikelets (× 4); 3, raceme, from rear, showing pedicels (× 4); 4, spikelet pair (× 6); 5, raceme joint, disarticulated (× 6), 2–5 from *McCallum Webster* T115. Drawn by Ann Davies. From F.T.E.A.

41. COELORACHIS Brongn.

By T.A. Cope

Coelorhachis Brongn. in Duperrey, Voy. Monde, Phan.: 64, t. 14 (1831). —Clayton in Kew Bull. **24**: 309–314 (1970).

Perennials, often tall. Basal leaf sheaths often laterally compressed; ligule a short membrane; leaf laminas linear, flat, rarely filiform. Inflorescence a single raceme, sometimes solitary and terminal but commonly aggregated into a spathate false panicle; racemes cylindrical or dorsiventrally flattened, the spikelets paired or occasionally in triplets of 2 sessile and 1 pedicelled; internodes clavate, often squatly so, shorter than the sessile spikelet, transversely articulated. Sessile spikelet convex on the back; callus truncate with a prominent central peg; inferior glume crustaceous or thinly coriaceous, nearly always winged towards the apex, 2-keeled in the inferior half, smooth, rugose or areolar, usually glabrous; superior glume awnless; inferior floret reduced to a hyaline lemma, with or without a palea; superior lemma entire and awnless. Caryopsis oblong, dorsally compressed. Pedicelled spikelet well developed and male, or vestigial, awnless; pedicel clavate or foliaceous, oblong or curved, sometimes auriculate at the summit.

A genus of c. 21 species; occurring throughout the tropics.

Pedicel auriculate at the summit · 1. *afraurita*
Pedicel not auriculate at the summit · 2. *lepidura*

1. **Coelorachis afraurita** (Stapf) Stapf in Prain, F.T.A. **9**: 80 (1917). —Stent & Rattray in Proc. & Trans. Rhodesia Sci. Assoc. **32**: 6 (1933). —Sturgeon in Rhodesia Agric. J. **51**: 143 (1954). —Jackson & Wiehe, Annot. Check List Nyasal. Grass.: 33 (1958). —Vesey-FitzGerald in Kirkia **3**: 109 (1963). —Simon in Kirkia **8**: 19, 52 (1971). —Clayton in F.W.T.A., ed. 2, **3**: 509 (1972). —Clayton & Renvoize in F.T.E.A., Gramineae: 842, fig. 197 (1982). TAB. **57**. Type from Mali.
    Rottboellia afraurita Stapf in Mém. Soc. Bot. France **8**: 98 (1908).

Caespitose perennial; culms up to 400 cm high, robust, erect. Basal leaf sheaths strongly laterally compressed and flabellate; leaf laminas 30–100 cm or more × 4–12 mm, conduplicate when dry. Racemes 2–7 cm long, dorsiventrally flattened with imbricate spikelets, embraced below by the spatheole, fastigiate in a copious false panicle. Sessile spikelet narrowly oblong; inferior glume 3–4.5 mm long, thinly coriaceous, smooth, faintly nerved, membranously winged above, the wings distinct from the body of the glume. Pedicelled spikelet 2.5–4 mm long, narrowly lanceolate, asymmetrically winged above; pedicel clavate, with a narrowly lanceolate auricle on one side towards the apex.

Zambia. N: Mbala Distr., Mbulu River, 1680 m, 2.iii.1955, *Siame* 618 (K; SRGH). **Zimbabwe**. C: Goromonzi Distr., Chinamora Res., east of Ngomakurira, 1580 m, 5.i.1967, *Simon* 959 (K). E: Chipinge Distr., Kabanga Dam, 26.ii.1972, *Gibbs Russell* 2556 (K). **Malawi**. N: Nkhata Bay Distr., Viphya, Mazamba, 1520–1830 m, 10.viii.1949, *Wiehe* 184 (K). C: Ntcheu Distr., near Masasa on Golomoti road, 1460 m, 13.vi.1950, *Wiehe* 588 (K). **Mozambique**. N: Sanga Distr., Unango, rio Malagui, c. 1100 m, 2.iii.1964, *Torre & Paiva* 10957 (K; LISC). T: entre Furancungo e Vila Coutinho, 25.vi.1956, *Myre & Balsinhas* 2508 (LISC).
Throughout tropical Africa. Growing in marshes and bogs and along the edges of streams, rivers and ponds; 1100–1830 m.

2. **Coelorachis lepidura** Stapf in Prain, F.T.A. **9**: 79 (1917). —Clayton & Renvoize in F.T.E.A., Gramineae: 842 (1982). Type: Mozambique MS: Kongone River, *Kirk* s.n. (K, holotype).
    Rottboellia lepidura (Stapf) Pilg. in Engler & Prantl, Nat. Pflanzenfam., ed. 2, **14e**: 139 (1940).

Caespitose perennial; culms up to 150 cm high, erect. Basal leaf sheaths laterally compressed; leaf laminas up to 30 cm × 2–7 mm, conduplicate when dry. Racemes 6–12 cm long, dorsiventrally flattened with imbricate spikelets, eventually exserted from the spatheole, gathered into a leafy false panicle. Sessile spikelet narrowly ovate-elliptic; inferior glume 3–4 mm long, thinly coriaceous, smooth, faintly nerved,

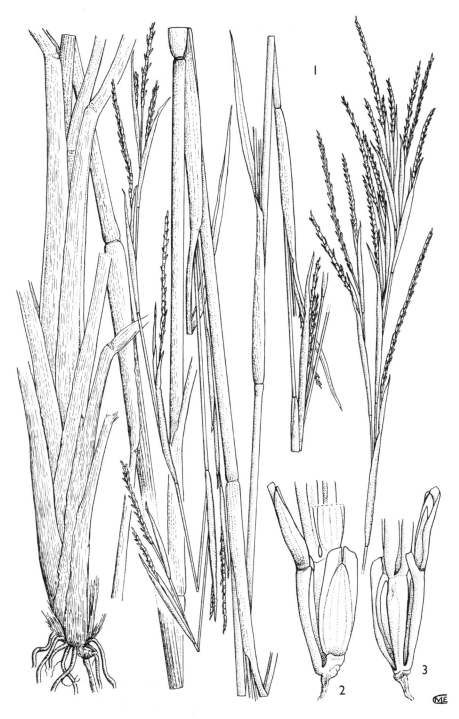

Tab. 57. COELORHACHIS AFRAURITA.   1, habit (× ⅔); 2, spikelet pair (× 9); 3, spikelet pair, from rear, showing auriculate pedicel (× 9), 1–3 from *Milne-Redhead & Taylor* 9839. Drawn by M.E. Church. From F.T.E.A.

membranously winged above, the wing merging with the body of the glume. Pedicelled spikelet 2.5–3 mm long, lanceolate, asymmetrically winged; pedicel slenderly clavate, not auriculate.

Mozambique. N: Monapo Distr., andados 4 km do Régulo Chihir para Itoculo, c. 140 m, 2.xii.1963, *Torre & Paiva* 9378 (LISC). Z: Namacurra, km 7 on road to Maganja da Costa (Vila da Maganja), c. 40 m, 25.i.1966, *Torre & Correia* 14095 (LISC). MS: Cheringoma Distr., Nhamissembe (Inhamissembe), 7–8 m, 14.vii.1972, *Ward* 7943 (K).
    Also in Kenya and Tanzania. Growing in floodplains below 200 m.
    Very similar to *C. capensis* Stapf from South Africa, differing by little more than the length of the inferior glume of the sessile spikelet (4.5–5 mm in *C. capensis*). The relationship between these two species would repay further investigation.

42. ROTTBOELLIA L.f.

By T.A. Cope

Rottboellia L.f., Nov. Gram. Gen.: 23 (1779); Suppl. Pl.: 114 (1781). *nom. conserv.*, non Scop. (1777).
Stegosia Lour., Fl. Cochinch.: 51 (1790).
Robynsiochloa Jacq.-Fél. in J. Agric. Trop. Bot. Appl. **7**: 406 (1960).

Annuals. Ligule a very short membrane; leaf laminas flat. Inflorescence a single raceme, either terminal or aggregated into a spathate false panicle; racemes cylindrical or slightly dorsiventrally flattened; internodes clavate, flattened or semi-cylindrical with sunken spikelets, wholly or partly fused to the adjacent pedicel. Sessile spikelet callus truncate, with prominent central peg or ridge; inferior glume coriaceous, broadly convex on the back, 2-keeled, smooth, narrowly winged at the apex; inferior floret male, with a hyaline lemma and well developed palea; superior lemma entire and awnless. Caryopsis ovate in face view, crescentic in side view. Pedicelled spikelet herbaceous, a little smaller than the sessile spikelet; pedicel oblong, flattened, often scarcely distinguishable from the internode.

A genus of 4 species; occurring in the Old World tropics, and introduced in the Caribbean region. *R. cochinchinensis* is becoming a serious weed in the tropics.

Sessile spikelet sunk in a semi-cylindrical segment formed from the fused pedicel and internode; callus circular with a prominent central knob ·········· 1. *cochinchinensis*
Sessile spikelet not sunken, the flattened or triquetrous pedicel and internode usually fused only in the inferior half, the keeled superior glume of the spikelet thrust between them above; callus elliptic with a low central ridge ·················· 2. *purpurascens*

1. **Rottboellia cochinchinensis** (Lour.) Clayton in Kew Bull. **35**: 817 (1981). —Clayton & Renvoize in F.T.E.A., Gramineae: 853 (1982). —Gibbs Russell et al., Grasses South. Africa [Mem. Bot. Surv. S. Africa No. 58]: 283 (1990). TAB. **58**. Type from Cochin Chine.
    Rottboellia exaltata L.f., Suppl. Pl.: 114 (1781), non (L.) L.f. (1779). —Stapf in Prain, F.T.A. **9**: 73 (1917). —Robyns, Fl. Agrost. Congo Belge **1**: 64 (1929). —Stent & Rattray in Proc. & Trans. Rhodesia Sci. Assoc. **32**: 6 (1933). —Sturgeon in Rhodesia Agric. J. **51**: 144 (1954). —Chippindall in Meredith, Grasses & Pastures S. Africa: 520 (1955). —Jackson & Wiehe, Annot. Check List Nyasal. Grass.: 56 (1958). —Hood, A Guide to the Grasses of Zambia: 73 (1967). —Simon in Kirkia **8**: 20, 53 (1971). —Clayton in F.W.T.A., ed. 2, **3**: 506 (1972). —Hall-Martin & Drummond in Kirkia **12**: 159 (1980). Type from India.
    Stegosia cochinchinensis Lour., Fl. Cochinch. **1**: 51 (1790).
    Rottboellia arundinacea A. Rich., Tent. Fl. Abyss. **2**: 444 (1851). Type from Ethiopia.

Tall annual up to 400 cm high, supported below by stilt roots. Leaf sheaths, especially the lower, painfully hispid; leaf laminas up to 45 cm × 20 mm, broadly linear. Racemes 3–15 cm long, cylindrical, glabrous, terminating in a tail of reduced spikelets, gathered into a spathate false panicle. Sessile spikelet oblong-elliptic, pallid; callus circular with a prominent central peg; inferior glume 3.5–5 mm long; superior glume boat-shaped with keel narrowly winged above, fitting into the concave internode. Pedicelled spikelet 3–5 mm long, narrowly ovate, herbaceous, green; pedicel shorter than the internode, the two fused throughout.

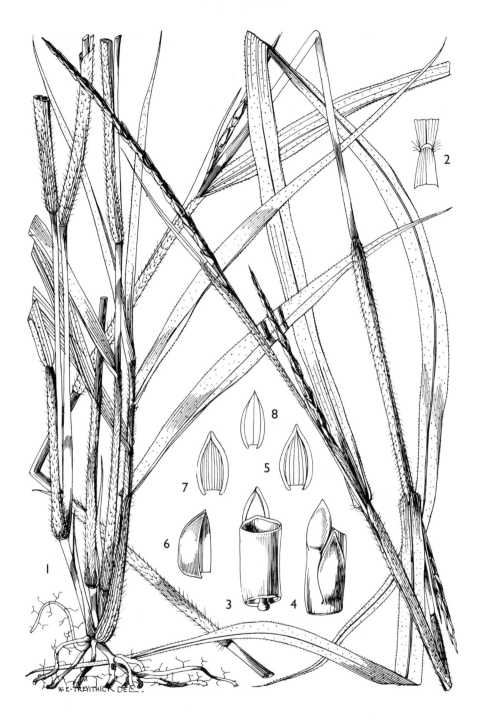

Tab. 58. ROTTBOELLIA COCHINCHINENSIS. 1, habit (× ²/₃); 2, ligule (× ²/₃); 3, rhachis internode (× 4); 4, spikelet pair (× 4); 5, inferior glume of sessile spikelet (× 4); 6, superior glume of sessile spikelet (× 4); 7, inferior glume of pedicelled spikelet (× 4); 8, superior glume of pedicelled spikelet (× 4). Drawn by W.E. Trevithick. From F.T.E.A.

Botswana. N: Central Distr., Mosetse River, 121 km on Francistown–Maun road, 1040 m, 9.iii.1961, *Vesey-FitzGerald* 3150 (SRGH). SE: Central Distr., Madibeng, 980 m, 11.ii.1959, *de Beer* 866 (SRGH). **Zambia**. B: Mongu Airport, 23.iii.1964, *Verboom* 1339 (K). N: Samfya, vi.1958, *Cooling* 136 (SRGH). W: Kitwe, 17.ii.1960, *Fanshawe* 5370 (SRGH). C: Lusaka Distr., Quien Sabe, 1100 m, 19.ix.1929, *Sandwith* 27 (K; PRE; SRGH). S: Livingstone, iv.1909, *Rogers* 7076 (K; SRGH). **Zimbabwe**. N: Mutoko Reserve, iii.1954, *Rodda* 19 (K; SRGH). W: Matobo Distr., Matopos Research Station, 1370 m, 24.iv.1954, *Rattray* 1645 (K; PRE; SRGH). C: Harare, Royal Salisbury Golf Course, 31.iii.1954, *Sturgeon* in *GHS* 57585 (K; PRE; SRGH). E: Mudzi Distr., Lawleys Concession, 21.ii.1954, *West* 3412 (K; SRGH). **Malawi**. N: Nkhata Bay Distr., 8 km east of Mzuzu at Roseveare's, 1220 m, 5.vi.1973, *Pawek* 6825 (K). S: Nsanje Distr., Makhanga (Makanga) Experimental Station, 80 m, 19.iii.1960, *Phipps* 2555 (K; PRE; SRGH). **Mozambique**. N: Namapa Distr., Estação Experimental de Namapa, 2.iii.1960, *Lemos & Macuácua* 47 (K; LISC; PRE; SRGH). Z: Mocuba, Posto Agrícola, 18.v.1949, *Barbosa & Carvalho* 2711 (SRGH). T: between Lupata and Tete, 1809, *Kirk* s.n. (K). MS: Chemba Distr., Chiou, Estação Experimental do C.I.C.A., 13.iv.1960, *Lemos & Macuácua* 100 (K; LISC; PRE; SRGH). M: Namaacha Distr., Goba (Estação C.F.), 20.iv.1949, *Myre & Balsinhas* 679 (SRGH).

Throughout the Old World tropics and introduced in the Caribbean region. Growing in shade in wooded grassland, along roadsides and an increasingly problematic weed of cultivation; 80–1400 m.

2. **Rottboellia purpurascens** Robyns, Fl. Agrost. Congo Belge 1: 66 (1929). Type from Dem. Rep. Congo.
> *Robynsiochloa purpurascens* (Robyns) Jacq.-Fél. in J. Agric. Trop. Bot. Appl. **7**: 406 (1960). —Simon in Kirkia **8**: 53 (1971). —Clayton in F.W.T.A., ed. 2, **3**: 506 (1972).

Robust aquatic annual up to 160 cm high, the basal part often immersed in water and rooting profusely at the nodes. Leaf sheaths glabrous or pilose with bulbous-based hairs; leaf laminas 20–45 cm × 3–12 mm, broadly linear. Racemes 4–10 cm long, slightly dorsiventrally flattened, glabrous, terminating in a tail of reduced spikelets, gathered into a spathate false panicle. Sessile spikelet oblong, pallid; callus elliptic with low central ridge; inferior glume 3–4 mm long; superior glume boat-shaped with a winged keel, this thrust between the internode and pedicel, sometimes conspicuously protruding. Pedicelled spikelet 5–7 mm long, ovate, herbaceous, green; pedicel shorter than the internode, the two fused only in the inferior part.

Zambia. B: Mongu Distr., central Barotse Floodplain at Namushakende, 26 km south of Mongu, 980 m, 8.ii.1955, *Hinds* 289 (K; SRGH). N: Samfya Distr., Lake Bangweulu, Mboyalubambe Is., 10 km south of Lake Chali, 1062 m, 6.iii.1996, *Renvoize* 5754 (K).

Also in Guinea, Sierra Leone and Dem. Rep. Congo. Growing in up to 2 m of water; 980–1100 m.

43. HETEROPHOLIS C.E. Hubb.

By T.A. Cope

Heteropholis C.E. Hubb. in Hooker's Icon. Pl. **36**: t. 3548 (1956).

Annuals or perennials. Ligule shortly membranous; leaf laminas flat. Inflorescence a single axillary raceme; racemes cylindrical or slightly dorsally compressed; internodes thickened, clavate, fused to the adjacent pedicel. Sessile spikelet ± embedded in the internode; callus truncate, with prominent central boss or peg; inferior glume crustaceous, broadly convex, 2-keeled or sharply involute along the sides, the back smooth, cancellate or areolate, winged at the apex; inferior floret reduced to a hyaline lemma or sometimes with a palea; superior lemma entire and awnless. Caryopsis ellipsoid, dorsally flattened. Pedicelled spikelet thinly coriaceous, about the same size as the sessile spikelet or smaller; pedicel oblong, flattened.

A genus of 4 species; in Africa to Indo-China and the Philippines.

Heteropholis sulcata (Stapf) C.E. Hubb. in Hooker's Icon. Pl. **36**: t. 3548 (1956). —Vesey-FitzGerald in Kirkia **3**: 109 (1963). —Simon in Kirkia **8**: 53 (1971). —Clayton & Renvoize in F.T.E.A., Gramineae: 849, fig. 201 (1982). TAB. **59**. Type from Dem. Rep. Congo.
> *Peltophorus sulcatus* Stapf in Prain, F.T.A. **9**: 59 (1917).

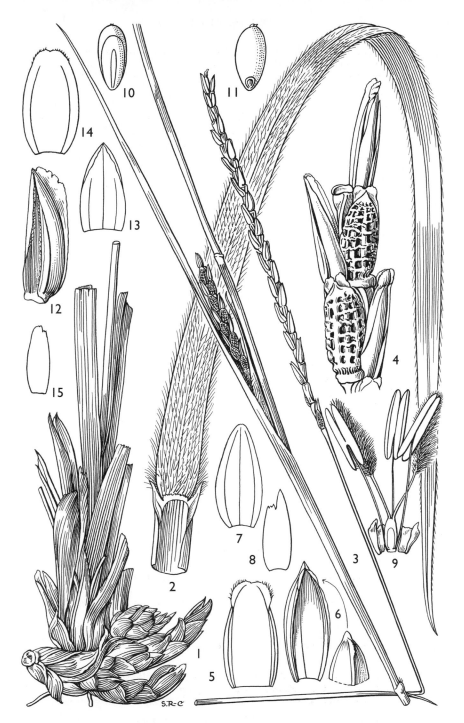

Tab. 59. **HETEROPHOLIS SULCATA**. 1, base of plant (× 1); 2, leaf (× 1), 1 & 2 from *Wiehe* N182; 3, racemes (× 1), from *Trapnell* 1706; 4, portion of raceme (× 6); 5–11, sessile spikelet: 5, inferior glume (× 8); 6, superior glume (× 8); 7, superior lemma (× 8); 8, its palea (× 8); 9, flower (× 8), 4–9 from *Shantz* 543; 10 & 11, caryopsis (× 6), from *Trapnell* 1522; 12–15, pedicelled spikelet: 12, inferior glume (× 8); 13, superior glume (× 8); 14, lemma (× 8); 15, palea (× 8), 12–15 from *Shantz* 543. Drawn by Stella Ross-Craig. From F.T.E.A.

Manisuris sulcata (Stapf) Dandy in J. Bot. **69**: 54 (1931). —Jackson & Wiehe, Annot. Check List Nyasal. Grass.: 47 (1958).

Perennial, arising from a short scaly rhizome; culms up to 120 cm high. Leaf laminas 16–60 cm × 3–10 mm, gradually tapering to a flexuous apex. Racemes 5–12 cm long, slightly compressed. Sessile spikelet broadly elliptic; inferior glume 2.5–4 mm long, sharply inflexed at the sides, strongly areolate on the back, pallid, with narrow wings on either side of the apex. Pedicelled spikelet narrowly oblong, 3–5 mm long, green or purplish, not areolate, the lateral keels of the inferior glume and the median keel of the superior glume narrowly winged; pedicel a little shorter than the internode.

Zambia. N: Mbala Distr., 13 km NW of Mbala (Abercorn), c. 1520 m, 11.iv.1961, *Phipps & Vesey-FitzGerald* 3059 (K; SRGH). W: Ndola, 1250 m, iii.1948, *Trapnell* 1986 (SRGH). C: Mkushi Distr., Munshiwemba, 12.i.1942, *Stohr* 736 (K; PRE). E: Lundazi Distr., 1370 m, 17.ii.1961, *Vesey-FitzGerald* 2975 (SRGH). **Malawi**. N: Nkhata Bay Distr., Nkhata Bay, 550 m, 20.ii.1961, *Vesey-FitzGerald* 2999 (SRGH).
Also in Tanzania and Dem. Rep. Congo. Growing in wooded grassland; 550–1520 m.

44. HACKELOCHLOA Kuntze

By T.A. Cope

Hackelochloa Kuntze, Revis. Gen. Pl. **2**: 776 (1891).

Annuals. Ligule a very short ciliate membrane; leaf laminas flat. Inflorescence a single axillary raceme, these numerous and aggregated into a spathate false panicle; racemes dorsally compressed; internodes flattened, oblong, cupuliform above, fused to the adjacent pedicel. Sessile spikelet globose; callus truncate, oblique, with a central boss; inferior glume hemispherical, crustaceous, rugose to cancellate, granular all over, wingless; superior glume partly adherent to the enclosing internode; inferior floret reduced to a hyaline lemma; superior lemma entire and awnless. Caryopsis oblate, slightly dorsally flattened. Pedicelled spikelet narrowly ovate, herbaceous, the inferior glume narrowly winged; pedicel broadly linear, scarcely distinguishable from the internode.

A genus of 2 species; one throughout the tropics, the other in India and Indo-China.
The genus is distinguished from *Heteropholis* by little more than the shape of the sessile spikelet.

Hackelochloa granularis (L.) Kuntze, Revis. Gen. Pl. **2**: 776 (1891). —Sturgeon in Rhodesia Agric. J. **51**: 144 (1954). —Chippindall in Meredith, Grasses & Pastures S. Africa: 523 (1955). —Jackson & Wiehe, Annot. Check List Nyasal. Grass.: 42 (1958). —Vesey-FitzGerald in Kirkia **3**: 108 (1963). —Hood, A Guide to the Grasses of Zambia: 72 (1967). —Simon in Kirkia **8**: 19, 52 (1971). —Clayton in F.W.T.A., ed. 2, **3**: 505 (1972). —Clayton & Renvoize in F.T.E.A., Gramineae: 849, fig. 200 (1982). —Gibbs Russell et al., Grasses South. Africa [Mem. Bot. Surv. S. Africa No. 58]: 173 (1990). TAB. **60**. Type from 'India Orientalis.'
Cenchrus granularis L., Mant. Pl. **2**: 575 (1771).
Manisuris granularis (L.) L.f., Nov. Gram. Gen.: 37 (1779). —Stapf in Prain, F.T.A. **9**: 57 (1917). —Stent & Rattray in Proc. & Trans. Rhodesia Sci. Assoc. **32**: 5 (1933).
Rottboellia granularis (L.) Roberty in Boissiera **9**: 79 (1960).

Coarse annual; culms up to 100 cm high, erect. Leaf laminas 2–15 cm × 4–12 mm, coarsely hispid, subamplexicaul. Racemes 5–15 mm long. Sessile spikelet c. 1 mm long and wide; superior glume narrowly crested at the apex. Pedicelled spikelet 1.5–2.5 mm long.

Botswana. N: Chobe Distr., near Mpandama-tenga (Pandamatenga), Zimbabwe frontier, 28.iii.1961, *Vesey-FitzGerald* 3364 (SRGH). **Zambia**. N: Mbala Distr., Mpulungu, 850 m, 12.iv.1959, *McCallum Webster* A306 (K; SRGH). C: near Sanje, west of Lusaka, 15°30'S, 27°52'E, 1110 m, 9.iii.1972, *Kornaś* 1368 (K). E: Lundazi Distr., Lukusuzi National Park, 12°46'S, 32°32'E, 800 m, 12.iv.1971, *Sayer* 1151 (K). S: Choma Distr., 5 km north of Mapanza, 1070 m, 18.iii.1956, *E.A. Robinson* 1373 (K; SRGH). **Zimbabwe**. N: Hurungwe Distr., north bank of R.

Tab. 60. HACKELOCHLOA GRANULARIS. 1, habit (× ⅔), from *Chevalier* 2280; 2, ligule (× 8); 3, raceme (× 6); 4, spikelet pair (× 16), 2–4 from *Rose Innes* 31406. Drawn by Victoria Goaman. From F.T.E.A.

Maura (Mauora), c. 610 m, 26.ii.1958, *Phipps* 910 (K; PRE; SRGH). W: Victoria Falls, 11.iii.1932, *Brain* 5586 (SRGH). C: Harare (Salisbury), Borrowdale, v.1927, *Eyles* 5862 (K; SRGH). E: Chimanimani Distr., Shinja East Farm, 12.ii.1955, *Crook* 564 (SRGH). **Malawi.** N: Kondowe–Karonga, 600–1800 m, vii.1896, *Whyte* s.n. (K). C: E.C.G.C. Station, Chitala, 31.iii.1950, *Wiehe* 486 (K). S: Nsanje Distr., Malawe Hills, west of Nsanje (Port Herald), 700 m, 23.iii.1960, *Phipps* 2632 (K; PRE; SRGH). **Mozambique.** N: Nampula, 26.iv.1946, *Morais* 22 (K; PRE). Z: Mocuba, Posto Agrícola, 16.v.1949, *Barbosa & Carvalho* 2642 (SRGH). T: Cahora Bassa Distr., Songo, 8.iii.1972, *Macêdo* 5025 (LISC). MS: Manica Distr., Bandula Forest, 700 m, 6.iv.1952, *Chase* 4428 (K; LISC; SRGH).

Throughout the tropics. Growing in open and wooded grassland, along roadsides and as a weed of alfalfa, cotton and maize; 400–1800 m.

45. OXYRHACHIS Pilg.

By T.A. Cope

Oxyrhachis Pilg. in Notizbl. Bot. Gart. Berlin-Dahlem **11**: 655 (1932).

Perennial. Ligule a very short membrane; leaf laminas involute. Inflorescence a single terminal or axillary raceme; raceme cylindrical, the spikelets borne singly on alternate sides of the rhachis; internodes semicylindrical, strongly oblique at base and apex. Spikelet sessile, sunk in the internode; callus obtuse; inferior glume coriaceous, broadly convex, smooth, wingless; inferior floret reduced to a hyaline lemma; superior lemma entire and awnless; palea small or suppressed. Caryopsis oblong-elliptic, dorsally compressed.

A genus of one species; occurring in tropical Africa and Madagascar.

There is no trace of either a pedicelled spikelet or a pedicel, but the structure of the sessile spikelet (particularly when there is a vestigial palea present in the superior floret) clearly indicates that it belongs with *Andropogoneae*. It is possible that the pedicel is present but is fused to and indistinguishable from the internode, a more extreme condition than is seen in the related genus *Ophiuros* (not in the Flora Zambesiaca area) in which the pedicel is just discernible.

Oxyrhachis gracillima (Baker) C.E. Hubb. in Hooker's Icon. Pl. **35**: t. 3454 (1947). —Simon in Kirkia **8**: 53 (1971). —Clayton in F.W.T.A., ed. 2, **3**: 506 (1972). —Clayton & Renvoize in F.T.E.A., Gramineae: 855, fig. 204 (1982). TAB. **61**. Type from Madagascar.

Rottboellia gracillima Baker in J. Linn. Soc., Bot. **22**: 533 (1887).

Oxyrhachis mildbraediana Pilg. in Notizbl. Bot. Gart. Berlin-Dahlem **11**: 655 (1932). Type from Tanzania.

Densely caespitose perennial; culms up to 60 cm high. Leaf laminas 5–30 cm long, filiform. Racemes 5–16 cm long, slender, long-exserted, purplish. Spikelet narrowly lanceolate, embedded in the internode; inferior glume 3–6 mm long, obtuse at the apex.

Zambia. N: Mporokoso Distr., 97 km SE of Mporokoso, 1430 m, 20.xii.1967, *Simon & Williamson* 1462 (K; SRGH). **Mozambique.** MS: Cheringoma Coast, Nyamanza Dambo, v.1973, *Tinley* 2907 (K; LISC; PRE).

Also in Sierra Leone, Cameroon, Tanzania and Madagascar. Swampy grassland and dambos; c. 40–1430 m.

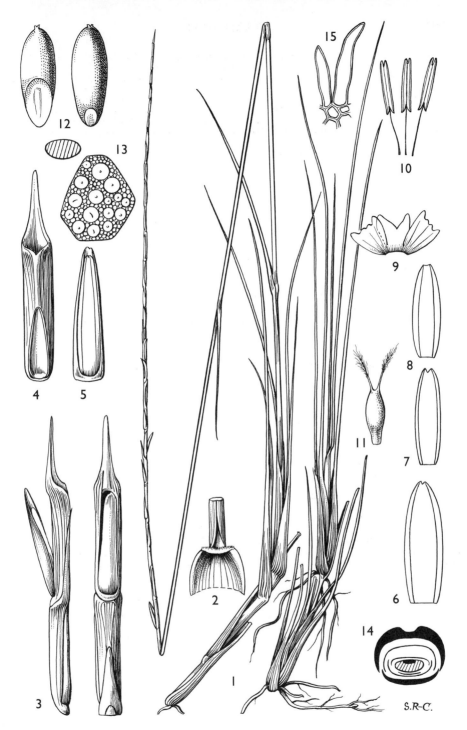

Tab. 61. OXYRHACHIS GRACILLIMA. 1, habit (× 1); 2, ligule (× 8); 3, portion of raceme, front and side views (× 8); 4, rhachis internode (× 8); 5, spikelet, viewed from inner side (× 8); 6, superior glume (× 8); 7, inferior lemma (× 8); 8, superior lemma (× 8); 9, its palea and lodicules (× 20); 10, stamens (× 8); 11, ovary; 12, grains; 13, cell with starch grains; 14, spikelet diagram; 15, leaf hairs. Drawn by Stella Ross-Craig. From F.T.E.A.

ADDENDUM

By T.A. Cope

Since publication of Volume 10, part 2 in 1999, a SABONET expedition to the Nyika Plateau in Malawi in 2000 found specimens of a species of *Eragrostis* new to the Flora Zambesiaca area. A modified portion of the key to species and a description of the species itself are provided below.

Group 9

8. Spikelets appressed and secund along the distinctly racemose primary branches · · · · · 8a
− Spikelets evenly distributed in the panicle on spreading pedicels · · · · · · · · · · · · · · · 10
8a. Anthers 2, 0.2–0.6(0.7) mm long; annual or short-lived perennial · · · · 82a. *schweinfurthii*
− Anthers 3, 0.7–1.3 mm long; caespitose perennial · 9
9. Basal leaf sheaths glabrous or thinly silky-pilose with white hairs · · · · · · · · · 83. *racemosa*
− Basal leaf-sheaths woolly-tomentose with off-white or yellowish hairs · · · · · · 88. *sclerantha*

82a. **Eragrostis schweinfurthii** Chiov. in Annuario Reale Ist. Bot. Roma **8**: 368 (1908). —Clayton in F.T.E.A., Gramineae: 231 (1974). Syntypes from Ethiopia.

Loosely caespitose annual or short-lived perennial; culms up to 55 cm tall, erect or geniculately ascending, unbranched, glabrous at the nodes, eglandular; leaf sheaths glabrous; ligule a line of hairs; leaf laminas 3–10 cm × 1–4 mm, linear, mostly flat, pilose with bulbous-based hairs, eglandular. Panicle 4–14 cm long, ovate to elliptic, the spikelets racemosely arranged on stiff pedicels 0.5–2 mm long, the lowermost primary branches not whorled, terminating in a fertile spikelet, glabrous or pubescent in the axils, stiff, eglandular. Spikelets 2.5–11 × 1.3–2 mm, narrowly ovate to narrowly oblong, lightly laterally compressed, 5–36-flowered, the florets disarticulating from below upwards, the rhachilla persistent; glumes subequal, boat-shaped, 0.8–1.4 mm long, reaching to about the middle of the adjacent lemmas, lightly keeled, narrowly ovate in profile, scabrid on the keel, glabrous on the flanks, acute at the apex; lemmas 1.3–1.8 mm long, lightly keeled, broadly ovate in profile, cartilaginous, the lateral nerves obscure, appressed to the rhachilla, those in opposite rows imbricate and concealing the rhachilla, dark green to olive or grey, glabrous, subacute to acute at the apex; palea persistent, glabrous on the flanks, the keels slender, wingless, scaberulous; anthers 2, 0.2–0.6(0.7) mm long. Caryopsis 0.5–0.6 mm long, subglobose to broadly ellipsoid.

Malawi. N: Rumphi Distr., Nyika Plateau, near Chelinda Camp at turn off to campsite, 27.iii.2000, *Smook* 10738 (K; PRE); Nyika Plateau, Chelinda Hill road at Kasaramba road turn off (10°28'S, 33°52'E), 31.iii.2000, *Smook* 10845 (K; PRE).

Also in east tropical Africa and tropical Arabia. In rolling montane grassland with patches of pine forest.

INDEX TO BOTANICAL NAMES